The Measurement of Roundwood

Methodologies and Conversion Ratios

The Measurement of Roundwood
Methodologies and Conversion Ratios

Matthew A. Fonseca

United Nations Economic Commission for Europe
Trade and Timber Branch
Geneva, Switzerland

CABI Publishing

CABI Publishing is a division of CAB International

CABI Publishing
CAB International
Wallingford
Oxfordshire OX10 8DE
UK

Tel: +44 (0)1491 832111
Fax: +44 (0)1491 833508
E-mail: cabi@cabi.org
Web site: www.cabi.org

CABI Publishing
875 Massachusetts Avenue
7th Floor
Cambridge, MA 02139
USA

Tel: +1 (617) 682-9015

A catalogue record for this book is available from the British Library, London, UK.

Library of Congress Cataloging-in-Publication Data

Fonseca, Matthew A.
The measurement of roundwood : methodologies and conversion ratios / by Matthew A. Fonseca.
p. cm.
Includes bibliographical references and index.
ISBN-13: 978-0-85199-079-8
ISBN-10: 0-85199-079-7 (alk. paper)
1. Forests and forestry--Mensuration. I. Title.

SD555.F62 2005
634.9'285--dc22

2005011482

ISBN-10: 0 85199 079 7
ISBN-13: 978 0 85199 079 8

Typeset by SPi Publisher Services, Pondicherry, India
Printed and bound in the UK by CPI Group (UK) Ltd, Croydon, CRO 4YY

Originally published in 2005. Transferred to POD in 2020.

Contents

Acknowledgements ix
Foreword *Harold E. Burkhart* xi
Abbreviations xiii
List of Tables xv
List of Figures xvii

1. INTRODUCTION 1

2. LOG SCALING 5
2.1 Basic Concepts, Commonalities and Differences of Log Scales 6
2.2 Cubic Measure and the Major Cubic Log Scaling Methods in Use 9
2.2.1 The USFS National Cubic Log Scale (USA) 10
2.2.2 BC Firmwood Scale (Canada) 15
2.2.3 Alberta Cubic Metre Scale (Canada) 18
2.2.4 The Ontario Cubic Method (Canada) 22
2.2.5 Swedish National Board of Forestry Log Scale (Sweden) 26
2.2.6 Russian Government Standard (Russia and members of the former USSR) 29
2.2.7 Cubage au Réel (France) 33
2.2.8 New Zealand 3-D and Mid-girth methods (New Zealand) 36
2.2.9 Brereton, ATIBT method (Africa, Oceania, South America, Asia, Japan) 38
2.2.10 Hoppus (Africa, Oceania, South America, Asia) 42
2.2.11 JAS Scale (Japan, Chile, East Asia, Oceania, Australia) 44
2.3 The Major Product Output Rules in Use 47
2.3.1 Scribner Short Log Rule (western USA) 48
2.3.2 Scribner Long Log Rule (northwestern USA, west coast Canada) 55
2.3.3 The Doyle Log Rule (central and eastern North America) 61
2.3.4 International ¼-Inch Log Rule (eastern North America) 67
2.4 Other Methods of Scaling 73
2.4.1 Weight scale 73
2.4.2 Stacked wood scale 78
2.4.3 Automated measurement systems (scanners, photo-cells) 80
2.5 Converting between Log Scaling Methods 83
2.5.1 Modelling conversion factors 83

2.5.2 Examples of using conversions 92
2.6 Sample Scaling 93
2.6.1 Determining sample size 94
2.6.2 Types of sample scaling 97
2.6.3 Population and subpopulations to be expanded (stratum) 99
2.6.4 Level of expansion 101
2.6.5 Expansion time window 102

3. MEASURING LOG YARD INVENTORIES AND MILL USAGE VOLUME 107
3.1 Basic Concepts 107
3.2 Methods of Measuring Log Yard Inventory 108
3.2.1 Stacked measure (deck factors) 108
3.2.2 Scaled inventory 110
3.2.3 Sample scaled inventory 111
3.2.4 Weight expanded inventory 112
3.2.5 Count (load or log) 112
3.2.6 Book estimated inventory 113
3.3 Calculating Mill Log Usage Volume 114
3.3.1 Measure the inventory and deliveries, and solve for the usage 114
3.3.2 Measure the production and solve for the usage by utilizing a recovery factor 114
3.3.3 Measure usage directly 115

4. MEASURING LOG QUALITY 117
4.1 Log Grading 117
4.1.1 Grading sawlogs and peelers 118
4.1.2 Grading chip logs 120
4.2 Log Manufacturing Quality 121

5. ROUNDWOOD WEIGHT AND GENERAL PHYSICAL PROPERTIES 125
5.1 The Variables that Determine Weight-to-Volume Ratios 125
5.1.1 Moisture content 125
5.1.2 Wood density 128
5.1.3 Bark volume and weight 129
5.1.4 Deducted defect volume 129
5.2 Conversions to and from Weight 129

6. METRICS OF LUMBER RECOVERY 131
6.1 Measuring Lumber Volume 133
6.1.1 Lumber board foot measure 133
6.1.2 Lumber cubic volume 138
6.2 Factors Affecting Lumber Recovery 139
6.2.1 Milling efficiency 139
6.2.2 Log characteristics 140
6.3 Recovery Trends by Log Size and Lumber Products Produced 145
6.3.1 Cubic scaled logs 151
6.3.2 Product output scaled logs 152

7. METRICS OF PLYWOOD/VENEER RECOVERY 159
7.1 Measuring Plywood and Veneer Volume 160
7.2 Factors Affecting Plywood and Veneer Recovery 160

7.2.1 Milling efficiency 160
7.2.2 Log characteristics 161
7.3 Plywood and Veneer Recovery Trends by Log Size and Scaling Method 164
7.3.1 Cubic scaled logs 164
7.3.2 Product output scaled logs 164

8. METRICS OF WOOD CHIPS AND OTHER RESIDUE RECOVERY FROM LOGS **169**
8.1 Units of Measure 169
8.2 Product Recovery 171
8.2.1 Chips 171
8.2.2 Sawdust 176
8.2.3 Shavings 177
8.2.4 Bark 177
8.2.5 Residual wood fibre-to-product ratios 180
8.2.6 Wood energy 181

REFERENCES **185**

Appendix 1 MEASURING LOG VOLUME **191**
Appendix 2 PHYSICAL PROPERTIES AND WEIGHT-TO-VOLUME DATA **225**
Appendix 3 GLOSSARY **255**

INDEX **263**

Acknowledgements

The author wishes to express his gratitude to the following persons who have contributed significant time, effort, expertise and knowledge to the making of this book:

Harry Bagley, Head of Quality Control, Plum Creek Timber Company, Columbia Falls, Montana, USA.
Ernie Bauer, Executive Director, Idaho State Board of Scaling Practices, Coeur d'Alene, Idaho, USA.
Lars Björklund, Swedish Timber Measurement Council, Tingvallavägen, Sweden.
Bill Cooper, Timber Scaling Supervisor, Alberta Forest Management Branch, Edmonton, Alberta, Canada.
Jim Crover, Scaling Policy Forester, Revenue Branch, BC Ministry of Forests, Victoria, British Columbia, Canada.
John Ellis, Group Technical Manager, Owens Cargo Company Ltd and Managing Director of Scaling Research International, Mount Maunganui, New Zealand.
Robert Frear, Technical Scaling Coordinator, Revenue Branch, BC Ministry of Forests, Victoria, British Columbia, Canada.
Adrian Whiteman, Forestry Officer, Planning and Statistics Branch, United Nations Food and Agriculture Organization, Rome, Italy.
Walter Zagrobelny, Provincial Measurement Supervisor, Wood Allocation and Measurement Section, Sault Ste Marie, Ontario, Canada.

Two individuals need to be singled out for the tremendous assistance and guidance that they have given:

Billy Dean, formerly the Manager of Measurements for Plum Creek Timber Company, also worked for the Inland Forest Resource Council and the Western Wood Products Association (WWPA) as a roundwood measurements expert. He spent many years dedicated to understanding the principles outlined in this book and communicating them to the forest sector at large and to the author of this book in particular. He is now retired and living on his ranch in Colorado.

Henry Spelter, Forest Economist, United States Forest Service Forest Products Laboratory, Madison, Wisconsin, USA. Mr Spelter was a great source of information and council. The author benefited significantly from Mr Spelter's experience and knowledge, not only regarding the subject matter in this book (of which he is a leading expert), but also about the process of producing a book.

Further assistance and expertise were given by Romeo T. Acosta, Lyne Bedard, Ernie Booth, Chris Bringloe, Mona Bynoe, Barber Cho, Clarence Hoot, Sterling James, Jerry Kingsbury, Godfrey Marshall, Volker Sasse and Caroline Stein.

Plum Creek Timber Company and its excellent staff deserve a great deal of thanks for providing a forward thinking environment and an excellent testing ground for learning the 'ins and outs' of roundwood metrics.

Finally, the author would like to thank his wife Cristina and daughters Tanya and Erin for their support and tolerance. Cristina, who is a librarian at the United Nations Office in Geneva, Switzerland, assisted in retrieving information and proof-reading. Much time was spent working on this publication which could have been family outings or a night at the cinema.

Disclaimer

The opinions and statements in this book are the author's and not those of present or past employers. Every effort was made to present accurate information from the best sources available, but this information should not be used in lieu of specific information sources or local knowledge when physical risk, issues of legality, or damages (financial or otherwise) could occur.

Note

A publication such as this is not possible without the contributions and expertise of many individuals. No doubt, there are many of you who have relevant information on roundwood metrics, which could be used to expand upon, clarify, and improve future revisions of this publication.

The author encourages any of you who wish to contribute information, make comments and suggestions, or have questions to contact him at: Roundwood@safe-mail.net

Foreword

Estimating the contents of roundwood is fundamental to forest products industries. Tree boles are irregular geometric solids with varying taper rates, making estimation of product quantities a formidable task. The wide array of products of possible interest – many of which involve converting highly irregular roundwood into regular dimensions – adds further complexity to the undertaking. Product specifications are subject to change, and manufacturing methods are variable and are undergoing constant evolution. Hence, it is not surprising that a bewildering span of metrics and methods for assessing roundwood has been promulgated.

While it would seem a relatively simple matter to estimate the probable product yield from roundwood, a wide variety of possible products must be considered. Furthermore, many primary forest products are bought and sold on a weight basis, as well as on volume measure, thus necessitating a thorough understanding of weight-to-volume relationships.

In the case of lumber, for example, the various dimensions of boards that may be produced from a log, variation in the equipment used in producing lumber, skills of various operators, and inherent variability in the logs makes for a multifarious estimation problem. In addition, the application of the log scaling procedures often varies by locality, with different definitions for obtaining scaling diameter and length. Scaling straight, sound logs is problematical, but many pieces have defects, which further complicates an already onerous job. Value depends on log quality or grade, as well as log volume and product quantity, thus adding yet another tangled layer to the knotty problem of assessing roundwood.

For many regions of the world, there is no industrial association or government agency with control over the measurement of roundwood. Therefore, it is essential that the various measurement standards and methods be clearly described and compiled in a readily available source. Because of the puzzling array of log rules, the difficulty of converting between different rules and units of measure and the scattered and incomplete documentation of the many assessment methods being applied by the forest products industry, pulling the material together in a logical and coherent manner under a single cover is a daunting endeavour.

Matthew A. Fonseca accepted the challenge of systematically summarizing a plethora of information on measuring roundwood with this volume entitled *The Measurement of Roundwood: Methodologies and Conversion Ratios*. In this ambitious effort, he summarizes the basic concepts, commonalities and differences of the principal log rules. He covers methods for measuring log yard inventories and mill usage volume, discusses measuring log quality, gives information on roundwood weight and general physical properties, details metrics of lumber recovery as well as plywood/veneer recovery, and ends with information on metrics of wood chips and other residue recovery from logs. This comprehensive volume on the metrics of roundwood includes coverage from major timber-producing regions of the world. With the increasing globalization of the forest products industry, having a ready reference to help sort through the maze of metrics and methods applied to assessing roundwood is certain to prove highly beneficial to foresters and forest products specialists.

Harold E. Burkhart
Department of Forestry
Virginia Polytechnic Institute and State University
Blacksburg, Virginia 24061, USA

Abbreviations

–	missing data
′	foot
″	inch
# *or* n	number
<	less than
>	greater than
√	square root
@	at
BC	British Columbia
BD	bone-dry
BDMT	bone-dry metric tonne
BDT	bone-dry ton
BDU	bone-dry unit
bf	board foot
BTU	British thermal unit
C	Celsius
ccf	cunit
cm	centimetre
CO_2	carbon dioxide
CT	computer tomography
CTL	cut-to-length log processor
CV	coefficient of variation
D	large-end diameter
d	small-end diameter
DBH	diameter at breast height
dia.	diameter
dm^3	cubic decimetre
F	Fahrenheit
EXP	antilog of natural logarithm
ft^2	square foot
ft^3	cubic foot
GJ	gigajoules
GTS	green target size

H	height
JAS	Japanese agricultural standard
kg	kilogram
L	length
lb	pound
LF	lineal foot
LL	long log
LN	natural logarithm
LRF	lumber recovery factor
m	metre
m^2	square metre
m^3	cubic metre
mbf	1000 board feet
MBTU	1000 BTU
MDF	medium density fibreboard
mc	moisture content
mm	millimetre
msf	1000 square feet $\frac{3}{8}''$ basis
MT	metric tonne
ns	not specified
OR	overrun
OSB	oriented strand board
PC	personal computer
PLE	prior load expansion
PNG	Papua New Guinea
Rt	revised t value
rw	random width
S	segment
SD	standard deviation
SE	sampling error
SG	specific gravity
SL	short log
spp	species
t	ton
UR	underrun
USFS	United States Forest Service
VRF	veneer recovery factor
W	width

List of Tables

Table 2.1	Russian regional taper and correction factors for bark	32
Table 2.2	Scribner log volume chart, lengths 1–20′	52
Table 2.3	Scribner Long Log Rule segment length and trim allowance chart	57
Table 2.4	Scribner Long Log volume chart, lengths 21–40′	58
Table 2.5	Doyle Log Rule volume chart	65
Table 2.6	International ¼″ Log Rule volume chart	71
Table 2.7	General guidelines for components of area of stacked roundwood	79
Table 2.8	Swedish National Board of Forestry stacked measure guidelines for pulp logs	80
Table 2.9	Volume index by length and small-end diameter class	85
Table 2.10	Revised *t* value	96
Table 5.1	Average physical characteristics of logs from some common North American tree species	126
Table 6.1	Typical distribution of fibre in the lumber manufacturing process	132
Table 6.2	Comparison of sawmill lumber volume as measured when rough-green, rough-dry, and finished-dry	133
Table 6.3	Example of lumber width yield in North American sizes by small-end diameter of log	135
Table 6.4	Board foot actual to nominal sizes and volumes by lumber product type	136
Table 6.5	Shrinkage rates of common North American trees	144
Table 6.6	Board mill, lumber and residual product recovery by small-end diameter of cubically scaled logs	146
Table 6.7	Stud mill, lumber and residual product recovery by small-end diameter of cubically scaled logs	147
Table 6.8	Dimension mill, lumber and residual product recovery by small-end diameter of cubically scaled logs	148
Table 6.9	Hardwood mill, lumber and residual product recovery by small-end diameter of cubically scaled logs	149

Table 6.10 Board mill, lumber recovery by small-end diameter of product output scaled logs 153
Table 6.11 Stud mill, lumber recovery by small-end diameter of product output scaled logs 154
Table 6.12 Dimension mill, lumber recovery by small-end diameter of product output scaled logs 154
Table 6.13 Hardwood mill, lumber recovery by small-end diameter of product output scaled logs 155
Table 7.1 Plywood recovery by small-end diameter of cubically scaled logs 165
Table 7.2 Plywood recovery by small-end diameter of product output scaled logs 166
Table 8.1 Bone-dry weight and volume conversions for selected tree species of North America 172
Table 8.2 Examples of residual product recovery by product type 175
Table 8.3 Physical characteristics and heat content of common North American tree species 178
Table 8.4 Composite panels and pulp recovery ratios 180
Table 8.5 Heating value of common North American tree species by volume 181

List of Figures

Fig. 2.1	Typical geometric shape of a log	7
Fig. 2.2	Examples of the Smalian cubic formula and the Huber cubic formula	9
Fig. 2.3	USFS diameter measurement methodology	11
Fig. 2.4	USFS taper distribution	12
Fig. 2.5	USFS National Cubic Log Scale gross volume determination	13
Fig. 2.6	BC Firmwood diameter measurement methodology	16
Fig. 2.7	BC Firmwood Scale gross volume determination	17
Fig. 2.8	Alberta Cubic Metre Scale diameter measurement methodology	19
Fig. 2.9	Alberta Cubic Metre Scale gross volume determination	21
Fig. 2.10	Ontario Cubic Method diameter measurement methodology	24
Fig. 2.11	Ontario Cubic Method gross volume determination	25
Fig. 2.12	Swedish National Cubic Log Scale diameter measurement methodology	27
Fig. 2.13	Swedish National Cubic Scale gross volume determination	28
Fig. 2.14	Russian Standard diameter measurement methodology	30
Fig. 2.15	Russian Standard gross volume determination	33
Fig. 2.16	Cubage au Réel diameter measurement methodology	34
Fig. 2.17	Cubage au Réel gross volume determination	35
Fig. 2.18	New Zealand 3-D diameter measurement methodology	37
Fig. 2.19	New Zealand 3-D gross volume determination	38
Fig. 2.20	Brereton diameter measurement methodology	40
Fig. 2.21	Brereton gross volume determination	41
Fig. 2.22	Hoppus gross volume determination	44
Fig. 2.23	JAS Scale diameter measurement methodology	46
Fig. 2.24	JAS Scale gross volume determination	47
Fig. 2.25	Product output diagram rule vs. formula-based rule	48
Fig. 2.26	Scribner scaling cylinder	49
Fig. 2.27	Scribner Short Log Rule gross volume determination	54
Fig. 2.28	Scribner Short Log Rule cull log example	54
Fig. 2.29	Scribner Long Log Rule diameter measurement methodology	56
Fig. 2.30	Scribner Long Log Rule taper distribution	57
Fig. 2.31	Scribner Long Log Rule gross volume determination	60

Fig. 2.32 Doyle Log Rule diameter measurement methodology 63
Fig. 2.33 Doyle Log Rule gross volume determination 64
Fig. 2.34 International ¼-Inch Log Rule 4′ scaling cylinder methodology (16′ scaling length) 68
Fig. 2.35 International ¼-Inch Log Rule gross volume determination 70
Fig. 2.36 Pounds per lineal foot 77
Fig. 2.37 Converting load A to Doyle bf using ‘pounds per lineal foot’ 78
Fig. 2.38 Converting load B to Doyle bf using ‘pounds per lineal foot’ 78
Fig. 2.39 Measurements to determine stacked measure volume 79
Fig. 2.40 Three-dimensional log scanning and images 81
Fig. 2.41 Net volume index by small-end diameter class 89
Fig. 2.42 Effects of log taper on converting between selected log scales 90
Fig. 2.43 Net volume index by small-end diameter class for product output (bf) rules 91
Fig. 3.1 Determining log deck volume using stacked measure 109
Fig. 3.2 Optimizer production report 116
Fig. 6.1 Effects on lumber recovery from taper in cubic and board foot scaled logs 143
Fig. 6.2 Lumber recovery by small-end diameter of log and mill type, actual log volume (BC Firmwood) 150
Fig. 6.3 Lumber recovery by small-end diameter of log and mill type, actual log volume (BC Firmwood) 150
Fig. 6.4 Cubic lumber recovery trends by log length 151
Fig. 6.5 Scribner Short Log Rule recovery by lumber product and log diameter 156
Fig. 6.6 Scribner Short Log Rule recovery trends by log length for selected diameters 156
Fig. 6.7 Scribner Long Log Rule recovery by lumber product and log diameter 157
Fig. 6.8 Scribner Long Log Rule recovery trends by log length for selected diameters 157
Fig. 6.9 Doyle Log Rule recovery by lumber product and log diameter 158
Fig. 6.10 International ¼-Inch Log Rule recovery by lumber product and log diameter 158
Fig. 7.1 Effects on plywood recovery from taper in cubic and board foot scaled logs 163
Fig. 7.2 Plywood recovery by small-end diameter (BC Firmwood) 166
Fig. 7.3 Plywood recovery by small-end diameter, product output scaled logs 167

1 Introduction

The ability to measure roundwood quantity and quality, and to predict product yields is of great importance to forest industries. In most areas, the cost of logs is the single largest operating expense that a forest product manufacturing company incurs. Despite the need for insight into roundwood metrics, it remains an area that is difficult to understand due to counterintuitive trends, complicated and inconsistent measurement logic, tremendous variability between regions, species, products, sizes, age classes, and the misnomers that exist regarding the units of measure. Further complicating one's ability to understand roundwood measurement is the proprietary nature of the industry: roundwood trade and conversion is highly competitive, and knowledge of roundwood metrics is often seen as a competitive advantage, not to be disseminated.

Beyond the obvious need for accurate volume and quality information on roundwood is the need to be able to use that information to make predictions, appraisals and forecasts, and conduct strategic analysis into core strengths and alternative scenarios that could provide improved efficiency and realization. This book has been written with the primary design to guide and provide reference to resource and manufacturing managers, professionals, analysts and log-scaling organizations. It is hoped that the data on product recovery will give the reader a good basis of understanding of the drivers that affect product recovery, and provide procedures and recommendations to implement or improve accounting procedures for measuring roundwood and product recovery.

Another intent of this book is to include a listing of comprehensive conversion ratios for log volume to weight, product volumes, and between various log-scaling methods. While there are few exact and universal conversions due to the variable physical characteristics of logs and inconsistencies in measurement methods, every effort has been made to obtain specific and accurate conversion ratios, and to present them with the necessary qualifiers.

In order to accomplish the stated aims, the following categories of roundwood metrics are covered:

Log scaling. Fifteen basic scaling methods utilized for the measurement of roundwood from various geographic regions around the world were chosen based on significance and availability of information. There is also an appendix table (Table A.1.B), which provides a brief description of ten more scaling methods. The key conventions used in these log scales are summarized, and relative volume trends compared to each other via a modelled log population in order to provide approximate conversion factors from one method of scaling to another. Because in many regions logs are scaled and traded by means other than physically measuring diameters and lengths, other methods of log scale are discussed, such as weight, stacked measure and the use of electronic scanning with necessary linkages provided to develop conversion ratios. As statistical sampling is a very common component used with log measurement systems, basic statistical methods and general guidance to methods and applications of different types of sample scale systems are discussed and diagrammed.

Log yard inventory management. Methods of measuring and managing log yard inventories are presented, discussed and tied together with the associated statistical methods. The use of weight, stacked measure, count, scaled and 'book determined' inventory methods are covered. Accounting procedures for reconciling inventory to usage and delivery volumes are discussed and illustrated with examples.

Measuring log grade and manufacturing quality. Relevant log characteristics which indicate the quality of products that can be produced are listed, and log manufacturing errors, which destroy value, are discussed in terms of accounting for these errors.

Roundwood weight and physical characteristics. Many systems of scaling, inventory and payments for transport and harvesting of timber, are dependent on weight. Weight-to-volume ratios are a key parameter for understanding costs and efficiency levels related to the conversion of trees into forest products. It is difficult to do meaningful realization analysis without understanding weight ratios, and this is especially true given the current trend toward purchasing wood and paying for timber harvest via weight. Weight is a key parameter along with bone-dry weight, which is critical to understanding value and making good allocation decisions for pulpwood and the manufacture of residual wood products. This book also includes a list of published weight-to-volume ratios and bone-dry weights for many worldwide commercial timber species in the form of raw logs. Bark content is shown for most major North American and some European timber species.

Product recoveries from roundwood. The key drivers that determine lumber, veneer, plywood, chip and residual wood fibre recovery from roundwood are discussed and diagrammed. Effects such as log size and characteristics, products manufactured, physical properties of wood, milling efficiency and log-scaling method used are quantified and reflected through graphs and tables, which show estimated recovery ratios within product lines, scaling methods and small-end diameter. The associated methods of measuring products are covered and quantified, including cubic content, board foot, surface measure and bone-dry measure. Product recovery from residual wood fibre such as composite panels, pulp and paper, and wood energy are covered briefly.

As stated earlier, every attempt has been made to present accurate information. But as with any information on variable subject matters such as the weight of wood, scaling method, conversion factors and recovery ratios, the reader needs to be aware that the listed data may not be reflective of every situation. It is best to obtain conversion ratios specific to the intended population; nevertheless, much of the content of this book exists because this is not always possible, nor is it always necessary.

Much of the matter in this book has been accumulated over many years by the author in his professional capacity, and much of it has come from published information on the subjects discussed. Official sources, such as Forest Services and Forest Ministries, and mensuration organizations provided much of the previously published information contained herein. There is undoubtedly a 'North American bias' in content as it is an important region in terms of the sheer size of the forest sector; it has many old and complicated systems of measure, and it is also where the author obtained his experience. Readers are encouraged to utilize the publications, papers and websites listed in the reference for further research.

Finally, it should be said that the forest products industry, while always accepting of new technology, is also steeped in tradition, and sometimes resistant to change. Many of the complexities contained in this book on understanding and managing roundwood are due to old systems of measurement that no longer apply well to the current state of forestry and the timber industry (particularly in North America). It is the author's hope that support for improved accuracy, simplification and harmonization of systems will eventually bring about an end to using systems, some of which have been utilized for more than 150 years and which no longer provide the users with much value.

2 Log Scaling

Log scaling is the process by which the gross and net content of timber is ascertained and expressed in an acceptable unit of measure. It may be expressed in cubic volume, predicted product output (e.g. bf), piece count, stacked measure or by weight. Normally, scaling is specifically associated with the measurement of volume, but it also provides a vehicle for measuring value via species, grade, dimensions and quality of log manufacturing. The purposes of log scaling are as follows:

1. As a unit of measure for transaction purposes.
2. As a gauge for work accomplished.
3. As a measure of inventory.
4. As a measure of mill efficiency (recovery).
5. As a predictor of output in finished products.

As much as 60–85% of the cost of producing wood products can be in the purchase of logs, so it is extremely important to fully understand the variables of the unit of measure used, and to control its consistency and accuracy. A typical log scaler measures a great deal of value in a year's time, and has the difficult task of obtaining accurate measurements while maintaining productivity. This situation can easily lead to inaccurate measurements, which are often biased toward an overstatement of volume. This happens, not because of personal bias, but because diameter measurements tend to be overstated and log defects tend to be missed in the production environment.

To ensure that the logs are scaled accurately, it is very important that the scalers are well trained and competent, have sufficient time, as well as good working conditions to do their scaling in. It is also important that scalers are regularly check-scaled, in order to measure their accuracy and correct any deficiencies.

To successfully manage a log conversion facility or market delivered logs, it is necessary to have accurate and consistent log scale, understand the nuances of log scale, be able to measure value, predict product output and establish accurate inventory methods. With the ongoing trend toward more globalization of roundwood trade, it is increasingly important to be able to work with a variety of scaling methods. Regardless of the unit of measure chosen for internal accounting, one may need to purchase and appraise logs based on a very different log-scaling method or by weight.

The desirable characteristics of a log-scaling system are accurate, consistent and relative scales through which one is able to measure volume and value, predict product output, establish accurate inventory and usage volumes, and to convert from one unit/method of measure to another (cubic volume to weight, weight to volume, scaled volume to cruised stand volume, etc.). Some methods are better than others for accomplishing these aims, but all can be managed and converted if one understands the variables and has enough data to work with. However, it is the author's opinion and experience that the cubic-based measures are the best for accomplishing the aims listed. Even when one is forced by regional acceptability or law to use other types of scaling systems, like weight or product output rules (Scribner, Doyle, International ¼-Inch, etc.), it is worthwhile determining and tracking cubic volume in parallel for 'internal' bookkeeping.

2.1 Basic Concepts, Commonalities and Differences of Log Scales

There is a wide variety of scaling methods utilized in the world. Choice of method is sometimes mandated by law or dictated by acceptability within a region, and often a company may utilize more than one method of scaling simultaneously on the same logs. Most scaling methods utilize length and diameter information to determine volume in cubic volumetric units, or in product output units (bf). Even simple weight scale (just weighing the logs) is typically converted to cubic or product output volumes based on conversions from sample scaling. There are two types of product output rules: diagram rules such as the Scribner rule, and rules based on mathematical formulas such as the Doyle log rule and the International ¼-Inch. Most of the world utilizes cubic measure, which measures all the wood fibre in cubic volumetric units and makes no assumption as to product output. Most scaling systems make deductions for defects (volume assumed not usable). Given that the same basic inputs (diameter, length and defects) are used in the different scaling methods, one would assume that there is a good level of uniformity and harmonization of methods; unfortunately, this is not often the case. In combination, the determination of diameters, length, taper, gross volume and defect can contribute to substantial discrepancies in volume determination between scaling systems even when the procedures appear very similar. To some degree, all of the scaling methods are arbitrary, and only through close examination can they be accurately assessed. Without analysing the actual log-scaling methods at this point, the major procedural differences causing disparities between the various log scales are as follows.

Diameter measurements. Many log-scaling methods require that diameter measurements be taken at both ends of the log while others only require one. The Scribner Long Log Rule only requires the small-end diameter for logs up to 41′ long (12.5 m). Other scaling methods also only utilize one diameter measurement, but taken in the middle of the log. There are also some rules that require three diameter measurements (both ends and the middle) for particularly long logs, or logs with a great deal of taper (in this situation a log is scaled as two log segments).

The diameter measurement itself may differ substantially from one log scale to the next. As already stated, some log-scaling methods utilize measurements taken outside bark (meaning that normally an assumed thickness of bark is subtracted), and circumference measurements that are converted to a diameter via π. Diameters converted from circumferences have a predisposed bias to be overstated because logs are not generally perfectly round; they all are at least somewhat elliptical. The more elliptical a log is, the more the volume will be overstated.

The rounding rules utilized for diameters can also create substantial differences. Both of the Scribner log rules use inches as a unit. Scribner Long Log truncates diameters (13.99″ rounds down to 13″), while Scribner Short Log and most other log scales use conventional rounding (13.51″ rounds up to 14″). To account for the elliptical nature of many diameters, most log scales request that two measurements be taken on each log end (narrow and perpendicular or narrow and wide axis), which also creates differences. For example, with both short and long log Scribner, an average that falls on the half inch is always rounded down (narrow measurement of 13″, right angle measurement of 14″ = 13.5″; rounds down to 13″); while with many other log scales, the average goes up or down to the closest even number.

Another reason that diameters create differences in log volume is because of the different methods of determining the large-end diameter for logs that are butt-cuts (first log in the tree taken above the stump). It is commonly accepted that the diameter taken at the butt-cut does not give a representative profile of the tree. Butt-cut logs are frustums of a neiloid in shape owing to the flare and buttressing that occurs from the ground level up to about 1.4 m (4.5′) above the ground or higher (Fig. 2.1). Methods for obtaining a representative butt diameter range from applying an assumed

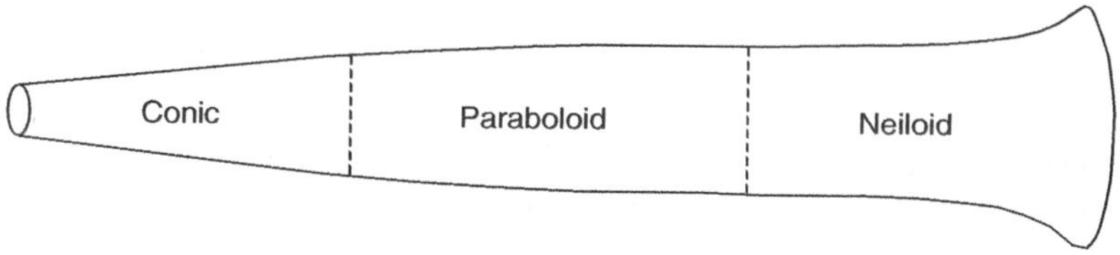

Fig. 2.1. Typical geometric shape of a log.

increase over the small-end diameter ('standard taper'), measuring the log over the bark above the butt-swell (and possibly estimating the diameter inside bark), or in the case of some scale methods, only using small-end or mid-point diameters (thus alleviating the need to obtain a butt-cut diameter).

Length measurements. Log lengths are measured much more consistently than diameters, but there are still differences. Some log-scaling methods include unmeasured additional length to allow for checking at the end of the log, shrinkage and unsquare cuts, while other log scales measure the entire length. This unmeasured length (trim allowance) can be as much as 7% of the log length (typically around 3%).

Log taper. Assumptions of how log taper is used to account for log shape have a great deal to do with the calculated volume. Most trees have similar shapes in that they tend to be frustums of a neiloid (concave) near the stump, paraboloid in the centre, and paraboloid or conic near the top (see Fig. 2.1). Without getting into the particulars of each log scale yet, it is important to understand that log scales may assume taper, measure the actual taper (with or without accounting for the actual profile of the stem) or ignore the taper altogether, and these differences can cause substantial volumetric disparities between scaling methods.

Gross volume determination. As stated earlier, the various scaling methods are expressed in either units of predicted output, such as the bf rules, or in cubic volumetric units (which make no output assumptions except in some cases where defects are concerned). The product output rules are generally based on assumptions that may have been accurate 100 years ago in a particular region and given particular products produced, but that may no longer be appropriate for predicting output based on current manufacturing standards and products. Of the product output rules listed in this publication, none correlate well with cubic scale, unless diameter and length data can be factored in, and even within the output rules, there is a great deal of variation. Further exacerbating these differences are the rounding rules for volume, which in some log rules can create huge differences in volume at key points. These abrupt and large changes (step functions) can be inadvertently or purposely manipulated to affect the volume assigned.

While volume determination for cubic is more consistent, there are differences in formulas that can also create substantial differences in the volume assigned. There are several different formulas for determining cubic volume, and they can all yield rather similar or different results depending, in great part, on the parameters of the logs, e.g. short, long, high taper, low taper, etc.

Defect deductions. There are many differences between scaling methods on what constitutes a deductible defect and procedures for determining the appropriate defect volume to deduct. There are three main ways for accounting for defect that are particular to the various scaling methods:

1. Ignore defects and just determine gross volume.
2. Deduct only for void, soft-rot and char.
3. Deduct anything that reduces the yield of primary products; this includes not only void, soft-rot and char but firmwood defects such as big knots, crook and sweep, checks, shake and twist.

2.2 Cubic Measure and the Major Cubic Log-Scaling Methods in Use

The methods listed below are some of the most common cubic log scales used around the world. There are more (some of which are described briefly in Table A.1.B), but the methods listed in this section are limited to those that the author was familiar with or on which information was readily available.

All but two of the cubic rules listed below are based on the formula 'area of a circle $= \pi \times$ radius^2'. The other methods listed are the JAS Scale, which utilizes a formula for the area of a square to calculate cubic volume, and Hoppus, which uses the area of one-fourth of the girth (circumference). There are two main variations of the 'area of a circle formulas' in use for log scaling: the Smalian formula and the Huber formula (Fig. 2.2).

The 0.7854 constant used in Fig. 2.2 is derived from $\pi \div 4$, which is the relative area of a circle to a square with the same dimensions, e.g. a circle with 20 cm diameter is 78.54% the area of a 20 × 20 cm square.

Smalian formula

Log volume = $((d^2 + D^2) \div 2) \times L \times 0.7854$.

Simply put, it measures the area of two cylinders, each for half the log length, one having the diameter of the small-end, the other having the diameter of the large-end.

Example: $[(0.25^2 + 0.31^2) \div 2] \times 5 \times 0.7854 =$ **0.311 m³**

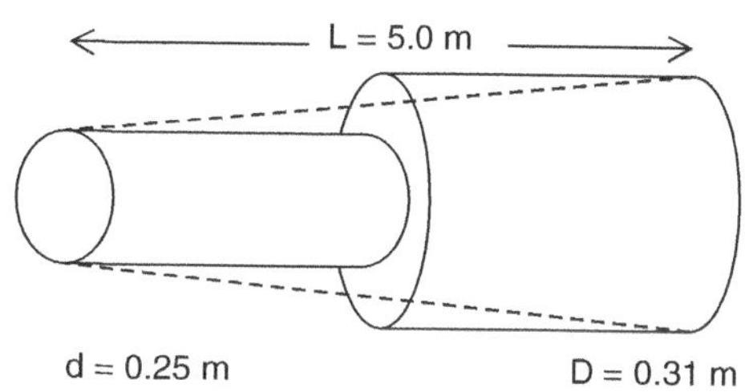

Huber formula

Log volume = $((d + D) \div 2)^2 \times L \times 0.7854$.

The Huber formula measures the area of a cylinder having the diameter equal to the mean of the small-end and large-end diameters or having a diameter equal to the midpoint.

Example: $0.28^2 \times 5 \times 0.7854 =$ **0.308 m³**

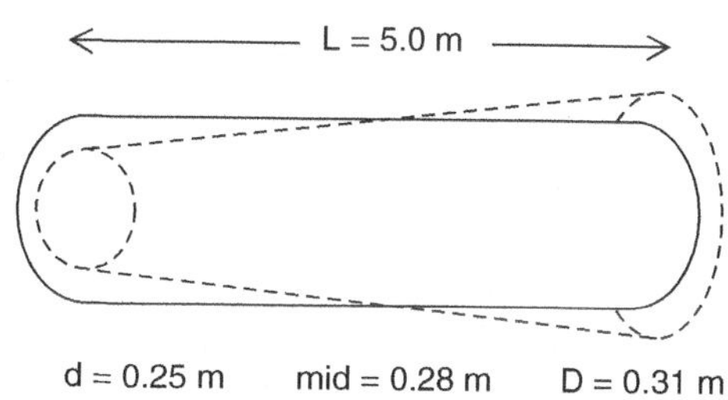

Fig. 2.2. Examples of the Smalian cubic formula and the Huber cubic formula.

Listed below is a brief summary of the principal methodology for some cubic log scales in use in the world today. The information on log scales contained herein should not be used in lieu of the actual scaling manuals.

2.2.1 The USFS National Cubic Log Scale (USA)

The benefits of measuring volume via a cubic log scale have been recognized for many years. Despite attempts by some in the United States Forest Service (USFS) to gain widespread acceptance and use of the cubic log method as far back as the 1920s, the product output rules ('bf measure') have maintained predominance in the USA. That is not to say that many companies and agencies in the USA have not used cubic measurement. Many companies, scaling bureaus and some agencies have used, and continue to use, cubic measurements for 'in-house' data and transactions.

It was not until 22 May 1991 that the USFS published the *National Forest Cubic Scaling Handbook*. Unfortunately, the timing was not good for obtaining universal acceptance of these guidelines, as they came during a time when the Forest Service was experiencing diminished influence amongst the timber community. This was due to a huge reduction in their timber sales programme (the spotted owl court decision and overall environmental agenda becoming politicized during these times), and because of tremendous pressure to improve accountability on removed volumes by switching from 'scaled sales' to other means such as 'lump sum' and 'ton rate'. Despite the slow acceptance rate in the USA by private companies and non-USFS agencies, some larger companies are using cubic log scale (notably in western USA) in parallel with bf measure. The USFS National Cubic Log Scale is normally reported in units of cubic feet (ft^3) or in units of 100 cubic feet (ccf), which is called a cunit. Listed below is a summary of the procedures for determining volume based on the *National Cubic Scaling Handbook* (USFS, 1991).

2.2.1.1 Diameter measurements

1. Measure diameters inside of bark and cambium layer.

2. Measure through the true centre of the log (ignore pith).

3. Take two measurements, narrow-way first, and right angle to the first measurement (the right angle measurement is generally, but not necessarily, the wide dimension).

4. Round diameters to the nearest inch; when one diameter falls exactly on the 0.5″, round it up; when both diameters fall exactly on the 0.5″, round one diameter up, and the other down.

5. Determine the average of both measurements (if the average is on the 0.5″, always round down).

6. Disregard bumps and depressions.

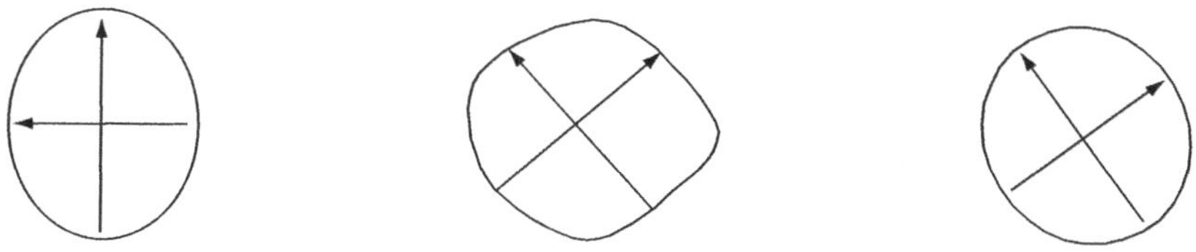

Narrow measurement = 8.84″
Right angle measurement = 9.62″
• 8.84 rounds to 9″
• 9.62 rounds to 10″
(9″ + 10″) ÷ 2 = 9.5″; round down to **9″**

Narrow measurement = 15.21″
Right angle measurement = 16.72″
• 15.21 rounds to 15″
• 16.72 rounds to 17″
(15″ + 17″) ÷ 2 = 16″; diameter is **16″**

Narrow measurement = 39.52″
Right angle measurement = 44.74″
• 39.52 rounds to 40″
• 44.74 rounds to 45″
(40″ + 45″) ÷ 2 = 42.5″; round down to **42″**

Fig. 2.3. USFS diameter measurement methodology.

7. Forked tops or swelled ends should be measured behind swell.
8. For butt-cut ends, measure diameter 4′ above the butt-end (which correlates with breast height, assuming a stump height of 6″).

Examples of diameter measurements and rounding conventions are shown in Fig. 2.3.

2.2.1.2 Length measurements

1. Measure short side to short side.
2. Recorded scaling length is actual length minus trim, recorded to the nearest foot. Maximum trim is normally 0.5′ per segment for logs 8–20′ in nominal length, 1′ for logs 21–40′ in nominal length, and 1.5′ for logs 41–60′ in nominal length. When the timber is from a stumpage sale, if the actual length minus trim allowance exceeds the whole foot, the recorded length is rounded up to the next higher foot in length (see Table A.1.C). If logs are purchased already cut, trim allowance is usually negotiated with insufficient trim allowance resulting in a log being scaled back to the next desired length multiple.
3. Maximum segment length is 20′ plus trim.
- Logs longer than 20′ are divided into two or more segments as necessary and scaled as individual log segments.
- Multisegment logs are divided evenly, if possible, into 2′ multiples, with the shortest segment(s) on the small-end of the log (in the event that the log cannot be divided evenly).
- Under no circumstances should there be more than a 2′ difference between the longest and shortest segment or more than one odd-length segment in any multisegment log (see Appendix Table A.1.C).

2.2.1.3 Taper distribution

1. One-segment logs (logs with a nominal length of 8–20′) are assumed to be conic, with a constant taper between the small-end and large-end.
2. Two-segment logs (logs with a nominal length greater than 20′ but less than 40′): total taper ÷ number of segments added to the small-end diameter of segment 1 = the small-end diameter of segment 2; if total

Two-segment log

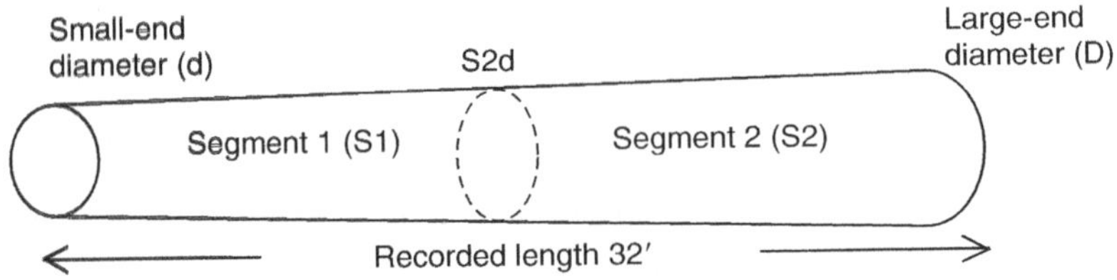

Even taper example: d = 12″; D = 14″

14 – 12 = 2″ of taper; 2 ÷ 2 segments = 1″; S2d = d + 1; S2d is **13″**

Odd taper example: d = 12″; D = 15″

15 – 12 = 3″ of taper; 3 must be raised to 4 to be divisible evenly by 2; 4 ÷ 2 segments = 2″; S2d = d + 2; S2d is **14″**

Note: The small-end diameter of segment 2 (S2d) is also the large-end diameter of segment 1 (S1D).

Fig. 2.4. USFS taper distribution.

taper is not evenly divisible by the number of segments, raise taper to next number evenly divisible by the number of segments.

3. Three-segment logs (logs with a nominal length of 40′–60′): use same procedure as for two-segment log to find small-end diameter of the middle segment; repeat procedure for the remaining two segments (as if the small-end segment did not exist) to find the small-end diameter of the butt segment (segment 3).

Figure 2.4 shows taper distribution for a two-segment log with both odd and even taper examples. Taper distribution can also be determined by using Table A.1.D.

2.2.1.4 Gross volume determination

The USFS National Cubic Log Scale uses the Smalian formula to determine log volume. To determine the volume of a log segment apply the formula:

$$\text{log volume} = [(\text{small-end diameter}^2 + \text{large-end diameter}^2) \times L] \times 0.002727,\ \text{round to nearest tenth of a cubic foot.}$$

Note that the constant in the above version of the Smalian formula is 0.002727 rather than the 0.7854 constant from Fig. 2.2; this is because there are 144 square inches in a square foot: $0.7854 \div 144 = 0.005454$. Because two diameters are used with the Smalian formula, $0.005454 \div 2 = 0.002727$.

Figure 2.5 shows an example of calculating gross volume. Volume can also be determined by using Table A.1.E, which is a half cylinder cubic foot volume chart.

2.2.1.5 Defect deduction rules

A defect is anything that causes a loss in volume of lumber. Stain is not considered a defect. The following is a list of deductible defects:

Two-segment log

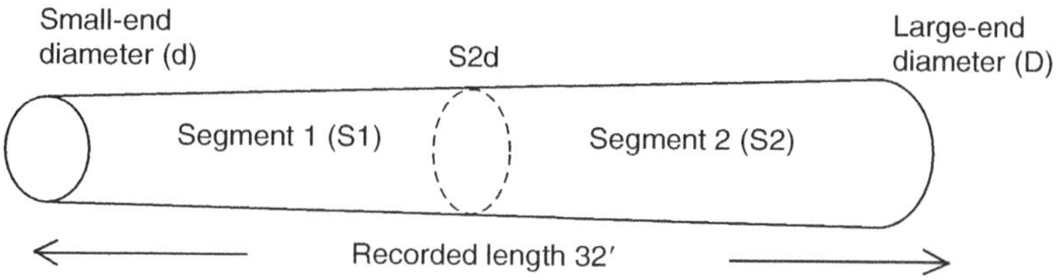

d = 15″

D = 18″

S2d (and S1D) are interpolated as 17

Segment 1 calculation: $(15^2 + 17^2) \times 16 \times 0.002727 =$ **22.4 ft³** (0.634 m³)

Segment 2 calculation: $(17^2 + 18^2) \times 16 \times 0.002727 =$ **26.7 ft³** (0.756 m³)

Total log volume = **49.1** ft³ (1.390 m³)

Fig. 2.5. USFS National Cubic Log Scale gross volume determination.

Bark seams	Crook (pistol grip butt)	Lightning scars	Conk rot
Breaks and splits	Crotch (forks)	Massed pitch	Heart rot
Burls	Fire scars	Pitch seams	Stump rot
Cat faces (basal scars)	Large knots (> 4.5″)	Pitch spangles	Sap rot
Heart checks	Knot clusters	Pitch rings	Twisted grain
Weather checks	Rotten knots	Stump pull	Worms and bug holes
Frost cracks	Metal	Shake	

There are four major methods for deducting for defects:

1. Squared area deduction.
2. Percentage deduction.
3. Length deduction.
4. Diameter deduction.

SQUARED AREA DEDUCTION. This is used primarily for interior defects that lend themselves to being partitioned into a square or rectangular area. The volume of a defect is calculated by the formula:

$$\text{defect deduction} = W'' \times H'' \times L' \div 144.$$

PERCENTAGE DEDUCTION. This is used for defects that go from the perimeter to the heart area of the log and can best be reckoned with by enclosing the defect in a 'pie-shaped' sector with a fractional representation for the length affected. This fraction is then applied to the length of the log segment, e.g. if a log segment has a scaling length of 16′, with one half of 6′ (½ of 6 = 3) of the log having a defect requiring a deduction, the net scale of the log is equal to that of a 13′-long log.

LENGTH DEDUCTION. This is normally used for defects that cause a loss of volume for all or some of the length affected. This deduction rule is commonly used in combination with the pie-cut method, e.g. ½ of an 8′ length reduction. The defect equals the difference between the volume as calculated from the gross length and the volume as calculated from the net length. Sweep (a gradual bow in the log) and crook (a sharp change of direction) deductions are taken by projecting a theoretically straight log through the straightest portion of the log segment and deducting for the volume that falls outside of the actual log segment. For example, a 16′ log with a crook affecting 6′ in length, of which ⅔ of the theoretical straight log falls outside of the actual log, would result in a 4′ length reduction: $4 \div 16 = 0.25$, so the defect volume is equal to 25% of the gross volume.

DIAMETER DEDUCTION. This is used for defects that occur in the perimeter of the log, such as rotten sapwood, surface checks and twist (excessive slope of grain). This deduction is made by establishing net diameters under the defective portion of the log. The defect deduction is the difference between the volume as calculated from the gross diameters and the volume as calculated from the net diameters.

2.2.1.6 Net volume and merchantability

Net volume is the gross volume minus defect. There is a point where a log is no longer considered merchantable and will be culled (giving it no net volume). This merchantability factor is designed to eliminate payment and/or mandatory utilization of logs that are not financially viable. The factor is based on the ⅓ sound rule from the Scribner bf rule. As the Scribner merchantability factor will be discussed in Section 2.3.1, we will not go into the particulars at this point. A good 'rule of thumb' is that a log will be culled when the net volume is less than 40% of the gross volume, but this is not always the case as the merchantability factor is actually based on net volume Scribner (generally ⅓ minimum net volume). A cull sawlog can be reclassified as a merchantable pulp log or specialty product log, depending on the purchase agreement. Figure 2.28 on page 54 shows a log that has a gross volume of 39.5 ft^3 and a rotten core that creates a defect deduction of 16.8 ft^3 (42.5% defect). As this log has less than 33⅓% net scale Scribner, it is a cull in cubic despite the percentage loss in cubic volume.

2.2.1.7 Pulp log scale

The USFS National Cubic Log Scale handles pulp log scaling in much the same way as listed above, with the following exceptions:

1. Log lengths used are actual (no trim allowance is given), with lengths rounded and recorded to the nearest foot.
2. Void, soft-rot and char are the only defects that can be deducted.
3. Round-shaped defects (such as centre rot) are deducted with a circular area formula rather than with squared defect (thus reducing the defective area by 21.46% vs. the same sawlog deduction).

2.2.2 BC Firmwood Scale (Canada)

Cubic log scale received its official start in British Columbia in 1945 via an amendment to the Forest Act, which introduced cubic scale as an improvement for usage and depletion of the forest inventory. As introduced, the scaling method, which at that time was called the 'BC Lumber Cubic Scale', was optional, and deducted for all defects that caused a loss of lumber volume (much like the current US National Cubic Log Scale). Volumes were presented in cubic feet or cunits (units of 100 ft^3). In the 1960s, the use of lumber recovery as a guide for defect deductions was dropped, and a system that only deducted for void, soft-rot and char was put in place, which was called the BC Firmwood Scale. In 1972, the government stopped authorizing the use of both the bf scaling rule and the 'BC Lumber Cubic Scale'. In 1978, the BC Firmwood Scale was made the only authorized log scale in British Columbia, and was converted to the metric system. The Yukon Territory uses essentially the same procedures as BC Firmwood. Listed below are the basic rules of the BC Firmwood Scale taken from the British Columbia *Scaling Manual* (British Columbia Ministry of Forestry, 1999).

2.2.2.1 Diameter measurements

1. Diameters are measured in units called 'rads', with each rad being 2 cm in length, e.g. a diameter of 20 rads is the equivalent of 40 cm, or put another way, having a radius of 20 cm (thus the name rad).
2. Measure through the true centre of the log (ignore pith).
3. Take two measurements at right angles to each other, through the narrow axis and the wide axis.
4. Disregard bumps and depressions.
5. Forked tops or swelled ends should be measured behind swell with line of taper projected toward the top and the appropriate reduction made to the diameter.
6. For butt-cut ends, measure diameter above the swelled end by callipering the log at a point just above the butt flare (theoretically inside the bark with a narrow and right angle measurement), and projecting the normal line of taper from the calliper point to the butt-end.
7. Measure diameters to the nearest rad diameter class, e.g. the diameter class of 18 rads includes all measurements from 35.0 to 36.99 cm; the diameter class of 19 rads includes all measurements from 37.0 to 38.99 cm.
8. Determine the average of both measurements; if the average is on the 0.5 rad, round to the closest even number, e.g. the average of 21 rads and 24 rads is 22.5, which rounds down to 22; the average of 21 rads and 22 rads is 21.5, which rounds up to 22.

Figure 2.6 is an example of diameter measurements using the BC Firmwood Scale.

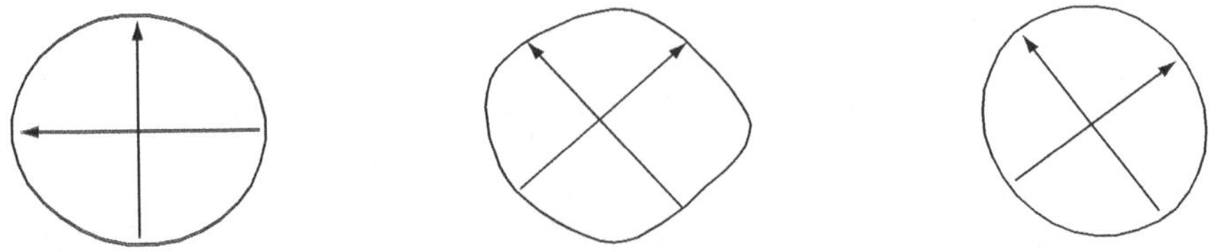

Narrow measurement = 22.5 cm
Right angle measurement = 24.4 cm
• 22.5 is equal to 11 rads
• 24.4 is equal to 12 rads
(11 + 12) ÷ 2 = 11.5; round to **12 rads**

Narrow measurement = 38.6 cm
Right angle measurement = 42.5 cm
• 38.6 is equal to 19 rads
• 42.5 is equal to 21 rads
(19 + 21) ÷ 2 = 20; diameter is **20 rads**

Narrow measurement = 100.4 cm
Right angle measurement = 113.6 cm
• 100.4 is equal to 50 rads
• 113.6 is equal to 57 rads
(50 + 57) ÷ 2 = 53.5; round to **54 rads**

Fig. 2.6. BC Firmwood scale diameter measurement methodology.

2.2.2.2 Length measurements

1. Lengths are measured in metres to the nearest tenth of a metre, e.g. 12.34 is recorded as 12.3 and 12.35 is recorded as 12.4.
2. Actual length is the scaling length (no unmeasured trim allowance is given).
3. Lengths are measured through the geometric centre of the log.
4. Logs that have a large-end diameter which is 1.5 times or larger than the small-end diameter may be required to be cut into two separate logs in order to improve the accuracy of the scale (the Smalian formula becomes less accurate as taper increases).

2.2.2.3 Taper distribution

1. Taper distribution is considered linear unless the large-end diameter is greater than 50% larger than the small-end diameter, e.g. 12 rads d and 20 rads D.
2. In the event that the large-end diameter oversteps the small-end diameter by 50% or more, the recommended procedure is to cut the logs into mill-preferred lengths and scale the individual segments (this procedure is generally implemented on sample scaled logs).

2.2.2.4 Gross volume determination

The BC Firmwood Scale uses the Smalian formula to determine log volume. To determine the volume of a log apply the formula:

$$m^3 = (\text{small-end diameter in rads}^2 + \text{large-end diameter in rads}^2) \times L \times 0.00015708\text{; round to three decimal places}$$

Figure 2.7 is an example of gross volume determination using the BC Firmwood Scale. Volume can also be determined by using Table A.1.F, which is a half cylinder volume chart.

2.2.2.5 Defect deduction rules

The BC Firmwood Scale only deducts for void, soft-rot and char (fibre loss defects). It should be noted, however, that the grading rules differ-

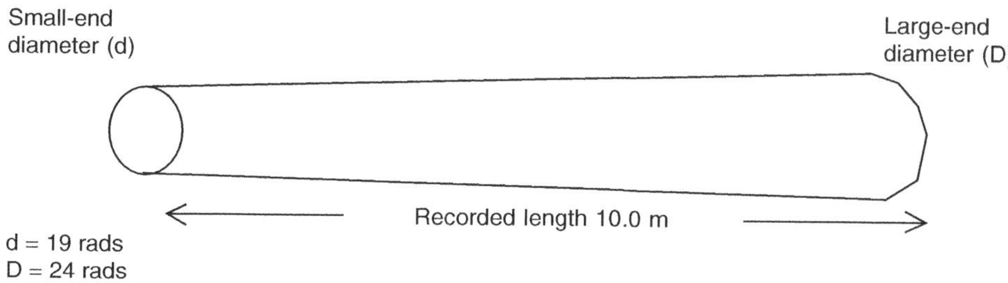

$$(19^2 + 24^2) \times 10 \times 0.00015708 = \mathbf{1.472\ m^3}\ (52.0\ ft^3)$$

Fig. 2.7 BC Firmwood Scale gross volume determination.

entiate between logs based on their suitability to make primary wood products such as lumber and plywood, and degree of lumber (or veneer) volume reducing defects such as sweep, crook, shake, check, spangle and twist are included in the grading criteria. The grading rules also handle void, soft-rot and char differently from the scaling procedures, e.g. the loss is considered to be much greater, as waste allowance is added to the defect size, and the physical limitations of producing primary products such as lumber and veneer are accounted for. In effect, the log scale only accounts for firm wood fibre, but the logs are virtually rescaled with end use in mind as part of the 'log-grading' procedures.

In general, depending on the actual numeric grade, the sawlog grades require a log to be > 50% or > 75% available for the manufacture of lumber, of which a minimum of 50% or more of the lumber will be merchantable ('strong general purpose lumber'). Peeler grades require 80% available volume for the manufacture of rotary peeled veneer. Utility grade requires > 50%, > 66⅔%, or > 75% (again depending on numeric grade) of the log volume available for the manufacture of lumber, of which > 35% is merchantable. The lumber reject grade (chip log) is a log that will not make a utility grade, but is better than a firmwood reject. A firmwood reject is a log where either the net volume is less than 50% of the gross volume due to rot, or where there is less than 1.2 m of merchantable length due to rot.

There are three major methods for deducting for defects:

1. Squared area deduction.
2. Length deduction.
3. Diameter deduction.

SQUARED AREA DEDUCTION. This is used primarily for interior defects that lend themselves to being partitioned into an area shaped like a square, rectangle, triangle, circle, semicircle, ring or portion of a ring. The defect volume is calculated by determining the cubic volume occupied by the defective area using the appropriate mathematical formula for the geometrical shape of the defective area.

LENGTH DEDUCTION. This is normally used for defects that can easily be enclosed in a fractional area for the length affected, e.g. ¼ of a 1.2 m length cut. The defect equals the difference between the volume as calculated from the gross length and the volume as calculated from the net length. In this case, ¼ of 1.2 is the same as a 0.3 m length reduction: assuming the log is 5.0 m in length, $0.3 \div 5 = 0.06$, so the defect volume is equal to 6.0% of the gross volume.

DIAMETER DEDUCTION. This is used for defects that occur in the perimeter of the log. This deduction is accomplished by establishing net diameters (under the defective portion of the log) and applying this reduction for the whole, or a percentage, of the log as needed. The defect deduction is the difference between the volume as calculated from the gross diameters and the volume as calculated from the net diameters.

2.2.2.6 Net volume and merchantability

Net volume is the gross volume minus defect. There is technically no minimum merchantability for logs in regard to volume, excepting that volume is not calculated for stem volume with a diameter of less than 10 cm. It should also be noted that logs containing less than 50% firm wood fibre are graded as grade ‘Z’, which means that there is no stumpage cost charged (even though there will be a volume reported).

2.2.2.7 Pulp log scale

Scaling procedures are the same for pulp and sawlogs as there is no differentiation between end use in the scaling procedures.

2.2.3 Alberta Cubic Metre Scale (Canada)

Apparently, the Alberta Cubic Scale was initiated in 1962 to facilitate the measurement of volume in a cubic manner. In 1979, it was modified to the Alberta Cubic Metre Scale in order to measure logs in cubic metres. In 1992, the log scale underwent some major revisions including the addition of taper equations to calculate taper and large-end diameters, and was adopted as the standard for Alberta. In 2000, a directive made use of the taper formulas option, and allowed the use of diameters taken at both ends of the log. This log scale is oriented toward end products as it allows unmeasured trim on log lengths and deducts for many firmwood defects that reduce the recovery of lumber and veneer.

As it would not be accurate to compare the Alberta Cubic Metre scaling procedures with other log scales utilizing taper equations specific to the timber types in Alberta (while the other log scales use the actual taper), comparisons and examples are via the two-diameter methodology. Besides the procedures outlined below, logs can be scaled in what is called tree length scaling, which is briefly discussed in Section 2.6.2.1.

The following information is based on the *Alberta Scaling Manual* (Alberta Land and Forest Service, 1992) and the 2004 Smalian Scale supplement (Alberta Scaling Manual, November, 2004).

2.2.3.1 Diameter measurements

1. Measure diameters inside of bark and cambium layer.
2. Measure through the true centre of the log (ignore pith).
3. Take two measurements at right angles to each other, through the narrow axis and the wide axis of the log end.
4. Disregard bumps and depressions.
5. Forked tops or swelled ends should be measured in a manner that reduces the diameter to what it would normally be without the unusual swell.
6. Butt-cut diameters are measured by disregarding the butt flare, projecting the normal line of taper and measuring at the representative coordinates.
7. Diameters are in 2-cm diameter classes; the even numbers are the midpoint of the class and the odd numbers are the boundaries, e.g. the 20-cm class includes all measurements from 19.0 cm to 20.99 cm.
8. Determine the average of both measurements; if the average is on the odd centimetre, round to the nearest even diameter divisible by four, e.g. $(18 + 20) \div 2 = 19$, round to 20; $(16 + 18) \div 2 = 17$, round to 16; $(16 + 20) \div 2 = 18$, diameter remains 18.

Figure 2.8 shows examples of diameter measurements using the Alberta Cubic Metre Scale.

2.2.3.2 Length measurements

1. Lengths are recorded in 0.2-m length classes (e.g. 4.6, 4.8, 5.0).
2. The recorded length is the actual length minus trim allowance, which is: 0.05 m for lengths 2.4 to 3.59 m; 0.10 m for lengths 3.60 to 4.79 m; and 0.15 m for lengths 4.8 m and over.

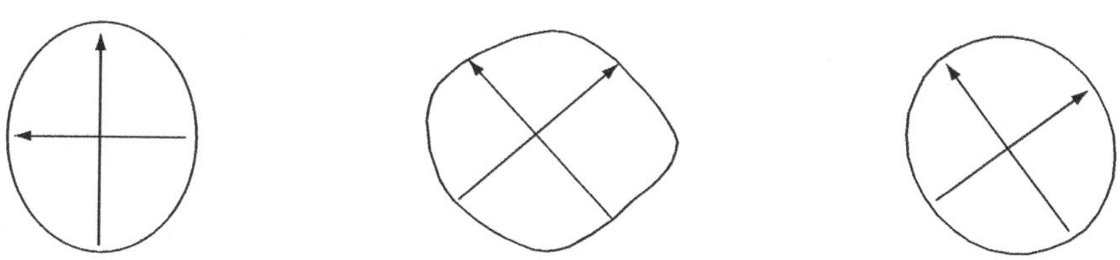

Narrow measurement = 22.5 cm
Right angle measurement = 24.4 cm
• 22.5 is equal to the 22 cm class
• 24.4 is equal to the 24 cm class
(22 + 24) ÷ 2 = 23; rounds to nearest even number divisible by 4, which is **24 cm**

Narrow measurement = 38.6 cm
Right angle measurement = 42.5 cm
• 38.6 is equal to the 38 cm class
• 42.5 is equal to the 42 cm class
(38 + 42) ÷ 2 = 40; diameter is **40 cm**

Narrow measurement = 100.4 cm
Right angle measurement = 113.6 cm
• 100.4 is equal to the 100 cm class
• 113.6 is equal to the 114 cm class
(100 + 114) ÷ 2 = 107; rounds to nearest number divisible by 4, which is **108 cm**

Fig. 2.8. Alberta Cubic Metre Scale diameter measurement methodology.

3. If the actual length minus trim allowance falls exactly on, or just under, an odd length class (e.g. 4.7, 4.9, 5.1), round down to the next 0.2-m length class; if it falls over the odd length class, e.g. 4.71, round up to 4.8. For example, if a log has a measured length of 5.25 m, 5.25 − 0.15 m = 5.1 m, which would round down to a recorded length of 5.0 m. If the log has a measured length of 5.26 m, the recorded length would be 5.2 m.
4. The maximum recorded segment length is 7.2 m plus trim; logs longer than 7.2 m are divided into two or more segments as needed (either theoretically by callipering or by physically cutting the log) and scaled as individual log segments.

2.2.3.3 Taper distribution
There are now two approved methods for determining log taper in Alberta:

1. The large-end diameter and thus taper of a log is determined by the following regression formula (with no differentiation between butt-cut and second-cut logs):

Butt or large-end diameter = [intercept + (A × small-end diameter) + (B × small-end diameter^2)] − [small-end diameter × (length ÷ 2.4)] + small-end diameter

	Coniferous	Broadleaved
Intercept	5.18862610	5.43503728
Factor A	0.73061559	0.67028964
Factor B	0.006046189	0.007636883

2. Measuring the large-end diameters utilizing the procedures outlined in section 2.2.3.1 and thus the actual taper can be determined.

2.2.3.4 Gross volume determination
The Alberta Cubic Metre Scale uses the Smalian formula to determine log volume with either one diameter (using the regression formula above) or with the use of the actual large-end diameter. The formula for determining log volume is:

$$m^3 = (\text{small-end diameter}^2 + \text{large-end diameter}^2) \times L \times 0.00003927,$$
round to three decimal places

Figure 2.9 is an example of volume calculation using the Alberta Cubic Metre Scale. Volume can also be determined by using Table A.1.F for use with the small-end and large-end diameters, or Table A.1.G for use with just the small-end diameter. The total volume for the logs in Fig. 2.9 would be 1.484 m^3 using the 'one diameter' method (log 1 has 0.660 m^3; log 2 has 0.824 m^3). Note that timber purchasers are required to stay with one method through the term of a timber sale, as the volumes could be seriously biased in either direction by using the 'one diameter' method on

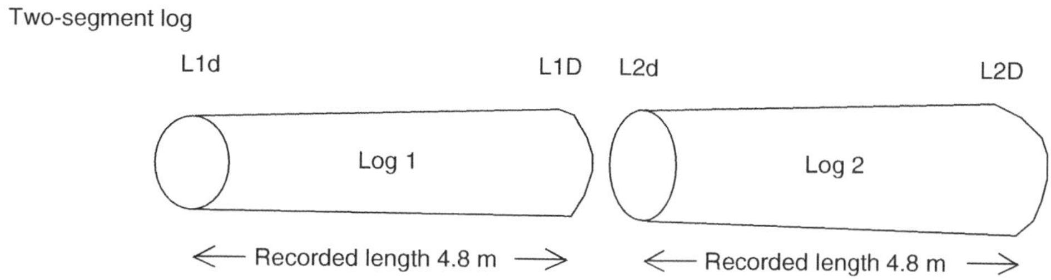

Log 1: d = 38 cm; D = 42 cm
Log 2: d = 42 cm; D = 48 cm

Log 1 calculation: $(38^2 + 42^2) \times 4.8 \times 0.00003927 = 0.605$ m^3 (21.4 ft^3)
Log 2 calculation: $(42^2 + 48^2) \times 4.8 \times 0.00003927 = 0.767$ m^3 (27.1 ft^3)

Total log volume = **1.372 m^3** (48.5 ft^3)

Fig. 2.9. Alberta Cubic Metre Scale gross volume determination.

high- or low-tapered logs, while using the 'two diameter' method (which measures actual taper) on logs with the opposite taper characteristics.

2.2.3.5 Defect deduction rules

With the exception of mechanical damage (from logging or mill yard handling), deductions are taken for defects that cause a loss of wood volume available for manufacturing. Firmwood stain is not considered a defect. The major defects that are deducted for are: heart rot, butt rot, sap rot, cat face, crook, sweep, shake, check, cracks, fork, crotch.

There are five major methods for deducting defects:

1. Squared or circular area deduction.
2. Percentage deduction.
3. Pie cut deduction (also known as reducing a fraction).
4. Length deduction.
5. Diameter deduction.

SQUARED OR CIRCULAR AREA DEDUCTION. This is used primarily for interior defects that lend themselves to being partitioned into an area shaped like a square, rectangle or circle. The defect volume is calculated by determining the cubic volume occupied by the defective area using the appropriate mathematical formula for the particular shape of the defective area. It is important to note that for interior rot defects, 2 cm is added to the diameter if circular in shape; if the defect is square or rectangular, add 2 cm to both the height and width. This additional waste allowance of 2 cm is not added to firmwood-type defects such as checks or shake.

PERCENTAGE DEDUCTION. This is commonly used to deduct for sweep and crook. Sweep and crook deductions are only considered for their effects within a 2.4-m-long section of a log segment (in other words the rule allows theoretical cutting of short logs 2.4 m long to eliminate crook or sweep). Sweep and crook each have a respective formula for calculating the percentage of volume loss, which are given below:

$$\%\ \text{loss from sweep} = (\text{maximum deflection} - 2) \div \text{top diameter} \times 100$$
$$\%\ \text{loss from crook} = (\text{crook deflection} \div \text{diameter of end with crook}) \times (\text{length affected} \div \text{log length}) \times 100$$

PIE CUT DEDUCTION (ALSO KNOWN AS REDUCING A FRACTION). This is used for defects that go from the perimeter to the heart area of the log, and can best be reckoned with by enclosing the defect in a 'pie-shaped' sector with a fractional representation for the length affected. For example, if a defect affects ¼ of the log for half the length of the segment, 12.5% of the gross volume is deducted.

LENGTH DEDUCTION. This is used for defects that have a basal area, which exceeds 50% of the basal area of the log end. The defect equals the difference between the volume as calculated from the gross length and the volume as calculated from the net length.

DIAMETER DEDUCTION. This is used for defects that occur in the perimeter of the log, such as rotten sapwood, surface checks and twisted grain. This deduction is made by establishing net diameters under the defective portion of the log. The defect deduction is the difference between the volume as calculated from the gross diameters and the volume as calculated from the net diameters.

2.2.3.6 Net volume and merchantability

Net volume is the gross volume minus defect. Any log that has a net volume less than 33⅓% of the gross volume, will be culled (meaning it will have a net volume of zero).

2.2.3.7 Pulp log scale

There is no distinction between pulp and sawlog scaling methods.

2.2.4 The Ontario Cubic Method (Canada)

The Ontario Cubic Method is used primarily for the measure and reporting of roundwood harvested from 'Crown' forests (publicly owned) within the province of Ontario. This log scale is not oriented toward end-product recovery; it does, however, allow deductions for some firmwood defects that reduce the recovery of lumber and veneer (namely heart check and shake). Deductions for crook, sweep, seams, mechanical damage or stain is not permitted, nor is there any unmeasured length allowed for trim. Like the BC Firmwood Scale, firmwood defects (like sweep and crook) are accounted for in the grading rules. Unlike the US National,

BC Firmwood or Alberta Cubic log scales, the Ontario Cubic Method uses the Huber formula for scaling most sawlogs (Smalian is used for the sample logs in tree length butt scale). The following procedures are taken from Ontario Ministry of Natural Resources (2000). There are three variations of the log scale that can be used, depending on the application (see Section 2.4.1 for information on Ontario weight scale):

1. Cubed fixed length (used on all conifers, poplar and white birch):
- Timber up to 5.7 m in length can be (optionally) scaled from one end of the log only when there is a statistically even distribution of large- and small-ends on the same side of the load.
- Timber 5.9 m long and longer has to be measured and scaled from both ends of the log.

2. Tree length butt measure is based on an average volume per stem and can be used on all conifers, poplar and white birch that are cut into tree length stems (see Section 2.6.2.1 for further explanation of the concept).

3. Cube grade (used on all commercial hardwoods except poplar, and on white birch).

The following rules are based on scaling logs as individual units.

2.2.4.1 Diameters

1. Measure diameters inside of bark and cambium layer.

2. Measure through the true centre of the log (ignore pith).

3. Take two measurements at right angles to each other, through the narrow axis and the wide axis of the log end.

4. Disregard bumps and depressions.

5. Forked tops or swelled ends should be measured in a manner that reduces the diameter to what it would normally be without the unusual swell.

6. Butt-cut diameters are measured by disregarding the butt flare, projecting the normal line of taper and measuring at the representative coordinates.

7. Diameters are in 2-cm diameter classes; the even numbers are the midpoint of the class and the odd numbers are the boundaries, e.g. the 20-cm class includes all measurements from 19.1 cm to 21 cm.

8. Determine the average of both measurements; if the average is on the odd centimetre, round to the nearest even diameter divisible by four, e.g. $(18 + 20) \div 2 = 19$, round to 20; $(16 + 18) \div 2 = 17$, round to 16; $(16 + 20) \div 2 = 18$, diameter remains 18.

9. Determine the average of the small-end and large-end diameter, and again, if the average is on the odd centimetre, round to the nearest even diameter divisible by four, e.g. $(38 + 44) \div 2 = 41$, round to 40; $(16 + 30) \div 2 = 23$, round to 24; $(76 + 88) \div 2 = 82$, diameter remains 82.

Figure 2.10 shows some examples of diameter measurements using the Ontario Cubic Method.

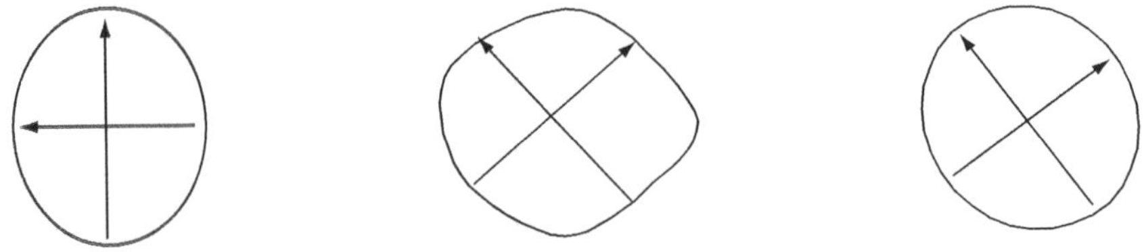

Narrow measurement = 22.5 cm
Right angle measurement = 24.4 cm
• 22.5 rounds to 22 cm
• 24.4 rounds to 24 cm
(22 + 24) ÷ 2 = 23; rounds to nearest even number divisible by 4, which is **24 cm**

Narrow measurement = 38.6 cm
Right angle measurement = 42.5 cm
• 38.6 rounds to 38 cm
• 42.5 rounds to 42 cm
(38 + 42) ÷ 2 = 40; diameter is **40 cm**

Narrow measurement = 100.4 cm
Right angle measurement = 113.6 cm
• 100.4 rounds to 100 cm
• 113.6 rounds to 114 cm
(100 + 114) ÷ 2 = 107; rounds to nearest number divisible by 4, which is **108 cm**

Fig. 2.10. Ontario Cubic Method diameter measurement methodology.

2.2.4.2 Length measurements

1. Measure short side to short side.
2. Lengths are recorded in odd, 0.2-m length classes with the class boundary being on the even tenths of a metre and the recorded length being on the odd tenths of a metre, e.g. a log that is recorded as being 5.7 m long includes lengths from 5.61 to 5.8 m in length.
3. No trim allowance is given.

2.2.4.3 Taper distribution

For the purposes of calculating volume, the log is considered a cylinder, having a diameter equal to the mean diameter of the small-end and large-end diameters of the log. In the case of 'one-end scaling', the diameter used is based on either the small-end or the large-end (again with the assumption that there will be an even distribution of small-end and large-end diameters on the same end of the load). Note that when using tree length butt measure, which uses the Smalian formula, taper is considered linear.

2.2.4.4 Gross volume determination

The Ontario Cubic Method uses the Huber formula to determine log volume. To determine the volume of a log segment, apply the formula:

$$\text{mean diameter}^2 \times L \times 0.00007854 = \text{log volume;}$$
round to three decimal places

Note: the mean diameter is replaced with the end diameter if using the 'one-end scaling method'.

Figure 2.11 is an example of using a formula to calculate the volume of a log with the Ontario Cubic Method. Volume can also be determined by using Table A.1.H, which is a full segment volume chart.

2.2.4.5 Defect deductions

As mentioned above, the Ontario Cubic Method does not deduct for crook, sweep, seams, mechanical damage or stain. Deductions are primarily made for void, soft-rot, and fibre separation defects such as checks and

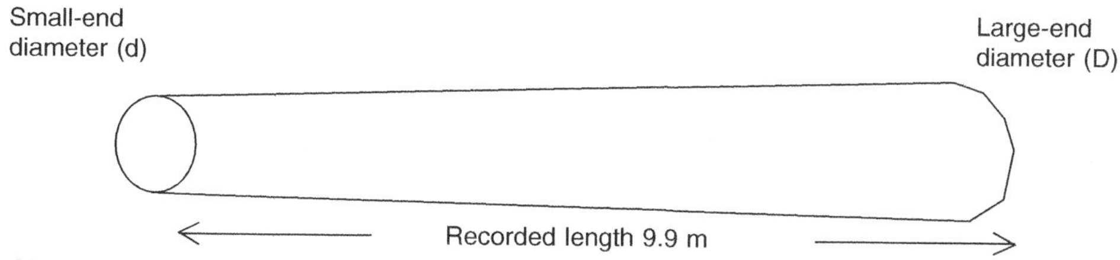

d = 38 cm

D = 48 cm (38 + 48) ÷ 2 = 43; 43 is rounded to nearest even number divisible by four, 44 cm

$$44^2 \times 9.9 \times 0.00007854 = \mathbf{1.505\ m^3}\ (53.1\ ft^3)$$

Fig. 2.11. Ontario Cubic Method gross volume determination.

shake (fibre separation defects only apply to white pine, red pine and eastern hemlock).

There are two methods for deducting for defects:

1. Diameter of defect deduction.
2. Diameter deduction.

DIAMETER OF DEFECT DEDUCTION. This is used primarily for interior defects that lend themselves to being partitioned into an area shaped like a square, rectangle or circle. The defect volume is calculated by determining the cubic volume of the defective area by converting all areas of defect into a diameter (circular area). This is done in two ways:

- For defects where the narrow dimension is greater than 50% of the wide dimension, e.g. 10 × 18, add the two numbers and divide by two to obtain the mean (in this case 14); if the mean is an odd number, raise or lower the diameter to the closest even centimetre class that is divisible by 4.
- If the defect is long and narrow (the narrow dimension is less than 50% the wide dimension), determine the square area of the defect, e.g. $8 \times 30 = 240$, find the square root of the number ($\sqrt{240} = 15.49$), and round to the nearest even number (in this case 16).

If the log is being scaled by the 'one end' method, the diameter of the defect is used for the entire length of the log (assuming that there will be a statistically even distribution of defective ends on a load). If, however, the log is 5.7 m or less, but both ends are measured, the rot is averaged for the length, e.g. the small-end diameter of the rot is 12 cm and the large-end diameter of rot is 20 cm, the rot will be considered to be 16 cm in diameter. If the same log had 16 cm of rot in the large-end but had no rot in the top, the rot would still be considered to run the full length of the log, but the diameter of the rot would be averaged with zero, e.g. 16 cm of rot in the butt and 0 cm in the top, the rot is considered to be 8 cm in diameter.

If the log has a recorded length of 5.9 m or greater, the defect is assumed to go into the log 2.5 m, unless it shows on the other end of the log, in which case the defect diameter is averaged over the length of the log. The formula for determining the defect volume is:

$$(\text{diameter of the defect}^2 \times \text{log length}^*) \times 0.00007854$$

*2.5 m if the log is 5.9 m in length or longer, and the rot only shows in one end.

DIAMETER DEDUCTIONS. This is used for rotten sapwood. It is accomplished by establishing a net diameter (under the defective portion of the log). The defect deduction is the difference between the volume as calculated from the gross diameters and the volume as calculated from the net diameters. It should be noted that percentage deductions and diameter of defect deductions are commonly reflected as diameter deductions via a chart, which equates percentage loss or squared area loss to a diameter deduction that matches closely the respective volume loss.

2.2.4.6 Net volume and merchantability

Net volume is the gross volume minus defect. Any log that has a net volume of less than 50% of the gross volume will be culled (meaning it will have a net volume of zero).

2.2.4.7 Pulp log scale

Scaling procedures are the same for pulp and sawlogs excepting that there is no deduction for heart checks or shake (firmwood defects).

2.2.5 Swedish National Board of Forestry Log Scale (Sweden)

The first Swedish timber measurement law was introduced in 1935, and the present version of the law was implemented in 1966. The present law states that ‘the measurement of timber in respect to coniferous sawlogs and pulpwood, which is intended to form the basis for calculating the value of the timber, has to be carried out according to regulations issued by the National Board of Forestry.’ The latest updates to the regulations were issued in 1999, *VMR 1-99,* and are based on recommendations from the Swedish Timber Measurement Council. The Timber Measurement Council is a non-governmental organization where buyers and sellers of timber in the Swedish market are equally represented. The current revision of the scaling procedures has been in effect since 1997. This scaling method has some end-product orientation, but primarily uses grading functions to indicate primary product recovery and quality.

The Swedish Log Scale utilizes only the small-end diameter and makes no allowance for taper when scaling sawlogs. However, when pulp logs are scaled, all of the fibre is accounted for. Most logs are now scaled via scanners (with scalers inputting bark thickness data, defect deductions and log grade). It is important to note that national statistics from Sweden are published in a corrected form, e.g. they apply a correction factor of approximately 1.21 to account for understating the actual sawlog volume. Listed below are the procedures of manually scaling sawlogs in Sweden taken from *VMR 1-99* (Swedish Timber Measurement Council, 1999). Traditionally logs in Sweden are processed into lengths of less than 6 m.

2.2.5.1 Diameter measurements

1. Diameters are taken inside the bark at the small-end of the log.
2. Take two measurements at right angles to each other, over the narrow axis and the wide axis of the log end. If the log is round, one measurement is permitted.
3. Diameter measurements are truncated and recorded in 1-cm diameter classes, e.g. 26.9 cm is rounded down to 26 cm.
4. Average the narrow and right angle measurement; if on the ½ centimetre, round to the nearest even number, e.g. a narrow measurement of 27 cm, and a perpendicular measurement of 28 cm, diameter is 28 cm.
5. Disregard bumps and depressions.
6. Forked tops or swelled ends should be measured in a manner that reduces the diameter to what it would normally be without the unusual swell.

Figure 2.12 shows some examples of diameter measurements using the Swedish National Board of Forestry Log Scale.

2.2.5.2 Length measurements

1. Lengths are recorded in truncated 0.1-m length classes, e.g. a measured length of 5.90–5.99 m is recorded as 5.9 m.
2. Lengths are measured at the shortest distance between the centres of the log ends.
3. Maximum log length is 6.0 m.

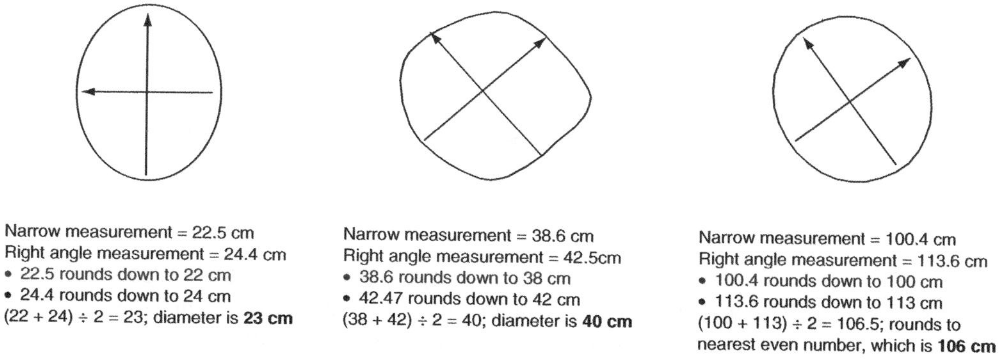

Fig. 2.12. Swedish National Cubic Log Scale diameter measurement methodology.

2.2.5.3 Taper distribution

For sawlogs, the Swedish Log Scale views the log as a cylinder having a diameter the same as the small-end of the log.

2.2.5.4 Gross volume determination

The formula used for sawlogs is based on the small-end only. In order to correct the effects of truncating the diameter and length measurements, volume is calculated for the midpoint of the diameter and length class, therefore 0.5 cm is added to the diameter, and 0.05 m added to the log length. The following formula is used to calculate gross volume for sawlogs:

$$(\text{small-end diameter} + 0.5)^2 \times (L + 0.05) \times 0.00007854 = \text{volume,}$$
round to at least two decimal places

Figure 2.13 shows an example of calculating gross volume using the Swedish National Board of Forestry Log Scale. Table A.1.I can be utilized to determine log volume scaled via the Swedish National Board of Forestry Log Scale.

2.2.5.5 Defect deduction rules

Firmwood stain, heart shake (heart check), twisted grain, wet wood and insect damage are not deducted. Deductions are taken for the following defects that cause a loss of wood volume available for manufacturing: spike knots, aniline wood, massed pitch, forest rot, shake, crookedness, mechanical damage, open scar.

There are two major methods for deducting defects:

1. Length deduction.
2. Diameter deduction.

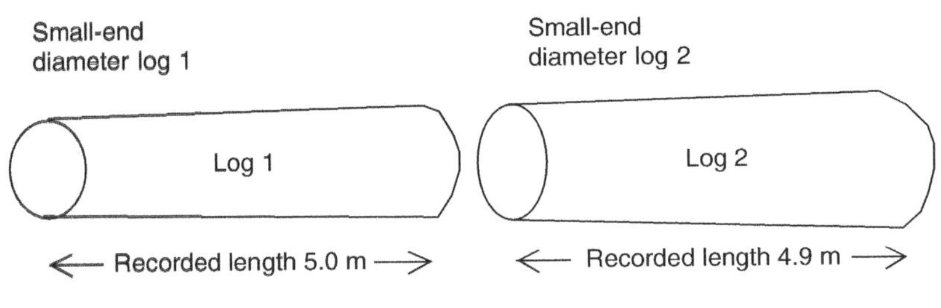

Log 1 d = 38 cm

Log 2 d = 42 cm

Log 1 calculation: $(38 + 0.5)^2 \times (5.0 + 0.05) \times 0.00007854 = 0.588\ m^3$ ($20.8\ ft^3$)

Log 2 calculation: $(42 + 0.5)^2 \times (4.9 + 0.05) \times 0.00007854 = 0.702\ m^3$ ($24.8\ ft^3$)

Total volume for both logs = **1.29 m³** ($45.6\ ft^3$)

Fig. 2.13. Swedish National Cubic Scale gross volume determination.

LENGTH DEDUCTION. This is used for defects that affect the interior of the log. Each log is divided into two semicircles, so it is possible to deduct the entire length, or 50% of the length affected by the defect. Deductions are made in units of a tenth of a metre.

DIAMETER DEDUCTION. This is levied against defects on the perimeter of the log that result in a reduced sawing yield in comparison with logs that do not contain the defect. A log with a defect that causes a diameter deduction exceeding 2 cm on logs with a small-end diameter less than 30 cm should be rejected; if the log is over 30 cm, the diameter deduction would have to exceed 3 cm before the log would be rejected.

2.2.5.6 Net volume and merchantability

Net volume is the gross volume minus defect. Any log that has a net length or diameter of less than the minimum agreed upon will be downgraded to pulp or rejected.

2.2.5.7 Pulp log scale

The Swedish National Board of Forestry Log Scale handles pulp log scaling in much the same way as listed above, with the following exceptions:

- Gross volume is determined using the following formula:

$$\{[(1-\alpha)\times(\text{small-end diameter}+0.5)^2]+[\alpha\times(\text{large-end diameter}+0.5)^2]\times(L+0.05)\}\times 0.00007854=\text{volume, round to at least two decimal places}$$

The constant α is taken from the following:

	Length class		
Small-end diameter	<3.5 m	3.5–4.49 m	>4.5 m
<14 cm	$\alpha = 0.485$	$\alpha = 0.485$	$\alpha = 0.485$
15–24 cm	$\alpha = 0.465$	$\alpha = 0.460$	$\alpha = 0.455$
>25 cm	$\alpha = 0.440$	$\alpha = 0.430$	$\alpha = 0.420$

- Void, soft-rot, char and very severe crook (rendering the log too crooked to be mechanically processed) are the only defects that can be deducted, and must not exceed half the volume of the log.

2.2.6 Russian Government Standard (Russia and members of the former USSR)

The scaling procedures below were taken from RD 13-2-3-97 (Kuritsyn *et al.*, 1997). These regulations were developed by the State Committee of the Russian Federation of Timber, Pulp and Paper and Woodworking Industry (Goskomlesprom of Russia). RD 13-2-3-97 is intended to guide the measurement of logs that are delivered for export.

2.2.6.1 Diameter measurements

1. Measure diameters inside of bark and cambium layer.
2. Measure through the true centre of the log (ignore pith).
3. Take two measurements at right angles to each other, through the narrow axis and the wide axis of the log end (one measurement is allowed for diameters 20 cm or less).
4. Disregard bumps and depressions.
5. Forked tops or swelled ends should be measured in a manner that reduces the diameter to what it would normally be without the unusual swell.
6. If measuring the large-end diameter on butt cut logs, measure at 0.5 m in from the butt end.
7. Diameters are rounded to the nearest centimetre (28.51 rounds to 29 and 28.49 rounds to 28). As an option, diameters over 14 cm can be rounded to nearest 2-cm increments.
8. Determine the average of both measurements, if the average is fractional, round to the nearest even diameter, e.g. $(19 + 20) \div 2 = 19.5$, round to 20; $(16 + 17) \div 2 = 16.5$, round to 16; $(16 + 20) \div 2 = 18$, diameter remains 18.
9. If using the Huber formula and utilizing the average of both end measurements, round to the nearest even number if the average is fractional, e.g. a small-end diameter of 20 and large-end diameter of 23 yield an average of 21.5, which is rounded up to 22.

Figure 2.14 shows some examples of diameter measurements using the Russian Standard.

2.2.6.2 Length measurements

1. Lengths are measured at the shortest distance between the centres of the log ends.
2. Lengths are rounded down to, and recorded in, 0.1-m length classes, e.g. a log that measures 4.71–4.8 m will be recorded as being 4.7 m long.

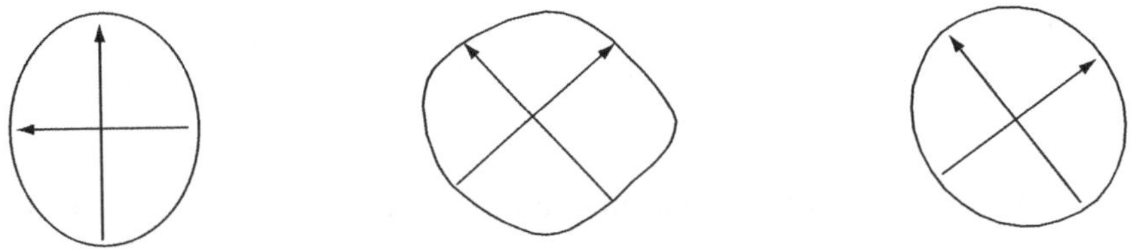

Narrow measurement = 22.45 cm
Right angle measurement = 24.44 cm
- 22.45 rounds to 22 cm
- 24.44 rounds to 24 cm

(22 + 24) ÷ 2 = 23; diameter is **23 cm**

Narrow measurement = 38.6 cm
Right angle measurement = 42.5 cm
- 38.6 rounds to 39 cm
- 42.47 rounds to 42 cm

(39 + 42) ÷ 2 = 40.5; diameter is **40 cm**

Narrow measurement = 100.4 cm
Right angle measurement = 113.6 cm
- 100.4 rounds to 100 cm
- 113.6 rounds to 114 cm

(100 + 114) ÷ 2 = 107; diameter is **107 cm**

Fig. 2.14. Russian Standard diameter measurement methodology.

2.2.6.3 Taper distribution

Taper can be accounted for in a number of different ways:

1. Utilize the actual taper by measuring both ends.
2. Measure the midpoint diameter thus ignoring taper (for use with the Huber formula).
3. Use Table 2.1 with average taper factors by species and region to establish the large-end diameter (for use with the Smalian formula) or midpoint diameter (for use with the Huber formula).

2.2.6.4 Gross volume determination

There are two primary methods (each with several variations) that are approved for determining gross volume. All will give similar answers except for the inherent differences in procedures for determining taper, and the differences that exist between using the Huber formula and the Smalian formula (which can be significant in small highly tapered wood). The basic gross volume determination methods and formulas are as follows:

1. Method of mid-diameter is based on the Huber formula:

   log volume in m^3 = midpoint diameter^2 $\times$ L $\times$ 0.00007854; round to three decimal places

The midpoint diameter used in the formula can be obtained in the following manners:

- Measure the actual midpoint diameter.
- An interpolated midpoint diameter obtained by averaging the small-end and actual large-end diameters.
- A midpoint diameter obtained by applying a 'standard' 1 cm of taper per metre for half of the log length and adding it to the small-end diameter and then applying the appropriate volume correction coefficient for the species and region from Table 2.1. The correction coefficient is normally applied to groupings of logs rather than to individual logs.
- A midpoint diameter obtained by applying the average taper factor for the species and region from Table 2.1 for half of the log length and adding it to the small-end diameter.

2. Smalian formula:

   m^3 = (small-end diameter^2 + large-end diameter^2) $\times$ L $\times$ 0.00003927; round to three decimal places

The large-end diameter used in the formula can be obtained in the following manners:

- An actual measurement on the large-end of the log if the log is a second cut.
- Callipering the log inside the bark at 0.5 m above the butt if the log is a butt-cut.
- Applying the species/region-specific average taper ratios from Table 2.1 to the log, and calculating the large-end diameter from the small-end diameter.

Table 2.1. Russian regional taper and correction factors for bark.

Species	Average taper of logs cm/m	Correction coefficients to calculate the volume		
		for taper	for bark	for taper and bark
		Northern zone of Karelia		
Spruce	1.150	1.059	0.903	0.956
Pine	1.149	1.040	0.932	0.969
Birch	0.872	0.969	0.847	0.821
		Middle zone of Karelia		
Spruce	1.194	1.053	0.908	0.956
Pine	1.142	1.019	0.939	0.957
Birch	0.947	0.977	0.864	0.844
		South-western zone of Karelia		
Spruce	1.043	0.986	0.909	0.896
Pine	1.120	0.995	0.950	0.945
Birch	0.918	0.969	0.865	0.838
		Western zone of Karelia		
Spruce	1.022	0.986	0.907	0.894
Pine	1.098	1.008	0.941	0.949
Birch	0.869	0.964	0.876	0.844
		Southern zone of Karelia		
Spruce	0.968	0.983	0.922	0.906
Pine	1.002	0.992	0.943	0.935
Birch	0.786	0.931	0.897	0.835
		Vologda district		
Spruce	0.947	0.977	0.913	0.891
Pine	1.072	1.018	0.939	0.956
Birch	0.861	0.973	0.879	0.855
Aspen	0.928	0.988	0.892	0.881
		Pechora area of the Komy Republic		
Spruce	1.186	1.051	0.902	0.948
Pine	1.097	1.009	0.922	0.930
Birch	1.086	1.018	0.867	0.882
		Irkutsk district – Ustj-Ilimsk Area		
Pine	1.010	1.010	0.930	0.939
Larch	1.109	1.061	0.867	0.920
Spruce	0.922	0.972	0.941	0.915
Fir	0.804	0.934	0.914	0.854
Cedar	0.941	0.966	0.934	0.902
Birch	0.822	0.951	0.858	0.816
Aspen	0.987	1.023	0.877	0.897
		Irkutsk district – Southern regions		
Pine	0.903	0.963	0.927	0.896
Larch	1.196	1.069	0.880	0.941
Birch	0.857	0.953	0.859	0.819
Aspen	0.812	0.934	0.887	0.824

Source: Kuritsyn *et al*., 1997.

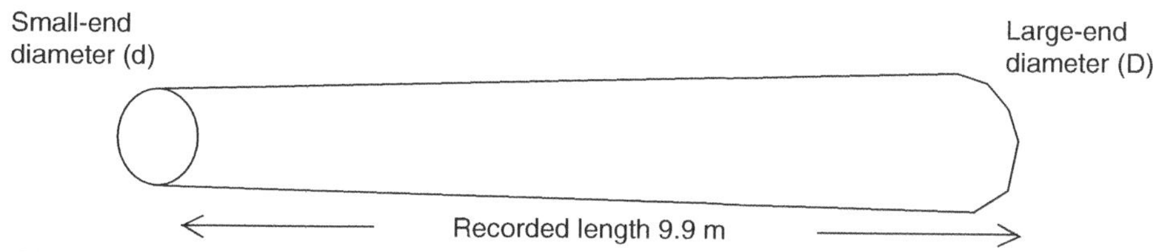

d = 38 cm
D = 46 cm

Example 1

Method of crosscut ends (Smalian formula): $(38^2 + 46^2) \times 9.9 \times 0.00003927 =$ **1.384 m³** (48.9 ft³)

Example 2

Method of mid-diameter using top measurement and standard 1 cm/m taper ratio for half the log length:

$38 + (1 \times L \div 2) = 42.95$ cm; $42.95^2 \times 9.9 \times 0.00007854 =$ **1.434 m³** (50.6 ft³)

Assuming the log is a cedar from the Irkutsk district, apply the 0.966 correction coefficient for taper from Table 2.1: $1.434 \times 0.966 =$ **1.385 m³** (48.9 ft³).

Note: If the top measurement is taken outside bark, the correction coefficient for taper and bark of 0.902 is used.

Fig. 2.15. Russian Standard gross volume determination.

Figure 2.15 shows two methods of calculating gross volume. Volume can also be determined from Table A.1.F if using the Smalian formula, or from Table A.1.H if using the Huber formula.

2.2.6.5 Defect deduction rules

Apparently, defect is accounted for by a grading structure that limits, amongst other things (such as length and diameter), the quantity of different types of defects by log grade. The grading structure varies by product classification and the purchase agreement. Customers can refuse to accept logs that contain defects outside of the agreed specifications.

2.2.6.6 Net volume and merchantability

As stated above, there are apparently no provisions to deduct volume; defect is handled via a grading system and rejecting logs which do not make the lowest grade.

2.2.6.7 Pulp log scale

Scaling procedures are the same for pulp and sawlogs.

2.2.7 Cubage au Réel (France)

This log scale is based on the Huber formula, and is oriented toward primary products in that it allows for trim allowance and makes deductions for defects that can reduce the recovery of primary wood products.

Procedures were taken from *Manuel d'exploitation forestière* (Association pour la rationalisation et la mécanisation de l'exploitation forestière, *et al.* 1994) and *Le Monde Forestier – Le cubage des bois* (Le Monde Forestier, 2004), both from the French Standard, NF B 53-020.

2.2.7.1 Diameter measurements

1. Measure through the geometric centre of the log at the midpoint of the log length.
2. Measurements are normally taken via callipers; or with a tape measure in circumference, outside the bark (with diameter inside the bark calculated using actual double bark thickness or regional averages based on species and size class).
3. If using callipers, take two measurements at right angles to each other through the narrow axis and the wide axis of the log end (one measurement is allowed for diameters under 19 cm).
4. Disregard bumps and depressions.
5. When averaging the two measurements, round to the nearest even centimetre if the average of the two measurements is fractional (if using a circumference tape, round to the nearest centimetre).

Figure 2.16 shows examples of diameter measurements using Cubage au Réel.

2.2.7.2 Length measurements

1. Measure short side to short side.
2. When logs are less than 6.25 m in length, lengths are recorded in truncated 0.25-m increments, e.g. a log that measures 4.98 m in length is recorded as being 4.75 m; for logs longer than 6.25 m, log lengths are recorded in truncated 0.50-m increments, e.g. 12.88 m is recorded as 12.5 m.

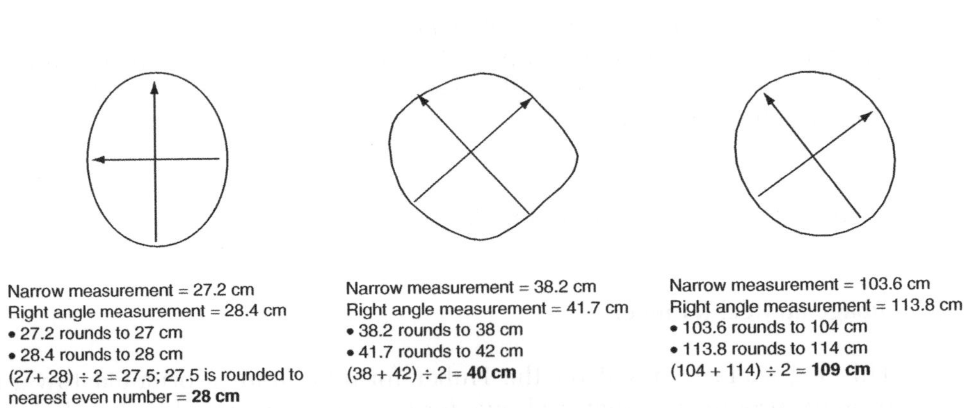

Fig. 2.16. Cubage au Réel diameter measurement methodology.

2.2.7.3 Taper distribution

Since Cubage au Réel uses the Huber formula, taper is not considered. For the purposes of calculating volume, the log is considered to be a cylinder having a diameter that corresponds to the diameter equidistant from both ends of the log.

2.2.7.4 Gross volume determination

Cubage au Réel uses the Huber formula to determine log volume. To determine the volume of a log segment, apply the formula:

$$\text{mid-diameter}^2 \times L \times 0.00007854 = \text{log volume};$$
round to three decimal places (Fig. 2.17)

If using the circumference, convert to diameter by dividing circumference by π (3.1416). Table A.1.H can be used to determine volumes for logs scaled via Cubage au Réel.

2.2.7.5 Defect deduction rules

1. Diameter deduction.
2. Length deduction.

DIAMETER DEDUCTION. This is used for defects that occur in the perimeter of the log, such as rotten sapwood, surface checks and twisted grain. This deduction is made by establishing net diameters under the defective portion of the log. The defect deduction is the difference between the volume as calculated from the gross diameters and the volume as calculated from the net diameters.

LENGTH DEDUCTION. This is normally used for defects that cause a loss of volume exceeding 50% of the volume for the length affected. The defect equals the difference between the volume as calculated from the gross length and the volume as calculated from the net length.

2.2.7.6 Net volume and merchantability

Net volume is the gross volume minus defect. No specific data were found regarding minimum merchantability requirements.

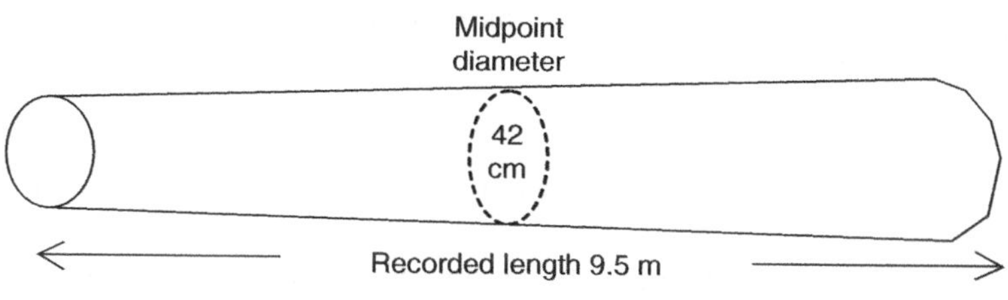

$42^2 \times 9.5 \times 0.00007854 = \mathbf{1.316\ m^3}$ (46.5 ft^3)

Fig. 2.17. Cubage au Réel gross volume determination.

2.2.7.7 *Pulp log scale*

No specific information on pulp log scale was found.

2.2.8 New Zealand 3-D and Mid-girth methods (New Zealand)

These two New Zealand scaling methods, as well as some others, are listed in *Procedures for the Measurement of Roundwood* (Ellis, 1994), which was the source for the procedures below. The 3-D method is unique in that the formula adjusts stem form (neiloid, parabolic, conic) based on taper, and thus allows the scaler to measure the butt-cut end of a log with normal flare.

2.2.8.1 *Diameter measurements*

1. Measure diameters inside of bark and cambium layer.
2. Measure through the true centre of the log (ignore pith).
3. Take two measurements at right angles to each other, through the narrow axis and the wide axis of the log end.
4. Disregard bumps and depressions.
5. Forked tops or swelled ends should be measured in a manner that reduces the diameter to what it would normally be without the unusual swell.
6. The large-end diameter on butt-cut logs is measured to include normal butt flare but exclude nodal swellings that deviate from the normal form of the log.
7. Diameters are rounded to the nearest centimetre (28.51 rounds to 29 and 28.49 rounds to 28).
8. Determine the average of both measurements, if the average is fractional, round to the nearest even diameter, e.g. $(19 + 20) \div 2 = 19.5$, round to 20; $(16 + 17) \div 2 = 16.5$, round to 16; $(16 + 20) \div 2 = 18$, diameter remains 18. Note: if a hand-held data recorder is used, the mean fractional diameter can be used as well.
9. When using the Mid-girth method, the mid-point diameter is either determined by using callipers, or taken outside the bark with a diameter tape. If using circumference, divide by π (3.14159). A double bark thickness is subtracted from the over-bark diameter.

Figure 2.18 shows some examples of diameter measurements using the New Zealand 3-D.

2.2.8.2 *Length measurements*

1. Lengths are measured at the shortest distance between the sawn log ends.
2. Lengths are recorded to the nearest 0.1-m length class with no trim allowance given, e.g. a log that measures 4.75–4.84 m will be recorded as being 4.8 m long. It can also be measured to the nearest centimetre.

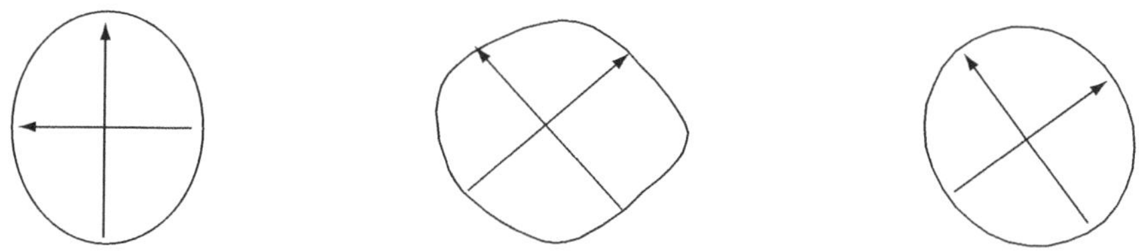

Narrow measurement = 22.45 cm Right angle measurement = 24.44 cm • 22.45 rounds to 22 cm • 24.44 rounds to 24 cm (22 + 24) ÷ 2 = 23; diameter is **23 cm**	Narrow measurement = 38.6 cm Right angle measurement = 42.47 cm • 38.6 rounds to 39 cm • 42.47 rounds to 42 cm (39 + 42) ÷ 2 = 40.5; diameter is **40 cm**	Narrow measurement = 100.4 cm Right angle measurement = 113.6 cm • 100.4 rounds to 100 cm • 113.6 rounds to 114 cm (100 + 114) ÷ 2 = 107; diameter is **107 cm**

Fig. 2.18. New Zealand 3-D diameter measurement methodology.

2.2.8.3 Taper distribution

The New Zealand 3-D method uses the amount of taper to assign log shape, e.g. butt flare is given to a log with a reasonable amount of taper, while a short log with little taper will be given a paraboloidal shape. Mid-girth assumes that a log is a cylinder having the diameter which occurs equidistant from each end of the log.

2.2.8.4 Gross volume determination

Gross volume is calculated with the following formulas:

1. 3-D method: m^3 = (EXP{1.944157 × LN(log length) + (0.029931 × small-end diameter) − 0.038675 + 0.884711 × LN[(large-end diameter − small-end diameter) ÷ log length]} + small-end diameter^2 × 0.07854 ×log length) ÷ 1000

Note: EXP = is the antilog of natural logarithm; LN = is the natural logarithm.

2. mid-girth method: m^3 = mid-diameter^2 × L × 0.00007854

Volume can also be determined via Table A.1.H if using the Mid-girth method, and Table A.1.J if using the 3-D formula (for a limited selection of diameters and lengths).

Figure 2.19 is an example of calculating log volume with the New Zealand 3-D and Mid-girth methods.

2.2.8.5 Defect deduction rules

There are three methods of making deductions:

1. Squared area deduction.
2. Length deduction.
3. Diameter deduction.

These deductions are only utilized when measuring the old growth indigenous logs as New Zealand plantation logs are considered free of defect.

SQUARED AREA DEDUCTION. This is applied by covering the defective area with a square or rectangle for the affected length. This defecting method is used

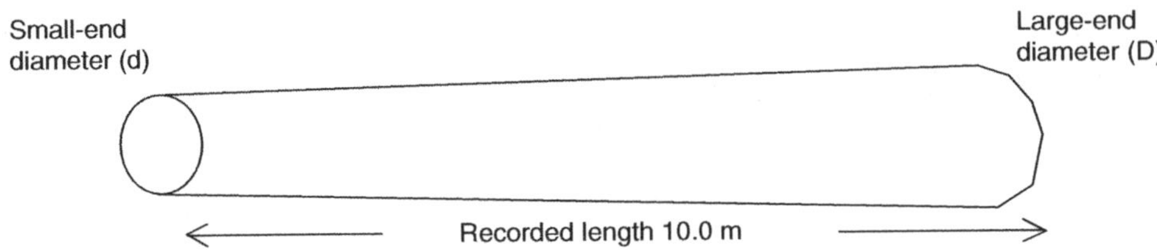

d = 38 cm
D = 50 cm (including butt flare)

3-D method

$\{EXP(1.944157 \times LN(10) + (0.029931 \times 38) \quad 0.038675 + 0.884711 \times LN\,[(50 \quad 38) \div 10]) + 38^2 \times 0.07854 \times 10\} \div 1000 = \mathbf{1.444\ m^3}$ (51.0 ft^3)

Mid-girth method

$42^2 \times 10.0 \times 0.00007854 = \mathbf{1.385\ m^3}$ (48.9 ft^3)

Fig. 2.19. New Zealand 3-D gross volume determination.

primarily for interior defects such as heart rot, shake, etc. The defect volume in $m^3 = W^{cm} \times H^{cm} \times L^{m} \div 10{,}000$.

LENGTH DEDUCTION. This is used for defects such as decayed log end, broken or split ends, etc. All of, or a percentage of, the length affected by the defect can be deducted. The defect volume is the difference between the volume as calculated by the gross length and the volume as calculated from the net length.

DIAMETER DEDUCTION. This is used for surface defects (knots, fluting, flanges, crook and sweep). When these defects occur, the mid-diameter is reduced for either the entire length or a portion of the length. The defect volume is the difference between the volume as calculated from the gross diameter and the volume as calculated from the net diameter.

2.2.8.6 Net volume and merchantability
Any log with more than 50% defect is rejected.

2.2.8.7 Pulp log scale
Scaling procedures are the same for pulp and sawlogs; except if using Mid-girth, only void, soft rot and char are deducted.

2.2.9 Brereton, ATIBT method (Africa, Oceania, South America, Asia, Japan)

Brereton is one of the most common methods for volume determination of tropical hardwood logs. It is used in many regions across the globe, and is generally reflected in cubic metres, but can be reflected in cubic feet, and has even been expressed as a bf volume in its original formulation (derived by multiplying the cubic feet by 12). Bernard Brereton invented this scale with intended use for coastal softwoods in western North America (Brereton, 1940); however, the method found its main following in the tropical regions of the world.

Procedures for this scaling method may vary significantly between one country or region to another. The variance occurs mainly in the rounding rules for diameter(s) and length, trim allowances (if any), the accounting of defects, and the occasional practice of measuring diameters under the sapwood (for species with undesirable sapwood characteristics). The following guidelines are based on the Papua New Guinea scaling handbook (Papua New Guinea Forest Authority, 1996), with procedural differences pointed out for the Japanese (Briggs and Flora, 1991), Philippine (Philippine Department of Environment and Natural Resources, no date) and the ATIBT (International Technical Association of Tropical Timber, 2003) versions of Brereton.

2.2.9.1 Diameter

1. Measure diameters inside of bark through the true centre of the log (ignore pith).

2. Take two measurements on each log end, through the wide axis and at a right angle to the wide axis (Japanese Brereton only; narrow and wide, which may not necessarily be at right angles).

3. Forked tops or swelled ends should be measured in a manner that reduces the diameter to what it would normally be without the unusual swell (this includes buttress and fluting at large-end).

4. Diameters are rounded down to nearest full centimetre, e.g. 28.00–28.99 = 28 cm.

- Philippine Brereton: round to nearest centimetre, e.g. 27.5–28.49 = 28 cm.
- Japanese Brereton: round down to the next even centimetre, e.g. 28.00 – 29.99 = 28 cm.

5. Determine average log diameter by averaging the four diameters (two from each end):

average diameter = (d1 + d2 + D1 + D2) ÷ 4

- Japanese and Philippine Brereton: find average of each log end, round to nearest even centimetre if fractional (only relevant with Philippine Brereton) and then average both of the log ends to obtain an average log diameter:

average diameter = (d + D) ÷ 2

Figure 2.20 shows examples of diameter measurements using Brereton.

2.2.9.2 Length measurements

Lengths are measured at the shortest distance between each end of the log and are recorded to the last full decimetre, e.g. 4.80–4.89 m is recorded as 4.8 m.

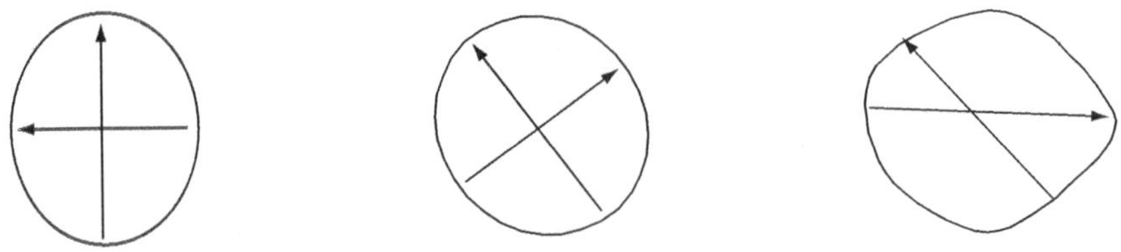

PNG Brereton, ATIBT
Narrow measurement = 22.6 cm
Wide measurement = 24.4cm
- 22.6 is rounded to 22 cm
- 24.4 is rounded to 24 cm

Note: end-diameters not averaged

Philippine Brereton
Narrow measurement = 12.7 cm
Wide measurement = 14.2 cm
- 12.7 is rounded to 13 cm
- 14.2 is rounded to 14 cm

(13 + 14) ÷ 2 = 13.5; round to **14 cm**

Japanese Brereton
Narrow measurement = 100.4 cm
Wide measurement = 113.6 cm
- 100.4 is rounded to 100 cm
- 113.6 is rounded to 112 cm

(100 + 112) ÷ 2 = **106 cm**

Fig. 2.20. Brereton diameter measurement methodology.

- Philippine Brereton: round to the nearest tenth of a metre, e.g. 4.75–4.84 m is recorded as being 4.8 m in length.
- Japanese Brereton: lengths are rounded down to, and recorded in 0.2-m length classes, e.g. a log measuring 4.8–4.99 m long is recorded as being 4.8 m in length.

2.2.9.3 Taper distribution

For the purposes of calculating volume, the log is considered to be a cylinder, having a diameter equal to the average of the diameters from each log end (four diameters for PNG and ATIBT, two diameters for Philippine and Japanese Brereton).

2.2.9.4 Gross volume determination

Brereton uses the Huber formula for determining log volume. To determine log volume, apply the following formula:

$$m^3 = (\text{interpolated mid-diameter in cm}^2 \times \text{length in metres}) \times 0.00007854$$

Figure 2.21 shows examples of calculating gross volume. Volume can also be determined by using Table A.1.H.

2.2.9.5 Defect deduction rules

The PNG and Philippine version of Brereton have well-documented defect deduction procedures, which are summarized below. The translated Japanese Brereton mentions defect deductions, but without clear instructions as to procedures. The ATIBT method acknowledges the need to reduce volumes in the case of log species prone to perishable sapwood and severe surface or end checking, but advise that the true gross volume be available, i.e. if one must take defect, show a gross volume and a net volume.

With the exception of preventable mechanical damage (from logging or mill yard handling), deductions are taken for defects that cause a loss of wood volume available for manufacturing. Firmwood stain is not

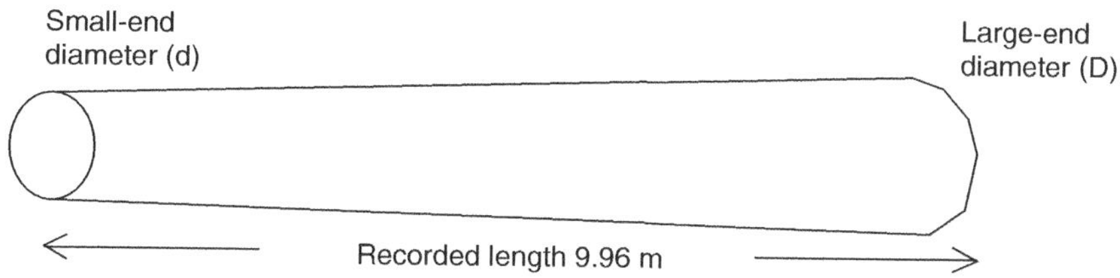

d1 = 37.7 cm
d2 = 39 0 cm

D1 = 45.9 cm
D2 = 47.2 cm

PNG Brereton and ATIBT methods

(37 + 39 + 45 + 47) ÷ 4 = 42;
interpolated mid-diameter = 42 cm; $42^2 \times 9.9 \times 0.00007854 =$ **1.372 m³** (48.4 ft³)

Japanese Brereton

d: (36 + 38) ÷ 2 = 37 D: (44 + 46) ÷ 2 = 45
(37 + 45) ÷ 2 = 41; interpolated mid-diameter = 41 cm; $41^2 \times 9.8 \times 0.00007854 =$ **1.294 m³** (45.7 ft³)

Philippine Brereton

d: (38 + 39) ÷ 2 = 38.5; round to 38 cm D: (46 + 47) ÷ 2 = 46.5; round to even = 46 cm
(38 + 46) ÷ 2 = 42; interpolated mid-diameter = 42 cm; $42^2 \times 10.0 \times 0.00007854 =$ **1.385 m³** (48.9 ft³)

Fig. 2.21. Brereton gross volume determination.

considered a defect. The major defects that are deducted for are: pipe, dry rot, termite holes, shake, oversize branch stubs, crook, sweep, burls.

There are four major methods for deducting for defects:

1. Area deduction.
2. Length deduction.
3. Percentage deduction.
4. Diameter deduction.

AREA DEDUCTION. This is used primarily for interior defects. The defect volume is calculated by determining the cubic volume occupied by the diameters and length of the defect. The formula is the same as is used for determining log volume. In the event that the interior defect does not go entirely through the log, volume is determined by assuming the defect goes through half the log length (if no indication is present), and by the following formula: defect deduction $= [(H^{cm} + W^{cm}) \div 2]^2 \times L^m \times 0.00007854$.

Philippine Brereton: interior defects are covered with a square or rectangle of sufficient size (with allowances made for tapering effects on defect) and multiplied by the length of the defect. The formula is: defect deduction $= H^{cm} \times W^{cm} \times L^m \div 10{,}000$.

LENGTH DEDUCTION. This is used for defects that can be best reckoned with by deducting all of, or a percentage of, the length affected. In PNG, this includes crook, burls or oversized knots. The defect volume equals the difference between the log volume as calculated from the gross and net lengths.

PERCENTAGE DEDUCTION. This is also not listed as being a valid method in PNG (although it is likely used in combination with the length deduction), but is used in the Philippines for defects confined to a sector or for deducting for sweep and crook.

DIAMETER DEDUCTION. This is also not listed as being a valid method in PNG. It is only referred to in the Philippine procedures. It is used for defects that occur in the perimeter of the log, such as rotten sapwood, surface checks and worm holes. This deduction is made by establishing net diameters (under the defective portion of the log). The defect deduction is the difference between the volume as calculated from the gross diameters and the volume as calculated from the net diameters.

2.2.9.6 Net volume and merchantability

The only reference to net volume and merchantability requirements in the PNG regulations is the following reference: 'it is expected that sections of a tree felled in the bush which will clearly not be merchantable will be docked off and left in the bush'. The Philippine procedures stipulate that logs are considered merchantable if over 1.83 m in length, having a diameter exceeding 30 cm, and having at least 33⅓% net volume for first-category logs (more valuable) and 50% net volume for second-category logs.

2.2.9.7 Pulp log scale

No specific information was found on pulp scale using Brereton.

2.2.10 Hoppus (Africa, Oceania, South America, Asia)

This is a very old system of measure and was once widely used in the UK and many former British colonies. It is still used in tropical regions of Asia, Africa and South America and is occasionally used in the UK, Australia and New Zealand. This log scale is also known as the 'quarter girth formula', or the Francon system (in its metric version). It was made popular around 1736 by the Englishman, Edward Hoppus (Honer, 1998). Hoppus gives 78.54% of Huber-derived cubic volume, given the same log dimension. This understatement of actual log volume has led some to conclude that the Hoppus formula includes assumptions as to lumber recovery. However, given that even modern sawmills do not achieve recovery ratios close to 78.54%, it may be more likely to conclude that the Hoppus formula was intended to just approximate the volume of a log, and simply dividing the girth (circumference) by four (thus the quarter girth moniker) would seem to reshape a circular log into a square timber, which might prove more tangible, volumetrically, to an 18th-century woodsman.

Hoppus is measured and reflected in both imperial (ft^3) and metric (m^3) units, and is occasionally reflected as superficial feet (a bf measure) and a unit called a Hoppus ton, which is a misnomer as it is a unit of volume and not a unit of weight. Further adding to the confusion is the fact that except for a few very low-density species or seasoned (dried) logs, a Hoppus ton would weigh substantially more than a ton (imperial or metric). The following procedures were put together based on Brereton (1940) and personal communications (G. Marshall, Georgetown, Guyana, 2004; Barber Cho, Yangoon, Myanmar, 2004).

2.2.10.1 Diameter measurements

1. Girth is generally used to determine volume, but diameter can also be used. Regardless of whether diameter or girth is used, the measurement is normally taken at the mid-point of the log length.
2. Measurements are normally taken with a girthing tape and measured in circumference. Over-bark measurements are used in some regions, but under-bark measurements are used as well (notably in Guyana) because the logging contractor debarks the log where the measurement will be taken. Girth measurements are usually rounded down to the nearest inch or centimetre.
3. If using callipers, take two measurements at right angles to each other through the narrow axis and the wide axis of the log, round each measurement to the nearest whole number. When rounding the pair of diameters, round to the even number if fractional.
4. A reduction is commonly made to the girth, diameter measurement or volume to approximate under-bark measurement, e.g. in Myanmar, 2″ is often subtracted from girth when measuring teak logs to account for bark (Brereton, 1940).
5. Disregard bumps and depressions.

2.2.10.2 Length measurements

1. Measure short side to short side.
2. Log lengths are rounded to the nearest foot or tenth of a metre, e.g. a log that is recorded as 32′ will have an actual length of 31.50–32.49′, or a log that is recorded as 10 m can have an actual length of 9.95–10.04 m.

2.2.10.3 Taper distribution

Taper is not considered or used in calculating Hoppus volume. For the purposes of calculating volume, the log is considered to be a cylinder, having a diameter that corresponds to the diameter equidistant from both ends of the log.

2.2.10.4 Gross volume determination

To determine log volume using imperial measurements (diameter or circumference), use the following formulas:

Hoppus ft^3 via girth = (mid-girth in inches $\div$ 4)^2 $\times$ length in feet $\div$ 144; round to the nearest tenth ft^3

Hoppus ft^3 via diameter = mid-diameter in inches^2 $\times$ length in feet $\times$ 0.004283; round to the nearest tenth ft^3

Hoppus superficial feet = Hoppus ft^3 $\times$ 12

Hoppus ton = Hoppus ft^3 $\div$ 50

To determine log volume using metric measurements (diameter or circumference), use the following formulas:

Hoppus m^3 via girth = (mid-girth in cm $\div$ 4)^2 $\times$ length in metres $\div$ 10,000; round to three decimal points

Hoppus m^3 via diameter = mid-diameter in cm^2 $\times$ length in metres $\times$ 0.000061685; round to three decimal points

Figure 2.22 shows an example of calculating volume via Hoppus. Table A.1.K gives volumes in Hoppus ft^3 and m^3 for various diameter and length combinations.

2.2.10.5 Defect deduction rules

No specific information for deduction rules was found.

2.2.10.6 Pulp log scale

No specific information on pulp scale was found.

2.2.11 JAS Scale (Japan, Chile, East Asia, Oceania, Australia)

The JAS Scale (Japanese Agricultural Standard) was apparently developed for use in measuring roundwood in Japan in the late 1940s. Rather than determining volume based on the area of a circle, JAS Scale determines volume based on the area of a square having the same cross dimensions as the diameter of the small-end of the log. In principle, this could tend to overstate the volume (a square having 27.3% more area than a

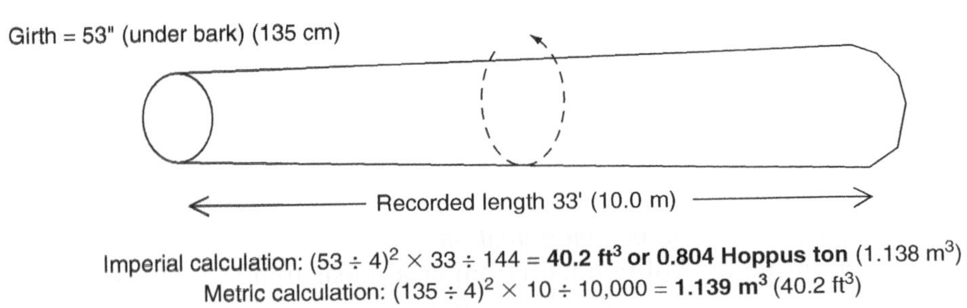

Fig. 2.22. Hoppus gross volume determination.

circle with the same cross dimensions), excepting that there are some compensating procedures, e.g. the small-end diameter is used (ignoring taper on logs 6 m long and less). There are also some compensating effects in the manner that diameters are measured and rounded. Information on JAS was taken from Ellis and Elliot (2001) and Briggs and Flora (1991).

2.2.11.1 Diameter measurements

1. Measure diameters inside of bark and cambium layer.
2. Measure the small-end of the log only.
3. Measure through the true centre of the log (ignore pith).
4. For logs less than 14 cm (nominally) on the small-end, take one measurement through the narrow axis.
5. For logs 14 cm and larger, take two measurements, through the narrow axis and the wide axis of the log end (narrow and at right angles in New Zealand and Australia).
6. Disregard bumps and depressions.
7. Forked tops or swelled ends should be measured in a manner that reduces the diameter to what it would normally be without the unusual swell.
8. Nominal diameters under 14 cm are truncated down to the nearest centimetre (13.9 rounds down to 13).
9. Diameters 14 cm and over are recorded in 2-cm diameter classes (14, 16, 18, 20) and are always truncated to the next lower even centimetre.
10. For elliptical logs 14 cm and larger, use the following rules to determine the scaling diameter:

- If the narrow measurement is less than 14 cm, diameter equals the narrow measurement.
- If the narrow measurement is between 14 cm and 39.99 cm, add 2 cm to the small-end diameter for each full 6 cm of difference between the narrow and wide dimensions, e.g. if the narrow measurement is 18 cm and the wide measurement is 23 cm, the diameter equals 18 cm.
- If the narrow measurement is 40 cm or greater, add 2 cm to the small-end diameter for each full 8 cm of difference between the narrow and wide dimensions, e.g. if the narrow measurement is 78 cm and the wide measurement is 93 cm, the diameter equals 80 cm.

Figure 2.23 shows examples of diameter measurements using JAS Scale.

2.2.11.2 Length measurements

1. Lengths are normally measured, rounded down to, and recorded in, 0.2-m length classes, e.g. 4.6, 4.8, 5.0. The exception to this rule are export

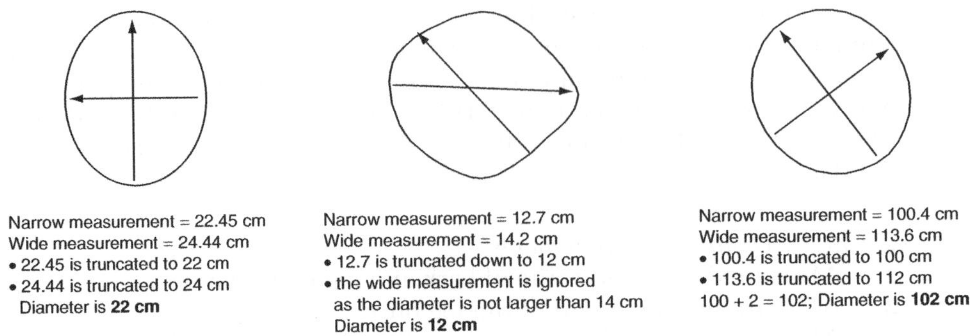

Fig. 2.23. JAS Scale diameter measurement methodology.

logs, which are measured in odd multiples providing that there is at least 0.1 m of trim allowance.

2. Lengths are measured at the shortest distance between each end of the log.

2.2.11.3 Taper distribution

1. If the scaling length is less than 6 m, taper is ignored.

2. If the scaling length is 6 m or greater, the following formula is used to increase the scaling diameter for volume calculation: (scaling length rounded down to a whole metre − 4) ÷ 2 = number added to small-end diameter to get the scaling diameter (do not round).

Note: this formula is included in the gross volume formula.

2.2.11.4 Gross volume determination

1. If the scaling length is less than 6 m: $m^3 = \text{scaling diameter}^2 \times \text{scaling length} \div 10{,}000$.

2. If the scaling length is 6 m or greater: $m^3 = \text{scaling diameter} + [(\text{scaling length rounded down to a whole metre} - 4) \div 2]^2 \times \text{scaling length} \div 10{,}000$.

3. Round all volumes to three decimal places, e.g. 0.387 m^3.

Figure 2.24 shows examples of calculating gross volume. Volume can also be determined by using Table A.1.L.

2.2.11.5 Defect deduction rules

Based on the translated text available, it appears that deductions are taken via length, percentage and a squared defect formula, but except for the squared defect formula, it is not clear as to the methodology.

SQUARED DEFECT FORMULA (USED FOR VOID AND SOFT ROT). This is applied using the same procedures to determine defect diameter (including factoring for elliptical

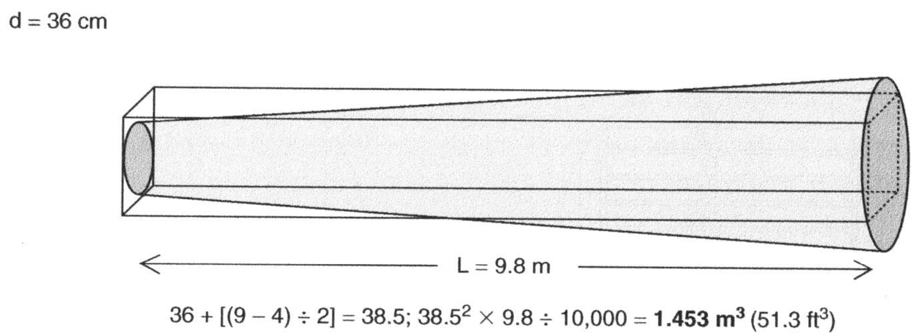

Fig. 2.24. JAS Scale gross volume determination.

defects) as is used to calculate log volume. Apply the following formulas to deduct for unsound or hollow areas:

1. When only one end of the log is rotten, the defective portion is calculated using the formula:

defect in m^3 = [diameter of defect^2 × (length ÷ 2)] ÷ 10,000; round to three decimal places

2. When both ends of the log are rotten, the deduction is calculated using the formula:

defect in m^3 = (average of defect diameters in centimetres at both ends of the log^2 × length) ÷ 10,000; round to three decimal places

2.2.11.6 Net volume and merchantability

Net volume is gross volume minus defect. No specific data was found regarding minimum merchantability requirements.

2.2.11.7 Pulp log scale

No specific information on pulp scale was found.

2.3 The Major Product Output Rules in Use

All of the product output rules listed in this section (and some of the rules listed in Table A.1.B) are reflected in volume units of bf. The bf measure is used in North America as a unit of measure for lumber volume, and in the case of product output rules, log volume. In principle, the bf is supposed to represent a piece of wood that has 0.0833 ft^3 (0.00236 m^3), and is 1″ thick × 1′ wide × 1′ long.

There are two types of product output rules, both of which attempt to predict the volume of lumber that a log will produce:

1. The diagram rule is based on diagrams of circles, with lumber hypothetically placed to maximize yield. There are assumptions made as to

Scribner Decimal C, Diagram Rule

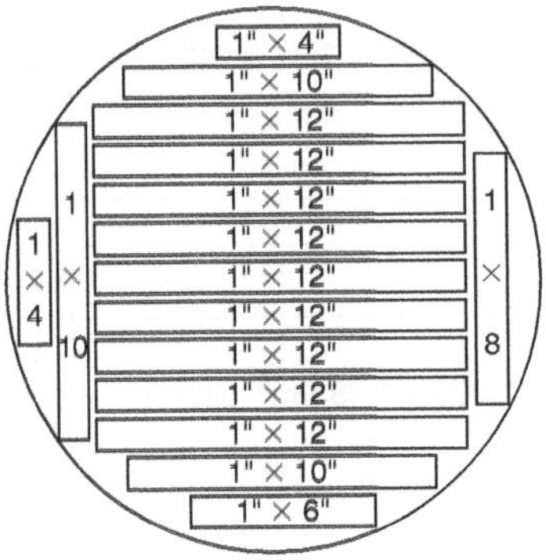

Number of boards	Board size inches	Bf per lineal foot of log
9	1 × 12	9.000
3	1 × 10	2.500
1	1 × 8	0.667
1	1 × 6	0.500
2	1 × 4	0.667

Total bf per lineal foot of log = 13.333

13.333 × 16′ = 213 bf (rounds to **210 b f**)

Doyle Log Rule Formula

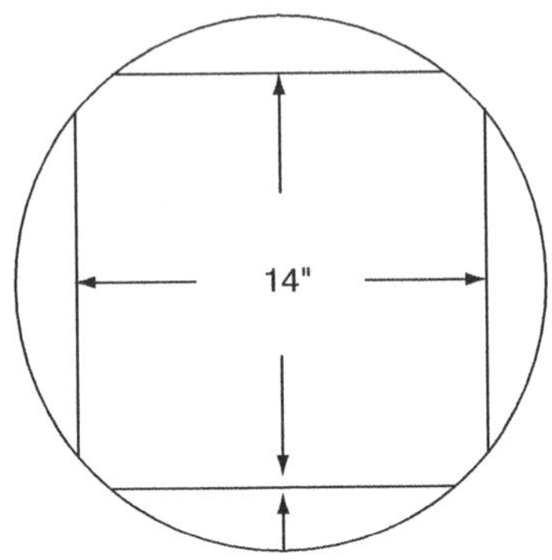

2" Slab allowance

(Small-end diameter – 4)^2 × (length ÷ 16) = Doyle bf

$(18 - 4)^2 = 196$

196 × (16 ÷ 16) = **196 bf**

Note: The devisor for determining board feet in the Doyle formula is 16 rather than 12. This has the same effect as allocating 25% of the cant area to saw kerf.

Fig. 2.25. Product output diagram rule vs. formula-based rule (16′ long 18″ small-end diameter log example).

specific lumber sizes, saw kerf, sawing practices and slab loss. Taper is ignored (the log is considered a cylinder) for the length of the log segment.

2. **The formula rule** also may make assumptions as to saw kerf and slab size, but utilizes a mathematical formula to calculate the hypothetical lumber yield, and some rules make allowances for taper.

Figure 2.25 shows examples of the two different approaches (diagram and formula log rules).

2.3.1 Scribner Short Log Rule (Western USA)

The Scribner log rule, originated in 1846, when it was invented and published by J.M. Scribner (Freese, 1973). Historically, the Scribner log rule has been the most widely used by the US Forest Service. Short Log Scribner also known as 'Eastside Scribner' is quite different from 'Long Log' Scribner in that, amongst other differences, the maximum scaling length of a log segment is 20′. This rule is based on diagrams of perfect circles, with 1″ thick boards of varying 2″ multiple widths, positioned in the circle to provide the best utilization, and allowing 0.25″ between sawing lines for kerf. The minimum size board used was a 1″ × 4″, and

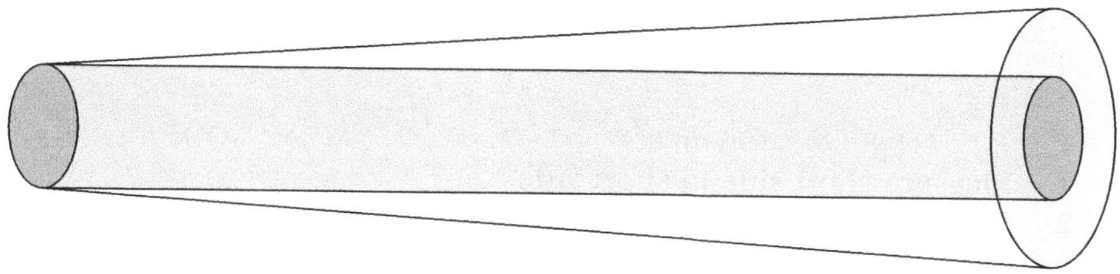

Fig. 2.26. Scribner scaling cylinder.

no wane was allowed. The rule assumes a log is a cylinder (no taper) within the log segment. Figure 2.26 illustrates the concept of the scaling cylinder. The procedures below are summarized from the *National Forest Log Scaling Rules* (US Forest Service, 1985).

It should be noted that the Pacific coast states of California, Oregon, Washington and Alaska use a revised volume chart (Table 2.2), which is slightly different from the volume table used in the rest of the USA. Table 2.2 also indicates where revised Scribner differs from the volumes used in the non-Pacific coast states (with a minus or a plus sign). Generally, the difference is not too significant (always + or −10 bf), unless the logs are weighted heavily toward 5–10″ d.

2.3.1.1 Diameter measurements

1. Measure diameters inside of bark and cambium layer on the small-end of the log only if scaling length is 20′ or less, and on both ends of the log if the scaling length is greater than 21′.
2. Measure through the true centre of the log (ignore pith).
3. Take two measurements, narrow-way first and at a right angle to the first measurement (the right angle measurement is generally but not necessarily the wide dimension).
4. Round diameters to the nearest inch; except when one diameter falls exactly on the ½″, round it up; when both diameters fall exactly on the ½″, round one diameter up and the other down.
5. Determine the average of both measurements (if the average is on the ½″, round down).
6. Disregard bumps and depressions.
7. Forked tops or swelled ends should be measured behind swell.
8. Methods of obtaining the small-end diameters of butt logs within multisegment logs vary by forest regions, but all are designed to mimic the true taper characteristics of a log. The methods for determining butt-cut diameter are as follows:

- Local species-specific taper tables (most oft-used method).
- Calliper the log 4.0′ in from the butt-end of the log.
- Callipering the log at the segment breaks to establish small-end diameter of log segments, thus eliminating the need for a butt diameter.

Figure 2.3 (on p. 11) is an example of the procedures used for diameter measurements.

2.3.3.2 Length measurements

1. Measure short side to short side.

2. Scaling length is actual length in feet minus trim; normally 6″ per segment for logs 8–20′ in nominal length; 12″ for logs 20–40′ in nominal length; and 18″ for logs 40–60′ in nominal length.

3. Maximum segment length is 20′ plus trim.

- Logs longer than 20′ are divided into two or more segments as necessary and scaled as individual log segments.
- Multisegment logs are divided as evenly as possible in 2′ multiples (if possible) with the shortest segment(s) on the small-end of the log.
- Under no circumstances should there be more than a 2′ difference between the longest and shortest segment, or more than one odd-length segment in any multisegment log (see Table A.1.C).

2.3.3.3 Taper distribution

1. All log segments are considered to be a cylinder having the diameter of the small-end of the log segment.

2. One-segment logs (logs with a nominal length of 8–20′); logs are assumed to be a cylinder having the diameter of the small-end of the log segment (taper can be completely ignored).

3. Two-segment logs (logs with a nominal length greater than 20′ but less than 40′): total taper ÷ number of segments, added to the small-end diameter = the small-end diameter of segment 2. If total taper is not evenly divisible by the number of segments, raise taper to next number evenly divisible by the number of segments (see Fig. 2.4 on p. 12).

4. Three-segment logs (logs with a nominal length of 40–60′), use same procedure as for two-segment logs to find the small-end diameter of the middle segment; repeat procedure for the remaining two segments (as if the small-end segment did not exist) to find the small-end diameter of the butt segment.

Taper distribution can also be determined by using Table A.1.D.

2.3.3.4 Gross volume determination

Volume is calculated by tallying the bf volume of the lumber for the respective log diameter and log segment length (see Fig. 2.25). The volume of a log segment is rounded to the nearest 10 bf. The formula for determining the bf content of lumber is: lumber volume = width in inches × thickness in inches × length in feet ÷ 12. All measurements used in this formula are nominal, e.g. a board that has a thickness of 1″, a width of 12″ and a length of 10′ is tallied as having 10 bf (1 × 12 × 10 = 120 ÷ 12 = 10). It should be noted that the actual finished size of a so-called 1″ × 12″ × 10′ is: 0.75″ × 11.25″ × 10′ (normally lengths are actual). As there is no formula that duplicates this process for all diameter and length combinations, a table is utilized for determining log segment volume (Table 2.2).

Figure 2.27 is an example of gross volume determination with the Scribner Short Log Rule.

2.2.3.5 Defect deduction rules

A defect is anything that causes a loss in volume of lumber. Stain is not considered a defect (see the list of deductible defects in Section 2.2.1 on p. 10 under defect deduction rules). There are four major methods for deducting for defects:

1. Squared area deduction.
2. Percentage deduction.
3. Length deduction.
4. Diameter deduction.

SQUARED AREA DEDUCTION. This is used primarily for interior defects that lend themselves to being partitioned into a square or rectangular area. The volume of a defect is calculated by the formula:

$$\text{defect deduction} = (W'' + 1) \times (H'' + 1) \times L' \div 15\text{, round to nearest 10 bf}$$

PERCENTAGE DEDUCTION. This is used for defects that go from the perimeter to the heart area of the log, and can best be reckoned with by enclosing the defect in a 'pie-shaped' sector with a fractional representation for the length affected. This fraction is then applied to the length of the log segment e.g. if a log segment has a scaling length of 16′, with one half of 6′ (half of 6 = 3) of the log having a defect requiring a deduction, the net scale of the log is equal to that of a 13′-long log.

LENGTH DEDUCTION. This is normally used for defects that cause a loss of volume for all or some of the length affected. This deduction rule is commonly used when the entire scaling cylinder is deducted or in combination with the percentage deduction (as above). The defect equals the difference between the volume as calculated from the gross length and the volume as calculated from the net length. Sweep and crook deductions are accounted for by projecting a theoretically straight cylinder, having a diameter the same as the small-end diameter of the log, through the straightest portion of the log segment, and deducting for the volume that falls outside of the actual log segment, e.g. if sweep affects 6′ of a log, of which ⅓ of the theoretic cylinder falls outside the log, the length deduction is 2′ (⅓ of 6 = 2).

DIAMETER DEDUCTION. This is used for defects that occur in the perimeter of the log, such as rotten sapwood, surface checks and twisted grain. This deduction is made by establishing net diameters (under the defective portion of the log). The defect deduction is the difference between the

Table 2.2. Scribner (revised) log volume chart, lengths 1–20′ (bf decimal C).

Small-end diameter of log segment in inches	Length of log segment in feet																			
	1	**2**	**3**	**4**	**5**	**6**	**7**	**8**	**9**	**10**	**11**	**12**	**13**	**14**	**15**	**16**	**17**	**18**	**19**	**20**
3	0	0	0	0	0	0	0	0	0	0	0	0	1	1	1	1	1	1	1	1
4	0	0	0	0	0	0	0	1	1	1	1	1	1	1	1	1	1	1	1	1
5	0	0	0	0	1	1	1	1	1	1	1	1	1	1	2–	2	2	2	2	2
6	0	0	0	0	1	1	1	1	1	1	1	1	2–	2–	2–	2	2	2	2	2
7	0	0	0	1	1	1	1	1	1	1	2	2	2	2	2	3	3	3	3	3
8	0	0	0	1	1	1	1	1	1	2	2	2	2	2	2	3	3	3	4–	4–
9	0	0	1	1	1	1	1+	2	2	2+	2+	3	3	3	3	4	4	4	5–	5–
10	0	1	1	1	2–	2	2	3	3	3	3	4–	4	4	5	6	6	6	7–	7
11	0	1	1	1	2	2	3–	3	3	4	4	4	5	5	6	7	7	8	8	8
12	0	1	1	2	2	3	3	4	4	5	5	6	6	7	7	8	8	9	9+	10
13	1	1	2	2	3	4	4	5	5	6	7	7	8	8	9	10	10	11	11+	12
14	1	1	2	3	4	4	5	6	6	7	8	9	9	10	11	11	12	13	14	14
15	1	2	3	4	4	5	6	7	8	9	10	11	12	12	13	14	15	16	17	18
16	1	2	3	4	5	6	7	8	9	10	11	12	13	14	15	16	17	18	19	20
17	1	2	3	5	6	7	8	9	10	12	13	14	15	16	17	18	20	21	22	23
18	1	3	4	5	7	8	9	11	12	13	15	16	17	19	20	21	23	24	25+	27
19	1	3	4	6	7+	9	10	12	13	15	16	18	19	21	22	24	25	27	28	30
20	2	3	5	7	9	10+	12	14	16	17	19	21	23	24	26	28	30	31	33	35
21	2	4	6	8	9+	11+	13	15	17	19	21	23	25	27	28	30	32	34	36	38
22	2	4	6	8	10	13	15	17	19	21	23	25	27	29	31	33	35	38	40	42
23	2	5	7	9	12	14	16	19	21	24–	26	28	31	33	35	38	40	42	45–	47
24	3	5	8	10	13	15	18	20+	23	25	28	30	33	35	38	40	43	45	48	50
25	3	6	9	11	14	17	20	23	26	29–	32	34	37	40	43	46	49	52	54	57
26	3	6	9	12	16	19	22	25	28	31	34	37	41	44	47	50	53	56	59	62
27	3	7	10	14	17	21	24	27	31	34	38	41	44	48	51	55	58	62	65	68

Small-end diameter of log segment in inches																				
28	4	7	11	15	18	22	25	29	33	36	40	44	47	51	55−	58	62	65	69	73
29	4	8	11	15	19	23	27	30+	34+	38	42	46	49	53	57	61	65	68	72	76
30	4	8	12	16	21	25	29	33	37	41	45	49	53	57	62	66	70	74	78	82
31	4	9	13	18	22	27	31	36	40	44	49	53	58	62	67	71	75	80	84	89
32	5	9	14	18	23	28	32	37	41	46	51	55	60	64	69	74	78	83	87+	92
33	5	10	15	20	24	29	34	39	44	49	54	59	64	69	73	78	83	88	93	98
34	5	10	15	20	25	30	35	40	45	50	55	60	65	70	75	80	85	90	95	100
35	5	11	16	22	27	33	38	44	49	55	60	66	71	77	82	88	93	98	104	109
36	6	12	17	23	29	35	40	46	52	58	63	69	75	81	86	92	98	104	110	115
37	6	13	19	26	32	39	45	51	58	64	71	77	84	90	96	103	109	116	122	129
38	7	13	20	27	33	40	47	53+	60	67	73	80	87	93	100	107	113	120	127−	133
39	7	14	21	28	35	42	49	56	63	70	77	84	91	98	105	112	119	126	133	140
40	8	15	23	30	38	45	53	60	68	75	83	90	98	105	113	120	128	135	143−	150
41	8	16	24	32	40−	48	56	64	72	79	87	95	103	111	119	127	135	143	151	159
42	8	17	25	34−	42	50	59	67	76	84	92	101	109	117	126	134	143	151	159	168
43	9	17	26	35	44−	52	61	70	78+	87	96	105	113	122	131	140	148	157	166	174
44	9	19	28	37	46	56	65	74	83	93	102	111	120	130−	139	148	157	167−	176	185
45	9	19	28	38	47	57	66	76	85	95	104	114	123	133	142+	152	161	171	180	190
46	10	20	30	40+	50+	59	69	79	89	99	109	119	129	139	149	159	168+	178	188	198
47	10	21	31	41	52	62	72	83	93	104	114	124	135−	145	155	166	176	186	197	207
48	11	22	32	43	54	65	76	86	97	108	119	130	140	151	162	173	184	194	205	216
49	11	22	34	45	56	67	79	90	101	112	124	135	146	157	168	180	191	202	213	225
50	12	23	35	47	58	70	82	94	105	117	129	140	152	164	175	187	199	211	222	234

Note: The volume numbers in this table are known as 'revised' Scribner, and are used in the states bordering the Pacific Ocean (California, Oregon, Washington and Alaska). The Scribner volumes utilized in the rest of the USA can be determined from this table by subtracting 1 from volumes with a minus (−) sign, and adding 1 to volumes with a plus (+) sign, e.g. the non-revised volume of a 13′–6″ log is 1 (10 bf). Decimal C signifies that volumes are shown at one-tenth their actual value, e.g. a value of 16 represents 160 bf.

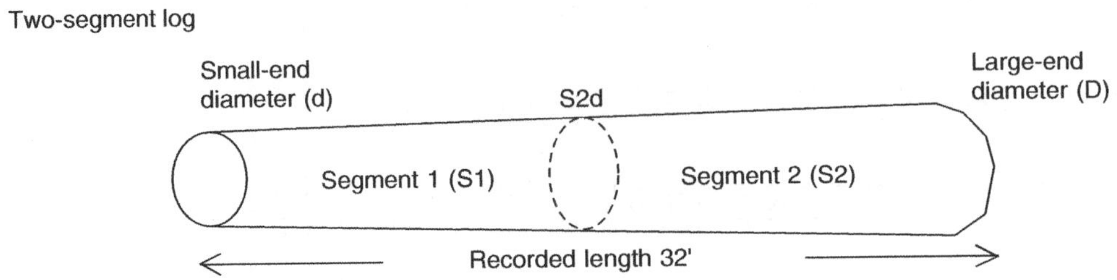

d = 15"
D = 18"
S2d is interpolated as 17"

Segment 1: 16' long, 15" d = 14 decimal C or **140 bf**
Segment 2: 16' long, 17" d = 18 decimal C or **180 bf**

Total log volume = **320 bf**

Fig. 2.27. Scribner Short Log Rule gross volume determination.

volume as calculated from the gross diameters and the volume as calculated from the net diameters.

2.2.3.6 *Net volume and merchantability*

Net volume is the gross volume minus defect. There is a point where a log is no longer considered merchantable and will be culled (giving it no net volume). This merchantability factor is designed to eliminate payment and/or mandatory utilization of logs that are not financially viable. In most regions, if the net scale of a log segment is less than 33⅓% of the gross volume, the log will be culled (having a net volume of zero). Figure 2.28 is an example of a cull log. It should be noted that a cull log can be reclassified as a merchantable pulp log or specialty product log (depending on the purchase agreement).

2.3.3.7 *Pulp log scale*

Since Scribner volumes are based completely on primary product (lumber) output (as it stood in the mid 19th century), it does not correlate very

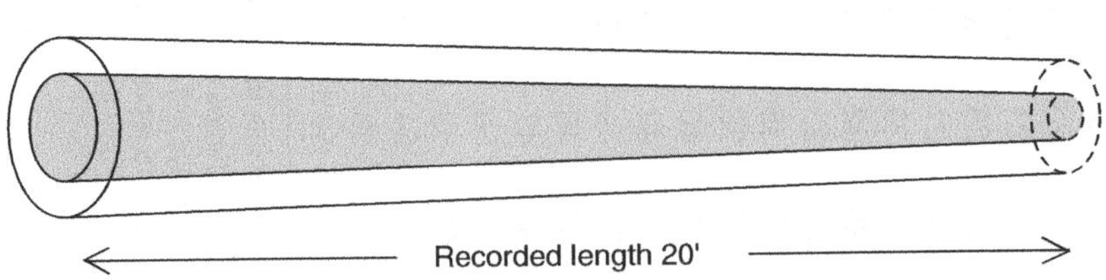

Log is 20' long, d of 18", gross = 270 bf; rotten core of wood is 10" in diameter on the small-end and 12" in diameter on the large-end (giving an average rot diameter of 11").

Defect volume; (11 + 1) × (11 + 1) × 20 ÷ 15 = 192 bf, rounded to 190 bf; since 190 exceeds 66.67% of the gross volume (190 ÷ 270 = 70.4%), the log is cull, and is assigned a net scale of **0 bf**.

Fig. 2.28. Scribner Short Log Rule cull log example.

well with actual lumber output given today's technology, let alone yield in usable wood fibre for the manufacture of chips, consequently it sees very little use in measuring pulp logs.

2.3.2 Scribner Long Log Rule (Northwestern USA, West Coast Canada)

The Scribner Long Log Rule is also known as 'Westside Scribner' as it is used almost exclusively west of the crest of the Cascade Mountains in northwestern USA. It is also used in southwestern Canada occasionally for exported logs. Conceptually, it is similar to Scribner Short Log, in that it is based on the same diagram rules produced by J.M. Scribner in 1846, and is based on a scaling cylinder. This rule did, however, evolve with many different interpretations and procedures; notably, log segments have a maximum scaling length of 40′ (thus the 'long log' moniker). Table A.1.N indexes the Scribner Long Log gross scale to Scribner Short Log gross scale. The following is based on *Official Log Scaling and Grading Rules* (Northwest Log Advisory Group, 1998).

2.3.2.1 Diameter measurements

1. Measure diameters inside of bark and cambium layer at the small-end of the log.
2. Measure through the true centre of the log (ignore pith).
3. Take two measurements, narrow-way first and at a right angle to first measurement (the right angle measurement is generally but not necessarily the wide dimension).
4. Disregard bumps and depressions.
5. Forked tops or swelled ends should be measured behind swell.
6. Measure diameters to the nearest inch truncating the measurement (11.0–11.99″ rounds to 11″).
7. Determine the average of both measurements (if the average is on the ½″, round down).

Figure 2.29 is an example of diameter measurements using the Scribner Long Log Rule.

2.3.2.2 Length measurements

1. Measure short side to short side.
2. Scaling length is actual length in feet minus trim; normally 6″ per segment for logs 8–20′ in nominal length; 12″ for logs 20–40′ in nominal length; 14″ for logs 42–50′ in nominal length, and 16″ for logs 51–60′.
3. Maximum segment length is 40′ plus trim.
   - Logs longer than 40′ are divided into two or more segments as necessary and scaled as individual log segments.
   - Multisegment logs are divided as evenly as possible in 2′ multiples (if possible) with the shortest segment(s) on the large-end of the log.

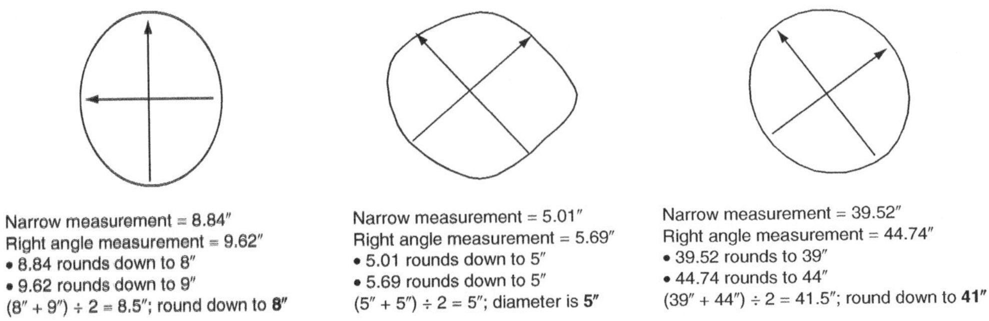

Fig. 2.29. Scribner Long Log Rule diameter measurement methodology.

- Under no circumstances should there be more than a 2′ difference between the longest and shortest segment, or more than one odd-length segment in any multisegment log (see Table 2.3).

2.3.2.3 Taper distribution

1. One-segment logs (logs with a nominal length of 8–40′); logs are assumed to be a cylinder having the diameter of the small-end of the log segment (taper is completely ignored).

2. Two-segment logs (logs with a nominal length greater than 40′ but less than 80′) allow 0.1″ of taper per lineal foot of nominal segment length. If taper is fractional, round down, e.g. a log that has a nominal length of 54′ is scaled as a 28′ top segment and a 26′ large-end segment, with 2.8″ of taper from the small-end of segment 1 to the small-end of segment 2, thus round 2.8″ down to 2″ (giving segment 2 a small-end diameter 2″ larger than segment 1 (Fig. 2.30).

2.3.2.4 Gross volume determination

The procedures for developing volumes are the same as for Scribner Short Log; volume is calculated by tallying the bf volume of the lumber for the respective log diameter and length of the segment (except with Scribner Long Log the scaling cylinder can be up to 40′ long in nominal length). The volume of a log segment is rounded to the nearest 10 bf. A table is utilized for determining log segment volumes (for lengths 20′ and less, use Table 2.2 on pp. 52–53; for lengths 21–40′, use Table 2.4). Note that Scribner Long Log always uses the revised volumes. Figure 2.31 is an example of gross volume calculation using Scribner Long Log Rule.

2.3.2.5 Defect deduction rules

A defect is anything that causes a loss in volume of lumber. Stain is not considered a defect unless associated with rot (see the list in Section 2.2.1 on p. 10 under defect deduction rules). There are two methods for deducting for defects:

1. Length deduction.
2. Diameter deduction.

Table 2.3. Scribner Long Log Rule segment length and trim allowance chart (feet).

Measured length	Recorded length	Segment length	Measured length	Recorded length	Segment length	
		Top			Top	Segment 2
8–8.5	8	8	35.1–36.0	35	35	
8.6–9.5	9	9	36.1–37.0	36	36	
9.6–10.5	10	10	37.1–38.0	37	37	
10.6–11.5	11	11	38.1–39.0	38	38	
11.6–12.5	12	12	39.1–40.0	39	39	
12.6–13.5	13	13	40.1–41.0	40	40	
13.6–14.5	14	14	41.1–42.2	41	21	20
14.6–15.5	15	15	42.3–43.2	42	22	20
15.6–16.5	16	16	43.3–44.2	43	22	21
16.6–17.5	17	17	44.3–45.2	44	22	22
17.6–18.5	18	18	45.3–46.2	45	23	22
18.6–19.5	19	19	46.3–47.2	46	24	22
19.6–20.5	20	20	47.3–48.2	47	24	23
20.6–22.0	21	21	48.3–49.2	48	24	24
22.1–23.0	22	22	49.3–50.2	49	25	24
23.1–24.0	23	23	50.3–51.2	50	26	24
24.1–25.0	24	24	51.3–52.3	51	26	25
25.1–26.0	25	25	52.4–53.3	52	26	26
26.1–27.0	26	26	53.4–54.3	53	27	26
27.1–28.0	27	27	54.4–55.3	54	28	26
28.1–29.0	28	28	55.4–56.3	55	28	27
29.1–30.0	29	29	56.4–57.3	56	28	28
30.1–31.0	30	30	57.4–58.3	57	29	28
31.1–32.0	31	31	58.4–59.3	58	30	28
32.1–33.0	32	32	59.4–60.3	59	30	29
33.1–34.0	33	33	60.4–61.3	60	30	30
34.1–35.0	34	34				

Note: Measured length converted from feet to inches and rounded to the nearest tenth of a foot.

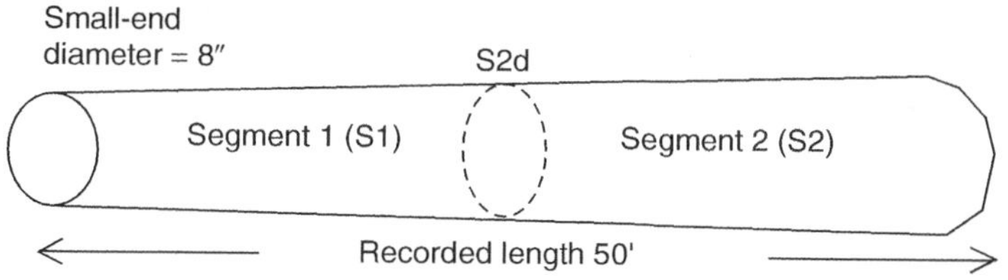

Two-segment log

With a recorded length of 50′, S1 = 26′ and S2 = 24′; the standard taper allowance is 0.1″ of taper per lineal foot, thus:

S2d = 26 × 0.10 = 2.6; 2.6 rounds down to 2″; 2″ + 8″ gives a scaling diameter of **10″**

Fig. 2.30. Scribner Long Log Rule taper distribution.

Table 2.4. Scribner Long Log Rule (revised) volume chart, lengths 21–40′ (bf decimal C).

Small-end diameter of log segment in inches	Length of log segment in feet																			
	21	**22**	**23**	**24**	**25**	**26**	**27**	**28**	**29**	**30**	**31**	**32**	**33**	**34**	**35**	**36**	**37**	**38**	**39**	**40**
3	1	1	1	1	1	1	1	1	1	1	1	1	1	1	1	1	1	1	2	2
4	1	1	2	2	2	2	2	2	2	2	2	2	2	2	2	2	3	3	3	3
5	2	2	2	3	3	3	3	3	3	3	3	3	4	4	4	4	4	4	4	4
6	3	3	3	3	3	3	3	3	4	4	4	5	5	5	5	6	6	6	6	6
7	3	4	4	4	4	4	4	5	5	5	5	6	6	6	6	6	7	7	7	7
8	4	4	4	4	5	5	5	5	5	6	6	7	7	7	8	8	8	8	9	9
9	5	5	6	6	6	6	7	7	7	7	7	9	10	10	10	10	11	11	11	12
10	7	8	8	9	9	9	10	10	10	11	11	12	13	13	13	14	14	14	15	15
11	9	9	10	10	10	11	11	12	12	13	13	14	15	15	16	16	17	17	18	18
12	10	11	11	12	12	13	13	14	14	15	15	16	16	17	17	18	18	19	19	20
13	13	13	14	15	15	16	16	17	18	18	19	19	20	21	21	22	22	23	24	24
14	15	16	16	17	18	19	19	20	21	21	22	23	24	24	25	26	26	27	28	29
15	19	20	20	21	22	23	24	25	26	27	28	28	29	30	31	32	33	34	35	36
16	21	22	23	24	25	26	27	28	29	30	31	32	33	34	35	36	37	38	39	40
17	24	25	27	28	29	30	31	32	33	35	36	37	38	39	40	42	43	44	45	46
18	28	29	31	32	33	35	36	37	39	40	41	43	44	45	47	48	49	51	52	53
19	31	33	34	36	37	39	40	42	43	45	46	48	49	51	52	54	55	57	58	60
20	37	38	40	42	44	45	47	49	51	52	54	56	58	59	61	63	65	66	68	70
21	40	42	44	46	47	49	51	53	55	57	59	61	63	65	66	68	70	72	74	76
22	44	46	48	50	52	54	56	58	61	63	65	67	69	71	73	75	77	79	81	84
23	49	52	54	56	59	61	63	66	68	71	73	75	78	80	82	85	87	89	92	94
24	53	55	58	61	63	66	68	71	73	76	78	81	83	86	88	91	93	96	98	101
25	60	63	66	69	72	75	77	80	83	86	89	92	95	98	100	103	106	109	112	115
26	66	69	72	75	78	81	84	87	91	94	97	100	103	106	109	112	116	119	122	125
27	72	75	79	82	86	89	92	96	99	103	106	110	113	116	120	123	127	130	133	137

Small-end diameter of log segment in inches																				
28	76	80	84	87	91	95	98	102	105	109	113	116	120	124	127	131	135	138	142	146
29	80	84	87	91	95	99	103	107	110	114	118	122	126	129	133	137	141	145	148	152
30	86	90	94	99	103	107	111	115	119	123	127	131	135	140	144	148	152	156	160	164
31	93	98	102	107	111	115	120	124	129	133	138	142	146	151	155	160	164	169	173	178
32	97	101	106	110	115	120	124	129	133	138	143	147	152	156	161	166	170	175	179	184
33	103	108	113	118	122	127	132	137	142	147	152	157	162	167	171	176	181	186	191	196
34	105	110	115	120	125	130	135	140	145	150	155	160	165	170	175	180	185	190	195	200
35	115	120	126	131	137	142	148	153	159	164	170	175	180	186	191	197	202	208	213	219
36	121	127	133	138	144	150	156	161	167	173	179	185	190	196	202	208	213	219	225	231
37	135	142	148	154	161	167	174	180	187	193	199	206	212	219	225	232	238	244	251	257
38	140	147	153	160	167	173	180	187	194	200	207	214	220	227	234	240	247	254	260	267
39	147	154	161	168	175	182	189	196	203	210	217	224	231	238	245	252	259	266	273	280
40	158	166	173	181	188	196	203	211	218	226	233	241	248	256	263	271	278	286	293	301
41	167	175	183	191	199	207	215	223	230	238	246	254	262	270	278	286	294	302	310	318
42	176	185	193	201	210	218	227	235	243	252	260	269	277	285	294	302	310	319	327	336
43	183	192	201	209	218	227	235	244	253	262	270	279	288	296	305	314	323	331	340	349
44	194	204	213	222	231	241	250	259	268	278	287	296	305	315	324	333	342	352	361	370
45	199	209	218	228	237	247	256	266	275	285	294	304	313	323	332	342	351	361	370	380
46	208	218	228	238	248	258	268	277	287	297	307	317	327	337	347	357	367	376	386	396
47	217	228	238	248	259	269	279	290	300	311	321	331	342	352	362	373	383	393	404	414
48	227	238	248	259	270	281	292	302	313	324	335	346	356	367	378	389	399	410	421	432
49	236	247	258	270	281	292	303	314	326	337	348	359	371	382	393	404	415	427	438	449
50	246	257	269	281	292	304	316	328	339	351	363	374	386	398	409	421	433	445	456	468

Note: For log lengths 20′ long and less, use Table 2.2 on p. 52 (Scribner Long Log always uses the revised volumes). Decimal C signifies that volumes shown are one-tenth their actual value, e.g. a value of 16 represents 160 bf.

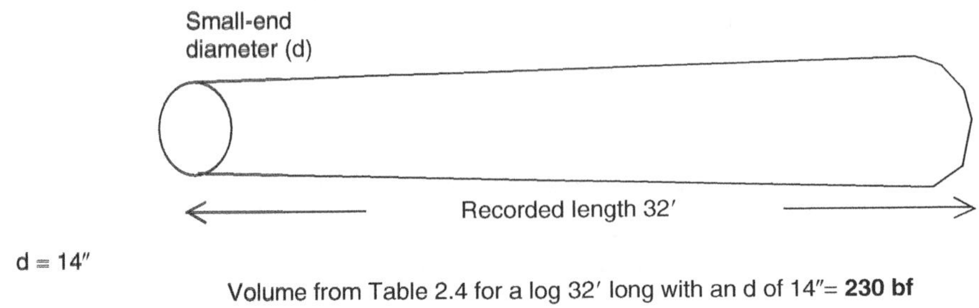

Fig. 2.31. Scribner Long Log Rule gross volume determination.

LENGTH DEDUCTION. This is normally used for defects that cause a loss of volume for all, or the majority of, the length affected. This deduction rule is commonly used in combination with a percentage, e.g. ½ of an 8′ length cut. The defect equals the difference between the volume as calculated from the gross length and the volume as calculated from the net length. Length deductions are also used for interior defects such as rot by estimating the percentage of the scaling cylinder affected by the interior defect as well as the length affected, and taking a corresponding length deduction. Sweep (a gradual bow in the log) and crook (a sharp change of direction) are deducted by projecting the scaling cylinder through the log segment to a point where the cylinder first leaves the log, and then redirecting the scaling cylinder until it leaves the log again, and so on. Deductions are taken for portions of the log that fall outside the redirected scaling cylinders or for portions of the log outside the scaling cylinder that have a recovered length of less than 8′ (which is the minimum product length for Scribner Long Log). A minimum of a 1′ length deduction is taken for each redirection of the scaling cylinder.

DIAMETER DEDUCTION. This is used for defects that occur in the perimeter of the log, such as rotten sapwood, surface checks and twisted grain (by establishing a net diameter under the defective portion of the log), and for interior defects if they are checks or ring separations. It should be noted that unlike the Scribner Short Log Rule, deductions for perimeter defects such as sap rot and surface checks, which occur outside the scaling cylinder, are not deducted. The defect deduction is the difference between the volume as calculated from the gross diameter and the volume as calculated from the net diameter. Diameter deductions are also used to account for the loss in volume from ring separations (shake, pitch, ring) and for heart check (heart shake, pitch check, rift crack). This is done by dividing the scaling cylinder (small-end diameter of the log segment) into three: inner third (heart), middle third (meat) and outer third (sap), and using the guidelines below to make a deduction:

Ring separation	Inner third (heart)	Middle third (meat)	Outer third (sap)*
½ ring in one end	deduct 1″ in diameter	deduct 1″ in diameter	deduct 1″ in diameter
½ ring in both ends	deduct 2″ in diameter	deduct 2″ in diameter	deduct 2″ in diameter
½–full ring in one end	deduct 1″ in diameter	deduct 1″ in diameter	deduct 2″ in diameter
½–full ring in both ends	deduct 2″ in diameter	deduct 2″ in diameter	deduct 4″ in diameter
Heart check/shake			
In one end	no deduction	deduct 1″ in diameter	deduct 2″ in diameter
In both ends	deduct 1″ in diameter	deduct 2″ in diameter	deduct 4″ in diameter

**Note*: For full rings that are less than 3″ from the outside of the radius, deduct entire area outside of the ring (net diameter is the ring diameter).

2.3.2.6 Net volume and merchantability

Scribner Long Log requires a minimum net volume of at least 33⅓% of the gross volume or the log will be culled (see Section 2.3.1.6).

2.3.2.7 Pulp log scale

This rule sees little use for measuring pulpwood, as it is very poor at measuring relative fibre volume.

2.3.3 The Doyle Log Rule (central and eastern North America)

The Doyle Log Rule is used primarily in the eastern half of North America (only occasionally outside the region), and appears to have originated in 1825 (Freese, 1973). Its use is especially favoured in southeastern USA. Unlike the Scribner rule, the Doyle Log Rule is based on a formula:

$$\text{bf volume} = (\text{small-end diameter in inches} - 4)^2 \times \text{length} \div 16$$

Thus this rule assumes a cylinder with a 4″ slab loss, and a 25% loss from shrinkage and saw kerf. Unfortunately, the formula is the only 'official' procedure of the Doyle Log Rule. There are no widely recognized and/or utilized standards or guidelines beyond this simple formula. The procedures for determining diameters, length, segmenting, taper distribution and defect deductions have as many different interpretations as there are users. The procedures listed are by no means an 'official' set of procedures; they are, however, an interpretation that is used, and in the author's opinion, 'sit in the middle ground' of the range of interpretations of this rule. The rule has been widely criticized as being very inaccurate in predicting product output. While it is true that this rule does a poor job of predicting product output (it grossly understates the relative recoverable product volume of small logs and even medium-sized logs to larger logs), it is not a bad rule for measuring value (especially for logs 9″ and larger). Doyle is seldom used for logs under 8″ on the small-end. Doyle understates typical lumber recovery by diameter class at a relative ratio to value by diameter class, which often allows a log purchaser to pay a single price per mbf (1,000 board feet), as opposed to having different price levels by size classes. This 'de facto grading mechanism', which appears to account for the higher manufacturing costs and lower value of products from small logs, is likely a fortunate coincidence; nevertheless, it is no doubt a good deal of the reason for Doyle's continued use and popularity. It is common for companies to utilize Doyle to purchase logs, and cubic or weight to track inventories and predict recovery.

2.3.3.1 Diameter measurements

Diameter measurements are sometimes measured over the bark, outside of just one bark thickness, or inside of bark. There is also no agreement as to the number of measurements utilized to establish the diameter of a log end either (to account for elliptical log ends); it can be one measurement or two. Listed below is just one interpretation:

1. Measure diameters inside of one bark and outside of the other bark (on average this increases the diameter by roughly 0.3″ over inside bark diameters for the Southern pines).
2. Measure the small-end of the log only if scaling length is 20′ or less, and both ends of the log if the scaling length is 21′ or greater in nominal length.
3. Measure through the true centre of the log (ignore pith).
4. Take two measurements, narrow-way first and at a right angle to first measurement (the right angle measurement is generally but not necessarily the wide dimension).
5. Round diameters to the nearest inch; except when one diameter falls exactly on the ½″, round it up; when both diameters fall exactly on the ½″, round one diameter up and the other down.
6. Determine the average of both measurements (if the average is on the ½″, round down).
7. Disregard bumps and depressions.
8. Forked tops or swelled ends should be measured behind swell.
9. Methods of obtaining butt-cut diameters vary, but all are designed to mimic the true taper characteristics of the log to the extent necessary to determine the diameter of the scaling cylinder of a log segment. The methods for determining butt-cut diameter are as follows:
- Local species-specific taper tables (assumed taper that is often 0.1″ per lineal foot).
- Callipering the log 4.0′ in from the butt-end of the log.
- Callipering the log at the segment breaks to establish small-end diameter of log segments thus eliminating the need for a butt diameter.

Figure 2.32 is an example of diameter measurements using the Doyle Log Rule.

2.3.3.2 Length measurements

1. Measure short side to short side.
2. Scaling length is actual length in feet minus trim; normally 6″ per segment for logs 8–20′ in nominal length; 12″ for logs 20–40′ in nominal length; and 18″ for logs 40–60′ in nominal length.

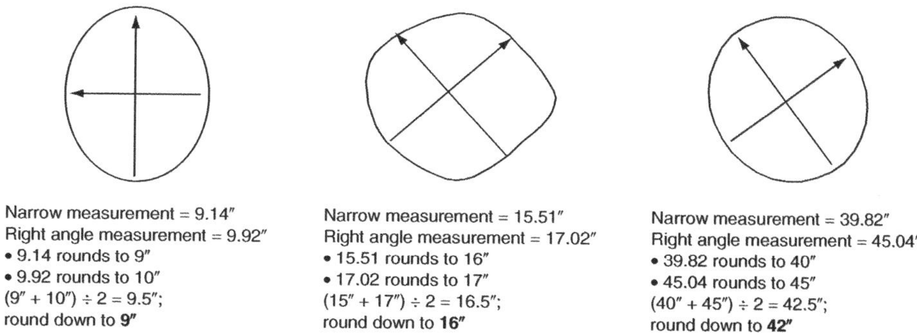

Fig. 2.32. Doyle Log Rule diameter measurement methodology.

3. Maximum segment length is 20′ plus trim.
   - Logs longer than 20′ are divided into two or more segments as necessary and scaled as individual log segments.
   - Multisegment logs are divided as evenly as possible in 2′ multiples (if possible) with the shortest segment(s) on the small-end of the log.
   - Under no circumstances should there be more than a 2′ difference between the longest and shortest segment or more than one odd-length segment in any multisegment log (see Table A.1.C).

2.3.3.3 Taper distribution

1. All log segments are considered to be a cylinder having the diameter of the small-end of the log segment.
2. One-segment logs (logs with a nominal length of 8–20′); logs are assumed to be a cylinder having the diameter of the small-end of the log segment (taper can be completely ignored).
3. Two-segment logs (logs with a nominal length greater than 20′ but less than 40′), total taper ÷ number of segments added to the small-end diameter = the small-end diameter of segment 2. If total taper is not evenly divisible by the number of segments, raise taper to next number evenly divisible by the number of segments (e.g. Fig. 2.4 on p. 12).
4. Three-segment logs (logs with a nominal length of 40–60′), use same procedure as for two-segment logs to find small-end diameter of large-end segment; repeat procedure for the remaining two segments (as if the large-end segment did not exist) to find the small-end diameter of the middle segment.

Total taper = difference between d and D. Taper distribution can also be determined by using Table A.1.D.

2.3.3.4 Gross volume determination

Volume is calculated by utilizing the Doyle formula for the respective log diameter and length of the log segment:

bf volume = (small-end diameter in inches − 4)^2 × length ÷ 16, round to the nearest bf

Two-segment log

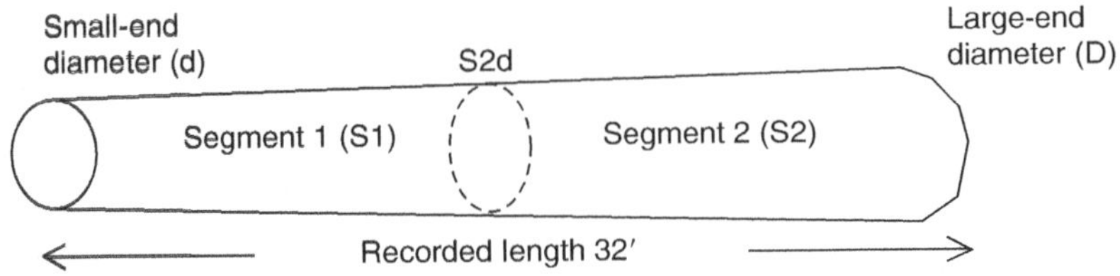

d = 15″
D = 18″
S2d (and S1D) are interpolated as 17

Segment 1: 16′ long, 15″ d $(15 - 4)^2 \times 16 \div 16 =$ **121 bf**
Segment 2: 16′ long, 17″ d $(17 - 4)^2 \times 16 \div 16 =$ **169 bf**

Total log volume = **290 bf**

Fig. 2.33. Doyle Log Rule gross volume determination.

As stated earlier, procedures for determining diameters vary greatly amongst the users of the Doyle Log Rule, but for the purposes of examples and comparisons in this publication, we will use the 'inside one bark' as this gives a common interpretation. An assumption of 0.3″ (7.6 mm) of bark thickness will be used for comparisons in Section 2.5. To use inside-bark diameter measurements with the 'one bark' interpretation of the rule, one can adjust the Doyle formula to:

bf volume = (small-end diameter in inches – **3.7**)^2 × length ÷ 16,
round to the nearest board foot[1]

Figure 2.33 is an example of determining the log volume with the Doyle Log Rule. Table 2.5 shows Doyle log volumes.

2.3.3.5 Defect deduction rules

A defect is anything that causes a loss in volume of lumber. Stain is not considered a defect unless associated with rot (see the list in Section 2.2.1 on p. 10 under defect deduction rules). There are four major methods for deducting for defects:

1. Squared area deduction.
2. Percentage deduction.
3. Length deduction.
4. Diameter deduction.

SQUARED AREA DEDUCTION. This is used primarily for interior defects that lend themselves to being partitioned into a square or rectangular area. The volume of a defect is calculated by the formula:

defect deduction = $(W'' + 1) \times (H'' + 1) \times L' \div 16$,
round to nearest board foot

[1] This change in the formula is used by some mills in Louisiana and Arkansas.

Table 2.5. Doyle Log Rule volume chart (board feet).

Small-end diameter of log segment in inches	Length of log segment in feet: 1	2	3	4	5	6	7	8	9	10	11	12	13	14	15	16	17	18	19	20
5	0	0	0	0	0	0	0	1	1	1	1	1	1	1	1	1	1	1	1	1
6	0	1	1	1	1	2	2	2	2	3	3	3	3	4	4	4	4	5	5	5
7	1	1	2	2	3	3	4	5	5	6	6	7	7	8	8	9	10	10	11	11
8	1	2	3	4	5	6	7	8	9	10	11	12	13	14	15	16	17	18	19	20
9	2	3	5	6	8	9	11	13	14	16	17	19	20	22	23	25	27	28	30	31
10	2	5	7	9	11	14	16	18	20	23	25	27	29	32	34	36	38	41	43	45
11	3	6	9	12	15	18	21	25	28	31	34	37	40	43	46	49	52	55	58	61
12	4	8	12	16	20	24	28	32	36	40	44	48	52	56	60	64	68	72	76	80
13	5	10	15	20	25	30	35	41	46	51	56	61	66	71	76	81	86	91	96	101
14	6	13	19	25	31	38	44	50	56	63	69	75	81	88	94	100	106	113	119	125
15	8	15	23	30	38	45	53	61	68	76	83	91	98	106	113	121	129	136	144	151
16	9	18	27	36	45	54	63	72	81	90	99	108	117	126	135	144	153	162	171	180
17	11	21	32	42	53	63	74	85	95	106	116	127	137	148	158	169	180	190	201	211
18	12	25	37	49	61	74	86	98	110	123	135	147	159	172	184	196	208	221	233	245
19	14	28	42	56	70	84	98	113	127	141	155	169	183	197	211	225	239	253	267	281
20	16	32	48	64	80	96	112	128	144	160	176	192	208	224	240	256	272	288	304	320
21	18	36	54	72	90	108	126	145	163	181	199	217	235	253	271	289	307	325	343	361
22	20	41	61	81	101	122	142	162	182	203	223	243	263	284	304	324	344	365	385	405
23	23	45	68	90	113	135	158	181	203	226	248	271	293	316	338	361	384	406	429	451
24	25	50	75	100	125	150	175	200	225	250	275	300	325	350	375	400	425	450	475	500
25	28	55	83	110	138	165	193	221	248	276	303	331	358	386	413	441	469	496	524	551
26	30	61	91	121	151	182	212	242	272	303	333	363	393	424	454	484	514	545	575	605
27	33	66	99	132	165	198	231	265	298	331	364	397	430	463	496	529	562	595	628	661
28	36	72	108	144	180	216	252	288	324	360	396	432	468	504	540	576	612	648	684	720
29	39	78	117	156	195	234	273	313	352	391	430	469	508	547	586	625	664	703	742	781
30	42	85	127	169	211	254	296	338	380	423	465	507	549	592	634	676	718	761	803	845
31	46	91	137	182	228	273	319	365	410	456	501	547	592	638	683	729	775	820	866	911

Continued

Table 2.5. *continued*

Small-end diameter of log segment in inches	Length of log segment in feet																			
32	49	98	147	196	245	294	343	392	441	490	539	588	637	686	735	784	833	882	931	980
33	53	105	158	210	263	315	368	421	473	526	578	631	683	736	788	841	894	946	999	1051
34	56	113	169	225	281	338	394	450	506	563	619	675	731	788	844	900	956	1013	1069	1125
35	60	120	180	240	300	360	420	481	541	601	661	721	781	841	901	961	1021	1081	1141	1201
36	64	128	192	256	320	384	448	512	576	640	704	768	832	896	960	1024	1088	1152	1216	1280
37	68	136	204	272	340	408	476	545	613	681	749	817	885	953	1021	1089	1157	1225	1293	1361
38	72	145	217	289	361	434	506	578	650	723	795	867	939	1012	1084	1156	1228	1301	1373	1445
39	77	153	230	306	383	459	536	613	689	766	842	919	995	1072	1148	1225	1302	1378	1455	1531
40	81	162	243	324	405	486	567	648	729	810	891	972	1053	1134	1215	1296	1377	1458	1539	1620
41	86	171	257	342	428	513	599	685	770	856	941	1027	1112	1198	1283	1369	1455	1540	1626	1711
42	90	181	271	361	451	542	632	722	812	903	993	1083	1173	1264	1354	1444	1534	1625	1715	1805
43	95	190	285	380	475	570	665	761	856	951	1046	1141	1236	1331	1426	1521	1616	1711	1806	1901
44	100	200	300	400	500	600	700	800	900	1000	1100	1200	1300	1400	1500	1600	1700	1800	1900	2000
45	105	210	315	420	525	630	735	841	946	1051	1156	1261	1366	1471	1576	1681	1786	1891	1996	2101
46	110	221	331	441	551	662	772	882	992	1103	1213	1323	1433	1544	1654	1764	1874	1985	2095	2205
47	116	231	347	462	578	693	809	925	1040	1156	1271	1387	1502	1618	1733	1849	1965	2080	2196	2311
48	121	242	363	484	605	726	847	968	1089	1210	1331	1452	1573	1694	1815	1936	2057	2178	2299	2420
49	127	253	380	506	633	759	886	1013	1139	1266	1392	1519	1645	1772	1898	2025	2152	2278	2405	2531
50	132	265	397	529	661	794	926	1058	1190	1323	1455	1587	1719	1852	1984	2116	2248	2381	2513	2645

PERCENTAGE DEDUCTION. This is used for defects that go from the perimeter to the heart area of the log, and can best be reckoned with by enclosing the defect in a 'pie-shaped' sector with a fractional representation for the length affected. This fraction is then applied to the length of the log segment, e.g. if a log segment has a scaling length of 16′, with one half of 6′ (½ of 6 = 3) of the log having a defect requiring a deduction, the net scale of the log is equal to that of a 13′ long log.

LENGTH DEDUCTION. This is normally used for defects that cause a loss of volume for all, or some of, the length affected. This deduction rule is commonly used when the entire scaling cylinder is deducted or in combination with the percentage deduction (as above). The defect equals the difference between the volume as calculated from the gross length and the volume as calculated from the net length. Sweep and crook deductions are accounted for by projecting a theoretically straight cylinder, having the same diameter as the small-end diameter of the log, through the straightest portion of the log segment and deducting for the volume that falls outside of the actual log segment, e.g. if sweep affects 6′ of a log, of which ⅓ of the theoretic cylinder falls outside the log, the length deduction is 2′ (⅓ of 6 = 2).

DIAMETER DEDUCTION. This is used for defects that occur in the perimeter of the log, such as rotten sapwood, surface checks and twisted grain. This deduction is made by establishing net diameters (under the defective portion of the log). The defect deduction is the difference between the volume as calculated from the gross diameters and the volume as calculated from the net diameters.

2.3.3.6 Net volume and merchantability

As with all aspects of Doyle log scaling procedures, merchantability requirements vary from one user to another. Some companies require a net scale of 33⅓% of the gross volume or the log will be culled, while others have no set standard at all (cull logs are determined at the discretion of the scaler).

2.3.3.7 Pulp log scale

As stated earlier, the Doyle Log Rule does a poor job of predicting product output (referring to lumber or veneer); it is even worse at measuring total fibre yield, and thus is of little use for scaling pulp logs.

2.3.4 International ¼-Inch Log Rule (eastern North America)

The International ¼-Inch Log Rule was published by Judson F. Clark in 1917 (Freese, 1973). Clark researched lumber recovery while working for the province of Ontario in Canada. The rule assumes a ¼″ sawkerf (originally it was developed with a ⅛″ kerf allowance), makes allowance

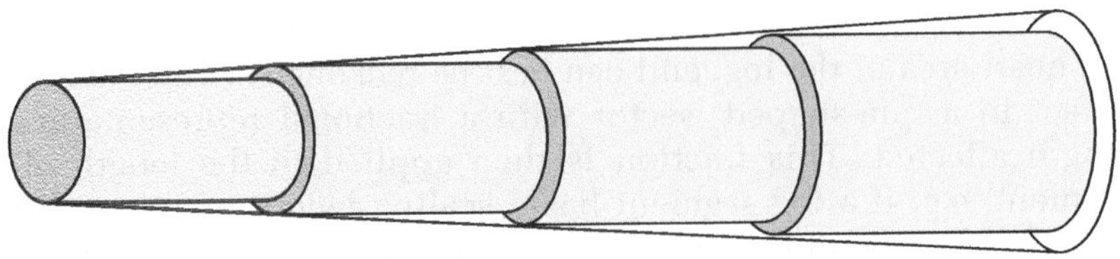

Fig. 2.34. International ¼-Inch Log Rule 4′ scaling cylinder methodology (16′ scaling length).

for 6.25% shrinkage and has a slab deduction that is 1.93″ thick and as wide as the diameter. Unlike most other primary product rules, the International ¼-Inch Rule accounts for taper in a log by increasing the diameter of the scaling cylinder ½″ every 4′ of nominal log length (Fig. 2.34). The last cylinder can be 4′ long or less when the log segment length is not a 4′ multiple, e.g. in the case of a 14′ segment (starting at the small-end of the log segment), the first three cylinders are 4′ long, and the fourth cylinder is 2′ long. This rule is used primarily in the eastern half of North America and appears to be especially popular in the New England and Quebec regions. It has also been adopted for use by the US Forest Service in some regions. The following rules are taken from *Wood Measurement Rules* (Maine Department of Agriculture, Food & Rural Resources, no date) and *National Forest Log Scaling Rules* (US Forest Service, 1985).

2.3.4.1 Diameter measurements

1. Measure diameters inside of bark and cambium layer.
2. Measure through the true centre of the log (ignore pith).
3. Take two measurements, narrow-way first and at a right angle to first measurement (the right angle measurement is generally but not necessarily the wide dimension).
4. Round diameters to the nearest inch; except when one diameter falls exactly on the ½″, round it up; when both diameters fall exactly on the ½″, round one diameter up and the other down.
5. Determine the average of both measurements (if the average is on the ½″, round down).
6. Disregard bumps and depressions.
7. Forked tops or swelled ends should be measured behind swell.
8. Methods of obtaining butt-cut diameters vary by regions, but all are designed to mimic the true taper characteristics of the log to the extent necessary to determine the diameter of the scaling cylinder of a log segment. The methods for determining butt-cut diameter are as follows:

- Assign 2″ of taper per segment (most often used).
- Calliper the log 4.0′ in from the butt-end of the log.
- Calliper the log at the segment breaks to establish small-end diameter of log segments thus eliminating the need for a butt diameter.

The procedures for establishing diameters with the International ¼-Inch Rule are the same as for the USFS National Cubic Log Scale (Fig. 2.3 on p. 11).

2.3.4.2 Length measurements

1. Measure short side to short side.

2. Scaling length is actual length minus trim; normally 6″ per segment for logs 8–20′ in nominal length; 12″ for logs 20–40′ in nominal length; and 18″ for logs 40–60′ in nominal length (see Table A.1.C).

3. Maximum segment length is 20′ plus trim.

- Logs longer than 20′ are divided into two or more segments as necessary and scaled as individual log segments.
- Multisegment logs are divided as evenly as possible in 2′ multiples (if possible) with the shortest segment(s) on the small-end of the log.
- Under no circumstances should there be more than a 2′ difference between the longest and shortest segment or more than one odd-length segment in any multisegment log (see Table A.1.C).

2.3.4.3 Taper distribution

1. Each log segment (logs with a nominal length of 8–20′) is considered to taper 0.125″ per lineal foot with the scaling cylinder readjusting to a new diameter every 4′, which is 0.5″ larger than the prior cylinder.

2. Two-segment logs (logs with a nominal length greater than 20′ but less than 40′), total taper ÷ number of segments added to the small-end diameter = the small-end diameter of segment 2. If total taper is not evenly divisible by the number of segments, raise taper to next number evenly divisible by the number of segments (Fig. 2.4 on p. 12).

3. Three-segment logs (logs with a nominal length of 40–60′), use same procedure as for two-segment log to find small-end diameter of large-end segment; repeat procedure for the remaining two segments (as if the large-end segment did not exist) to find the small-end diameter of the middle segment.

Taper distribution can also be determined by using Table A.1.D.

2.3.4.4 Gross volume determination

Volume is calculated for each 4′ section of scaling cylinder based on its small-end diameter using the following formula:

$$\text{bf} = (0.199 \times \text{diameter}^2) - (0.642 \times \text{diameter});$$

round the sum of the 4′ cylinders to the nearest 5 bf

Note: For cylinders that are less than 4′ in length (as would occur in a log segment that is not a multiple of 4′), use the following formula adjustment: volume from formula × (length of cylinder ÷ 4).

For example, if the formula yields 43.6 bf and the final cylinder length is 2′: $43.6 \times (2 \div 4) = 21.8$ bf.

Two-segment log

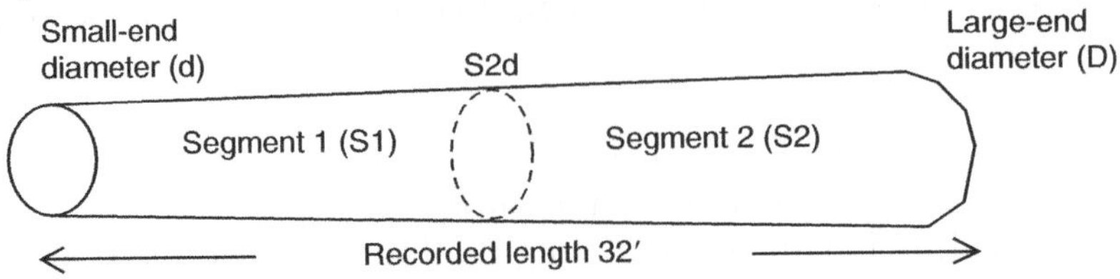

d = 15″
D = 18″
S2d is interpolated as 17″

Segment 1

1st 4′ cylinder: (0.199 × 15) (0.642 × 15) = 35.1 bf
2nd 4′ cylinder: (0.199 × 15.5^2) (0.642 × 15.5) = 37.9 bf
3rd 4′ cylinder: (0.199 × 16^2) (0.642 × 16) = 40.7 bf
4th 4′ cylinder: (0.199 × 16.5^2) (0.642 × 16.5) = 43.6 bf

Total = 157.3; rounds to 155 bf

Segment 2

1st 4′ cylinder: (0.199 × 17^2) (0.642 × 17) = 46.6 bf
2nd 4′ cylinder: (0.199 × 17.5^2) (0.642 × 17.5) = 49.7 bf
3rd 4′ cylinder: (0.199 × 18^2) (0.642 × 18) = 52.9 bf
4th 4′ cylinder: (0.199 × 18.5^2) (0.642 × 18.5) = 56.2 bf

Total = 205.4; rounds to 205 bf

Total gross volume of log = **360 bf**

Fig. 2.35. International ¼-Inch Log Rule gross volume determination.

Figure 2.35 is an example of calculating log volume using the International ¼-Inch Log Rule. As the formula is rather cumbersome to use in the field, volume is normally determined via a volume table (Table 2.6).

2.3.4.5 Defect deduction rules

A defect is anything that causes a loss in volume of lumber. Stain is not considered a defect unless associated with rot (see the list in Section 2.2.1.5 on p. 12 under *Defect deduction rules*). There are four major methods for deducting for defects, all of which are rounded to the nearest 5 bf:

1. Squared area deduction.
2. Percentage deduction.
3. Length deduction.
4. Diameter deduction.

SQUARED AREA DEDUCTION. This is used primarily for interior defects that lend themselves to being partitioned into a square or rectangular area. The volume of a defect is calculated by the formula:

$$\text{defect deduction} = (W'' + 1) \times (H'' + 1) \times L' \div 16, \text{ round to nearest 5 bf}$$

PERCENTAGE DEDUCTION. This is used for defects that go from the perimeter to the heart area of the log, and can best be reckoned with by enclosing the defect in a 'pie-shaped' sector with a fractional representation for the length affected. This fraction is then applied to the length of the log segment, e.g. if a log segment has a scaling length of 16′, with one half of 6′ (½ of 6 = 3) of the log having a defect requiring a deduction, the net

Table 2.6. International ¼-Inch Log Rule volume chart (board feet).

Small-end diameter of log segment in inches	Length of log segment in feet																			
	1	**2**	**3**	**4**	**5**	**6**	**7**	**8**	**9**	**10**	**11**	**12**	**13**	**14**	**15**	**16**	**17**	**18**	**19**	**20**
3	0	0	0	0	0	0	0	0	0	0	0	0	0	0	0	0	0	5	5	5
4	0	0	0	0	0	0	0	0	0	5	5	5	5	5	5	5	5	10	10	10
5	0	0	0	0	0	5	5	5	5	5	5	10	10	10	10	10	15	15	15	15
6	0	0	0	5	5	5	5	10	10	10	10	15	15	15	20	20	20	25	25	25
7	0	5	5	5	5	10	10	10	15	15	15	20	20	25	25	30	30	35	35	40
8	0	5	5	10	10	10	15	15	20	20	25	25	30	35	35	40	40	45	50	50
9	5	5	10	10	15	15	20	20	25	30	30	35	40	45	45	50	55	60	65	70
10	5	5	10	15	15	20	25	30	35	35	40	45	50	55	60	65	70	75	80	85
11	5	10	15	15	20	25	30	35	40	45	50	55	65	70	75	80	85	95	100	105
12	5	10	15	20	25	30	40	45	50	55	65	70	75	85	90	95	105	110	120	125
13	5	15	20	25	30	40	45	55	60	70	75	85	90	100	105	115	125	135	140	150
14	10	15	25	30	40	45	55	65	70	80	90	100	105	115	125	135	145	155	165	175
15	10	20	25	35	45	55	65	75	85	95	105	115	125	135	145	155	170	180	190	205
16	10	20	30	40	50	60	75	85	95	110	120	130	145	155	170	180	195	205	220	235
17	10	25	35	45	60	70	85	95	110	125	135	150	165	175	190	205	220	235	250	265
18	15	25	40	55	65	80	95	110	125	140	155	170	185	200	215	230	250	265	280	300
19	15	30	45	60	75	90	105	125	140	155	175	190	205	225	240	260	280	295	315	335
20	15	35	50	65	85	100	120	135	155	175	195	210	230	250	270	290	310	330	350	370
21	20	35	55	75	95	115	135	150	175	195	215	235	255	280	300	320	345	365	390	410
22	20	40	60	80	105	125	145	170	190	215	235	260	285	305	330	355	380	405	430	455
23	25	45	70	90	115	140	160	185	210	235	260	285	310	335	360	390	415	440	470	495
24	25	50	75	100	125	150	175	205	230	255	285	310	340	370	395	425	455	485	515	540
25	25	55	80	110	135	165	195	220	250	280	310	340	370	400	430	460	495	525	560	590
26	30	60	90	120	150	180	210	240	275	305	335	370	400	435	470	500	535	570	605	640
27	30	65	95	130	160	195	225	260	295	330	365	400	435	470	505	540	580	615	655	690
28	35	70	105	140	175	210	245	280	320	355	395	430	470	505	545	585	625	665	705	745
29	35	75	110	150	185	225	265	305	345	385	425	465	505	545	585	630	670	715	755	800

Continued

Table 2.6. *continued*

	Length of log segment in feet																				
30	40	80	120	160	200	245	285	325	370	410	455	495	540	585	630	675	720	765	810	855
31	45	85	130	170	215	260	305	350	395	440	485	530	580	625	675	720	770	820	870	915
32	45	90	135	185	230	280	325	375	420	470	520	570	620	670	720	770	820	875	925	980
33	50	100	145	195	245	295	345	395	450	500	555	605	660	715	765	820	875	930	985	1040
34	50	105	155	210	260	315	370	425	480	535	590	645	700	760	815	870	930	990	1050	1105
35	55	110	165	220	280	335	390	450	510	565	625	685	745	805	865	925	990	1050	1110	1175
36	60	115	175	235	295	355	415	475	540	600	665	725	790	855	915	980	1045	1110	1180	1245
37	60	125	185	250	315	375	440	505	570	635	700	765	835	905	970	1040	1105	1175	1245	1315
38	65	130	195	265	330	400	465	535	605	670	740	810	880	955	1025	1095	1170	1240	1315	1390
39	70	140	210	280	350	420	490	565	635	710	780	855	930	1005	1080	1155	1235	1310	1385	1465
40	75	145	220	295	370	445	520	595	670	745	825	900	980	1060	1140	1215	1300	1380	1460	1540
41	75	155	230	310	385	465	545	625	705	785	865	950	1030	1115	1195	1280	1365	1450	1535	1620
42	80	160	245	325	405	490	575	655	740	825	910	995	1085	1170	1260	1345	1435	1525	1615	1700
43	85	170	255	340	430	515	600	690	780	865	955	1045	1135	1230	1320	1410	1505	1600	1690	1785
44	90	180	270	355	450	540	630	725	815	910	1005	1095	1190	1290	1385	1480	1575	1675	1775	1870
45	95	185	280	375	470	565	660	755	855	955	1050	1150	1250	1350	1450	1550	1650	1755	1855	1960
46	100	195	295	390	490	590	690	790	895	995	1100	1200	1305	1410	1515	1620	1725	1835	1940	2050
47	100	205	305	410	515	620	725	830	935	1040	1150	1255	1365	1475	1585	1695	1805	1915	2025	2140
48	105	215	320	430	535	645	755	865	975	1090	1200	1310	1425	1540	1655	1765	1885	2000	2115	2230
49	110	225	335	445	560	675	790	900	1020	1135	1250	1370	1485	1605	1725	1845	1965	2085	2205	2325
50	115	235	350	465	585	705	820	940	1060	1185	1305	1425	1550	1675	1795	1920	2045	2170	2300	2425

scale of the log is equal to that of a 13′ long log. The defect is equal to the difference between the gross volume and the net volume. Sweep (a constant bow in the log) is deducted by the following formula: percent deduction from gross volume = (maximum deflection in inches per log segment − 2) ÷ small-end diameter.

LENGTH DEDUCTION. This is normally used for defects that cause a loss of volume for all of, or some of, the length affected. This deduction rule is commonly used when the entire scaling cylinder is deducted or in combination with the percentage deduction (as above). The defect equals the difference between the volume as calculated from the gross length and the volume as calculated from the net length. Crook deductions are accounted for by projecting a theoretically straight log through the straightest portion of the log segment and deducting for the volume that falls outside of the actual log segment, e.g. if sweep affects 6′ of a log, of which ⅓ of the theoretic cylinder falls outside the log, the length deduction is 2′ (⅓ of 6 = 2).

DIAMETER DEDUCTION. This is used for defects that occur in the perimeter of the log, such as rotten sapwood, surface checks and twisted grain. This deduction is made by establishing net diameters (under the defective portion of the log). The defect deduction is the difference between the volume as calculated from the gross diameters and the volume as calculated from the net diameters.

2.3.4.6 Net volume and merchantability

In most regions, if the net scale of a log segment is less than 33⅓% of the gross volume, the log will be culled (having a net volume of zero).

2.3.4.7 Pulp log scale

Like all primary product rules, the International ¼-Inch Rule is not very well adapted for measuring volume or value of pulp logs, and consequently is seldom utilized for this purpose.

2.4 Other Methods of Scaling

2.4.1 Weight scale

One of the most widely used methods of measuring roundwood is weight. As would be expected, there is a strong correlation between weight and volume. In many regions, weight is used to establish roundwood volume. Weight is also used to extrapolate volume (as well as size and grade) of logs from sample scaling. Weight is commonly used as the unit of measure for payment of transportation and harvesting of timber, and is utilized as a means of regulating trucking weight limits. Weight-scales are often easily accessible to logs in transport; commonly being located adjacent to transportation routes as well as at many roundwood processing facilities. The specialized trucks for transporting logs are often equipped with load cells under the bunks, which can very accurately measure the weight of logs. Even the equipment utilized for loading, unloading, and moving logs in the woods or the mill can now be equipped with scales that operate on static hydraulic pressure or load cells (for non-hydraulic equipment).

On top of the accessibility of weight, and fast data processing, weight is universally tangible, and is relatively inexpensive, especially in comparison with the cost of scaling loads via log-by-log measurements, which not only entails the labour cost of the scaler, but also log handling costs (heavy equipment and operators), and the associated log damage which occurs from handling. The relative low cost of using weight to determine volume is especially useful for low valued roundwood such as pulp logs, firewood and small saw timber (where the log by log measurement costs are the highest). Weight can also be utilized in higher valued logs to varying degrees dependent on the uniformity of the timber, the accuracy levels needed, and the amount of control needed for log manufacturing quality.

Normally, the weight of the logs is either directly converted to volume (m^3, ft^3, cords, steres, etc.), using established conversion factors, or weight is used as a vehicle for extrapolating volume via sample scaling. Weight can be utilized entirely as the final unit of measure in itself, but this is seldom done, as upstream and downstream processes (standing inventories, harvesting costs, timber taxes, government reporting, comparing product recoveries, etc.) typically require volumetric units, and log weight varies over time. In this section, we will focus on the use of weight directly to determine volume. The use of weight to extrapolate volume via sample scaling will be discussed in Section 2.6, and roundwood weight and physical properties will be covered in Chapter 5.

To extrapolate volume from weight, the normal procedure is to divide the weight by a standard conversion factor. For example, in parts of the southeast USA, 'Southern pine' (loblolly pine, short leaf pine, long leaf pine and slash pine) are directly converted using 70.7 lb per ft^3 gross and 74.0 lb per ft^3 net (this conversion assumes 4.5% defect) based on the use of the USFS National Cubic Log Scale.[2] For example, a load of logs is delivered having a net weight of 54,000 lb: $54{,}000 \div 74 = 729.7$ net ft^3 or 7.297 ccf.

In eastern Canada, weight is commonly converted directly to volume using fixed weight-to-volume factors set by the provincial forest service. The following is based on information taken from Ontario Ministry of Natural Resources (2000). In the province of Ontario the majority of the publicly owned stumpage is sold via 'mass scaling' (volume converted

[2] This number is derived from conversions used by some private timberland companies in the Southeast USA. It uses 71.8 lb per ft^3 and assumes 4.5% defect, which gives a 68.6 lb per ft^3 gross scale, based on actual length (no trim allowance). Assuming 3.0% additional unmeasured trim allowance (as would approximate the USFS National Cubic Log Scale conventions), the gross weight factor changes to 70.7 lb per ft^3 gross and 74.0 lb per ft^3 net.

from weight). The provincial forest service annually checks weight-to-volume ratios by species, and publishes a set of fixed gross volume conversion ratios to be used province-wide. The extrapolated gross volume for each timber sale agreement is converted to a net volume based on a cutting site specific factor for undersize volume and defect. Multiple species with very different weight-to-volume ratios can also be accommodated in this method as the weight conversions are weighted by species (determined on the entire sale area) and the appropriate adjustments from undersize and defects made at a species level. Example 2.1 demonstrates this methodology.

Weight scale can fulfill the objectives listed at the beginning of Chapter 1: to obtain accurate, consistent scale through which one is able to measure volume and value, predict product output, establish accurate inventory and usage volumes, and to convert from one unit/method of measure to another. As stated earlier, weight also has the advantage of being a relatively inexpensive method of accounting for log volumes. The relative accuracy and controls needed vary a great deal and are dependent on many factors, but in general, the following guidelines should be considered as to the suitability of weight for determining volume:

1. **Homogeneity of the timber.** The weight-to-volume ratios of timber vary a great deal by species, heartwood to sapwood ratios, wood density, seasonality, bark retention, time between cutting and weighing, types and amount of defects, weather conditions, and whether cut from a living or a dead tree. These variables will be discussed at further length in Chapter 5.
2. **Unit of measure utilized.** The correlation between weight and cubic volumes is quite good. However, the correlation between weight and

Example 2.1.

Load net weight 28,000 kg	Sale composition species %	Fixed kg/m^3		Undersize factor	Defect factor
Eastern hemlock	32%	1009	32% × 1009 = 322.88	0.9%	6.5%
Eastern larch	51%	994	51% × 994 = 506.94	4.1%	2.9%
Balsam fir	17%	791	17% × 791 = 134.47	3.9%	6.4%
			Weighted kg/m^3 = 964.29		

Load volume: 28,000 ÷ 964.29 = 29.037 m^3

		Gross m^3	Undersize m^3	Defect m^3	Net m^3
Eastern hemlock	32% × 29.037 =	9.292	0.084	0.604	8.604
Eastern larch	51% × 29.037 =	14.809	0.607	0.429	13.772
Balsam fir	17% × 29.037 =	4.936	0.193	0.316	4.428
	Extrapolated load total	29.037	0.883	1.349	26.804

some of the product output rules, such as the bf rules, is generally not very good unless the timber is uniform.

3. Log manufacturing quality control. Since logs are not individually scrutinized and measured, it is more difficult to control log manufacturing variables such as log lengths, diameters, defect removal or minimization, and overall manufacturing quality unless sample loads are taken.

4. Level of size and grade data needed. Timber can vary a great deal in value due to size and grade. As mentioned above, logs are not being scrutinized at an individual level.

2.4.1.1 Pounds per lineal foot

If there is a need in gaining log size information, average diameter data can be approximated at a composite level without log by log measurements if the following formula is applied (assuming the variables are accurately reflected):

Step 1. Net weight of load in pounds [or kilograms] ÷ (total number of logs × average log length in feet [or metres]) = pounds per lineal foot [or kilograms per lineal metre].

Step 2. Pounds per lineal foot [or kilogram per metre] ÷ average gross pounds per cubic foot [or kilogram m^3] = average square foot [or m^2] end area of log.

Step 3. Average midpoint log diameter in inches [or centimetres] $=\sqrt{}$(average ft^2 [or m^2] end area ÷ 0.7854) × 144 [or 10,000 for metric system].

Figure 2.36 shows examples of using the 'pounds per lineal foot' method for determining log size, and as shown in Figs 2.37 and 2.38, it can even be used to convert to other means of scaling that do not correlate well when converted to volume by weight alone (such as Doyle).

Based on the calculations from Fig. 2.36, a composite average log can be constructed by using average taper. Assuming the above logs have an average taper of 0.1″ per lineal foot (0.833 cm per m).

Load A would have a top diameter of (using imperial): (16.35 − (27.0 ÷ 2 × 0.1) = **15.0″** (37.5 cm); and a bottom diameter of: (16.35 + (27.0 ÷ 2 × 0.1) = **17.7″** (44.3 cm). The average log in load A is:

L = 27′, d = 15.0″, D = 17.7″ (L = 8.23 m, d = 38.1 cm, D = 45.0 cm)

To convert load A to Doyle: reduce length by 3% to account for normal trim allowance and increase the diameters by 0.3″ to account for over 'one bark' measurement, calculate volume using the Doyle formula, and multiply by the number of logs: (27 × 0.97) = 26.19 L; 15.0 + 0.3 = 15.3″ d; 17.7 + 0.3 = 18.0″ D. Because the log is over 20′ divide into two segments, but divide exactly in two rather than via the standard segmenting procedures: 26.19 ÷ 2 = 13.095. Figure 2.37 shows the procedures for calculating the volume of each segment, log and the load via the Doyle Log Rule.

Load B would have a top diameter of (using metric): (24.52 − (5.24 ÷ 2)) × 0.833 = **22.34 cm** (8.79″); and a bottom diameter of: 24.52 + ((5.24 ÷ 2) × 0.833) = **26.7 cm** (10.51″). The average log in load B is:

Load A

Net weight = 54,000 lbs (24,490 kg), average log length 27.0' (8.23 m), log count = 20, average lbs per ft^3 for Southern yellow pine = 68.6 (1098.9 kg/m^3), total load volume 7.87 ccf (22.29 m^3)

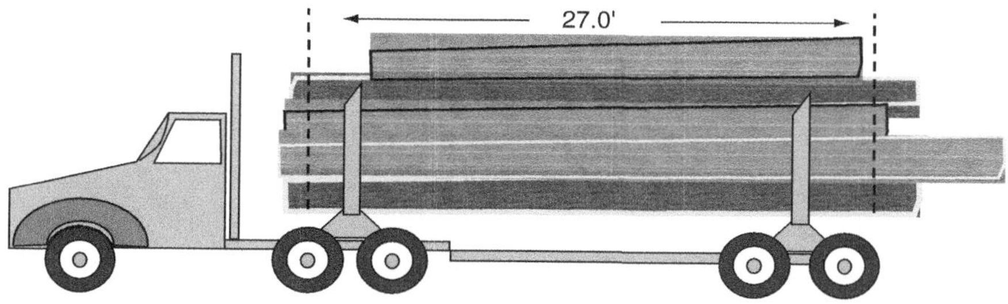

Imperial

54,000 ÷ (20 × 27) = 100 lbs per LF; 100 ÷ 68.6 = 1.458; (1.458 ÷ 0.7854) × 144 = 267.31; √267.31 = **16.35"**

Metric

24,490 ÷ (20 × 8.23) = 148.8 kg per m; 148.8 ÷ 1098.9 = 0.1354; (0.1354 ÷ 0.7854) × 10,000 = 1724; √1724 = **41.5 cm**

Load B

Net weight = 54,000 lbs (24,490 kg), average log length (16.8 + 17.6) ÷ 2 = 17.2' (5.24 m), log count = 90, average lbs per ft^3 for Southern yellow pine = 68.6 (1098.9 kg/m^3), total load volume 7.87 ccf (22.29 m^3)

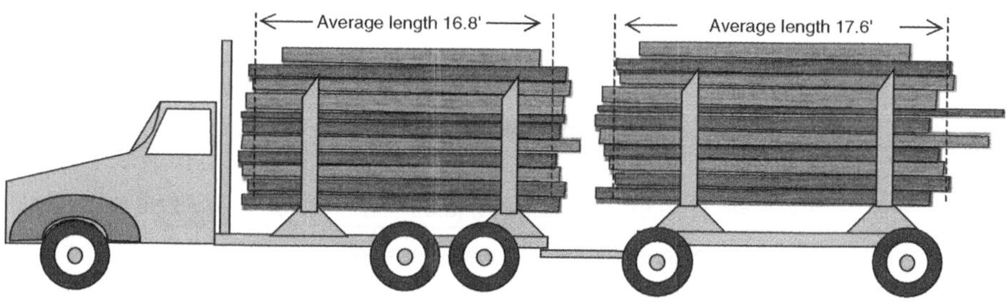

Imperial

54,000 ÷ (90 × 17.2) = 34.88 lbs per LF; 34.88 ÷ 68.6 = 0.508; (0.508 ÷ 0.7854) × 144 = 93.14; √93.14 = **9.65"**

Metric

24,490 ÷ (90 × 5.24) = 51.90 kg per m; 51.90 ÷ 1098.9 = 0.04723; (0.04723 ÷ 0.7854) × 10,000 = 601.35; √601.35 = **24.52 cm**

Fig. 2.36. Pounds per lineal foot.

$$L = 5.24\,\text{m},\ d = 22.07\,\text{cm},\ D = 26.44\,\text{cm}\ (L = 17.2',\ d = 8.79'',\ D = 10.51'')$$

To convert load B to Doyle: reduce length by 3% to account for normal trim allowance and increase the used diameters by 0.3″ to account for over 'one bark' measurement, calculate volume using the Doyle formula, and multiply by the number of logs: $(17.2 \times 0.97) = 16.68\,L$; $8.79 + 0.3 = 9.09''$ d. Because the log is less than 20′, it is scaled in one segment and only the small-end diameter is needed for the volume calculation. Figure 2.38 shows the procedures for calculating the volume of log and the load via the Doyle Log Rule.

As a generalization, weight as a direct form of log scale lends itself to operations where there is good control over log manufacturing

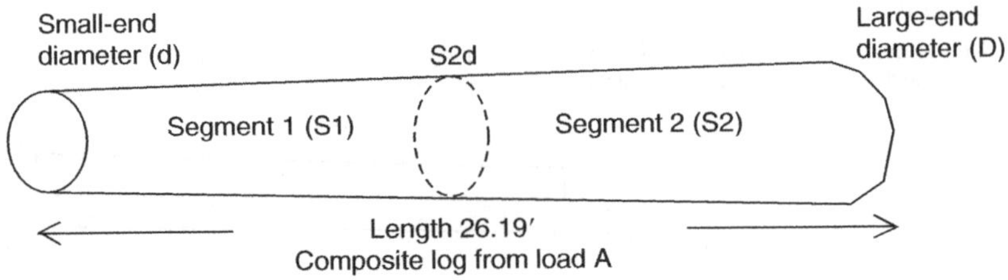

d = 15.3″
D = 18.0″
S2d (and S1D) = 16.65″

Segment 1: 13.095′ long, 15.30″ d; $(15.30 - 4)^2 \times 13.095 \div 16 =$ **105 bf**
Segment 2: 13.095′ long, 16.65″ d; $(16.65 - 4)^2 \times 13.095 \div 16 =$ **131 bf**

Total log volume = **236 bf Doyle;** total load gross volume = 236 × 20 logs = **4720 bf**

Fig. 2.37. Converting load A to Doyle bf using 'pounds per lineal foot'.

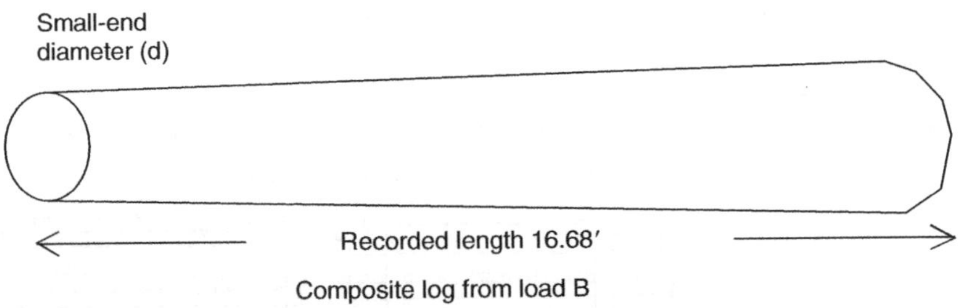

d = 9.09″

Segment 1: 16.68′ long, 9.09″ d; $(9.09 - 4)^2 \times 16.68 \div 16 =$ **27.0 bf**

Total log volume = **27.0 bf** Doyle; total load gross volume = 27.0 × 90 logs = **2430 bf**

Fig. 2.38. Converting load B to Doyle bf using 'pounds per lineal foot'.

specifications, logistics, quality and utilization guidelines. It also works best where variability of timber is not high (even aged stands, without much defect, single species strata or species with similar weight-to-volume ratios) and where cubic measure is utilized.

2.4.2 Stacked wood scale

When roundwood (occasionally split wood in the case of firewood and some specialty products) is stacked, the dimensions of the stacked wood can be measured and a volume derived. The height, width and length of the stack is measured and multiplied to determine the cubic area occupied by the stack of wood. For example, a stack of wood which has a height of 4′ (1.219 m), a width of 4′ (1.219 m), and a length of 8′ (2.438 m) occupies a cubic area of 128 ft^2 or 3.623 m^3. Two common units of stacked wood measure are the cord and the stere:

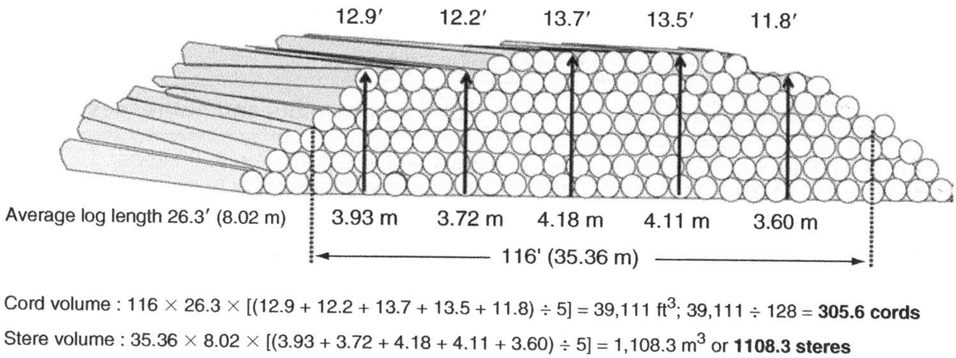

Fig. 2.39. Measurements to determine stacked measure volume.

1. **Cord** measure is used in North America. A cord is defined as being a unit of measure of stacked wood, occupying a space of 128 ft^3, thus the formula is: Cord $= H' \times W' \times L' \div 128$
2. **Stere** is typically used in regions outside of North America and is defined as being a unit of measure of stacked wood occupying the space of 1 m^3, thus the formula is: Stere $=$ Hm $\times$ Wm $\times$ Lm.

Figure 2.39 is an example of determining the volume of stacked wood with cords or steres.

Converting to solid wood. It is important not to confuse stacked cubic volume with solid wood cubic volume. There is obviously air space between the stacked logs and usually bark as well. The percentage of solid wood in stacked measure normally ranges from 60% to 70% of solid wood cubic volume (occasionally 50–80%), depending on the size of the timber, presence of bark and its thickness, straightness, presence of limbs, amount of taper, buttressing, the presence of unsound wood, the tightness of the stacking, and to what degree wood is deducted. It is best to obtain 'timber type' specific solid wood factors by doing periodic tests of measuring stacked volume against actual solid wood volumes from scaling the logs piece by piece. In lieu of doing that, however, the following factors shown in Table 2.7 (general guideline) and Table 2.8 (more specific guidelines) can be used for approximation.

Table 2.7. General guidelines for components of area of stacked roundwood.

Solid wood	66.67%
Bark	11.46%
Air	21.87%
Total	100.00%

Source: Ontario Ministry of Natural Resources, 2000.

Table 2.8. Swedish National Board of Forestry stacked measure guidelines for pulp logs.

Starting value 60%

Average log diameter under bark	%
<8.99 cm (<3.54″)	−4
9–9.99 cm (3.55–3.93″)	−3
10–10.99 cm (3.94–4.33″)	−2
11–11.99 cm (4.34–4.72″)	−1
12–12.99 cm (4.73–5.11″)	0
13–13.99 cm (5.12–5.51″)	+1
14–14.99 cm (5.52–5.90″)	+2
15–15.99 cm (5.91–6.30″)	+3
16–16.99 cm (6.31–6.69″)	+4
17–17.99 cm (6.70–7.08″)	+5
18–19.99 cm (7.09–7.87″)	+6
20–22.99 cm (7.88–9.05″)	+7
23–26.99 cm (9.06–10.63″)	+8
> 27 cm (> 10.64″)	+9

Quality of stacking	%
Very good	0
Good	−1
Fair	−2
Bad	−3 to −4
Very bad	−5 to −7

Trimming of knots and buttresses	%
Very well trimmed, few buttresses	0
A few knot stumps and buttresses	−1
Many knot stumps and buttresses	−2
Many large knot stumps, clusters and buttresses	−3 to 4
Very bad trimming	−5 to 7

Crookedness	%
Straight softwood, straight aspen	0 to −1
Crooked softwood, straight deciduous wood	−2 to 3
Very crooked softwood, crooked deciduous wood	−4 to 5
Very crooked deciduous wood	−6 to 7
Extremely crooked deciduous wood	−8 to 12

Bark volume	%
Debarked wood	+7
More than ⅔ thin bark	+2
More than ⅓ thin bark	+1
Normal conifer	0
Softwood with thick bark, normal deciduous bark	−1
More than 50% deciduous wood with thick bark	−2

Source: VMF Nord, 1999.

2.4.3 Automated measurement systems (scanners, photo-cells)

Scanning technology has been in use for quite a long time. Set-works for veneer lathes, head rigs, and canters have long used scanners to measure the log in order to position a log properly for peeling, sawing or chipping off slabs. Harvesting equipment utilized in logging operations, e.g. cut-to-length systems (CTLs), and log processors (delimbing, bucking), also are being fitted with mechanical and/or laser measuring devices. One of the residual benefits of this scanning process is that log volumes can be calculated from the scanned measurements. Scanners are currently finding use for measuring log deliveries (notably in the Nordic countries), and even more common is their use to measure usage going into the mill.

In many cases the volume from the scanner is factored to take into account hidden defects and/or differences in the gross volume determination, e.g. scanner usage $\times$ 0.92 = volume minus defect, butt flare, and trim allowance. This approach works quite well once factors are determined for a particular stratum.

The current generation of scanning technology uses multipoint lasers to take very accurate measurements of a log profile, and with today's high powered and quick computer processors, build a three dimensional model of the log (Fig. 2.40). This technology has not only allowed for tremendous improvements in the optimization of product recovery, but it

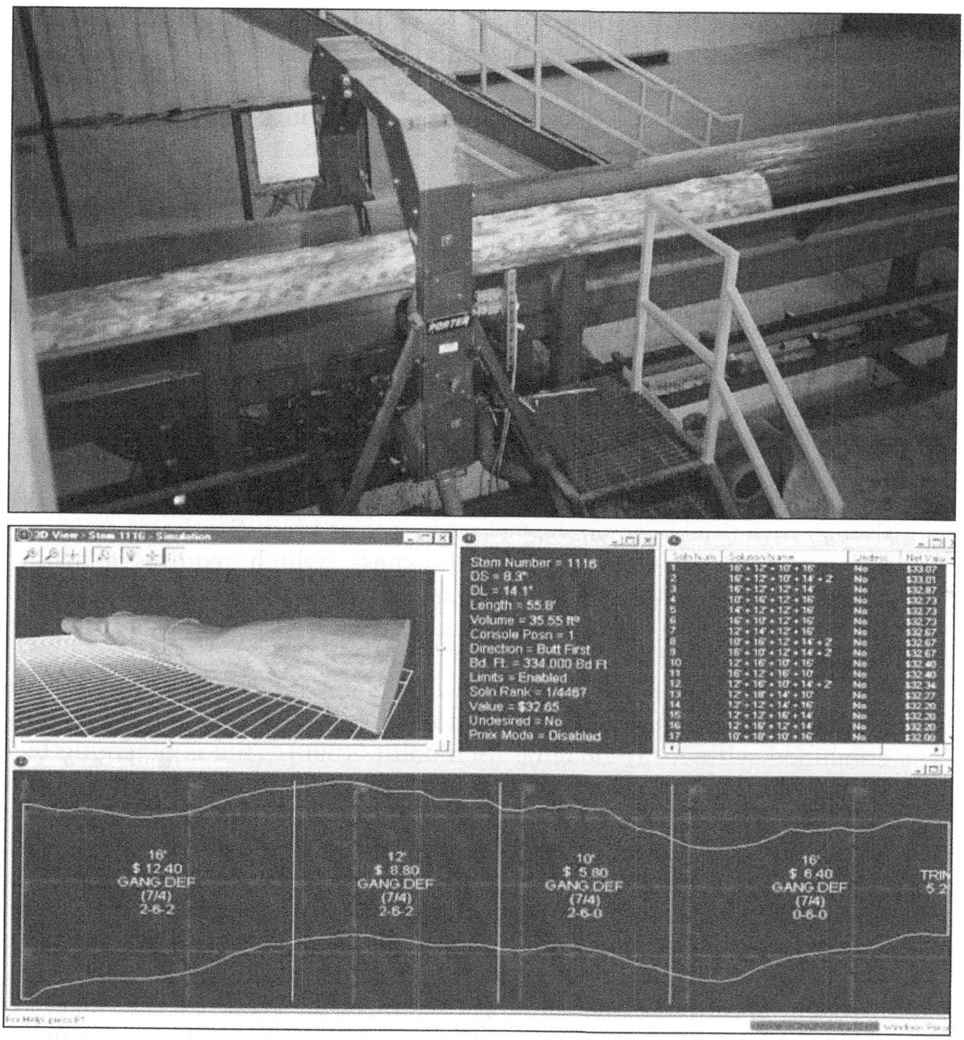

Photo courtesy: Porter Engineering, Richmond, British Columbia, Canada.

Fig. 2.40. Three-dimensional log scanning and images.

has the ability to measure, to a very high degree of accuracy, the log volume put through the scanner. This includes externally visible defects such as sweep, crook and voids which can all be readily recognized and measured via the three-dimensional image.

Of course, there are limitations to these systems which have thus far limited wide scale moves to utilize scanners for anything other than the intended purposes of optimization of product recovery, and the side benefit of measuring usage volume in the mill or production from logging equipment. One issue is cost; scanners are a sizeable investment, and not every operation can afford one. Furthermore, scanners exist primarily for product optimization purposes, which means that they are installed at the conveying systems leading into the mill (ahead of the bucking station), and just ahead of the primary log breakdown centres (head rig, canter, lathe). These locations are not conducive to the proper accounting of incoming logs. Generally, it is necessary for logs to have the volume and purchasing parameters (grade, species, size, source, logging contractor, trucker) determined for payment and accounting purposes, and then the logs placed into the log yard inventory for future use.

Given the need for accounting of incoming logs, most operations would need a separate dedicated scanning line in order to preserve the integrity of the data needed for purchasing and receiving timber, which of course is added expense without the primary benefits of this equipment (optimization of product recovery).

Further limitations exist because optical scanners cannot see log grade nor can they account for internal defects (rot, heart checks, shake, etc.), and logs put into storage in inventory should have bark (in order to prevent wood degradation from stain and splits). While logs can be scanned with bark, the accuracy of the measurements is reduced. Although bark thickness can be factored, species identification is an important parameter in factoring bark thickness, which requires a manned operation with someone imputing species and likely grade and internal defect information (somewhat of a challenge in a production line environment and added expense).

Excepting the factor of cost, all of the above limitations can be overcome through the use of computer tomography (CT) technology which is combined with X-ray imaging (Schmoldt, 1996). This is emerging technology, but it is already proven to effectively find, measure and map internal defects, as well as determine bark from wood. Coupled with laser scanning, there is tremendous potential for this technology. Again though, this technology will find its primary application in optimization equipment, with use for log scaling being a secondary focus.

It remains to be seen if the majority of log deliveries will someday be measured and recorded by scanners, but currently, most deliveries in Sweden are already measured via scanners for gross volume determination. Many mills are already utilizing three-dimensional scanning for calculating mill usage, and it is just a matter of time before CT technology finds its way into production facilities, especially where values are high.

2.5 Converting between Log Scaling Methods

The need for good conversions has become more important recently given the large volume of interregional roundwood trade as well as the desire for harmonized roundwood volume statistics. Obtaining accurate conversions from one method of log scale to another can be problematic. As has already been discussed, methodologies, procedures and formulas vary a great deal between the different log-scaling methods, and these variations seldom remain consistent between size classes, timber types, and quantity and types of defect. Furthermore, the variables that determine volume ratios (log lengths, diameters, taper, etc.) are occasionally manipulated to maximize or minimize volumes (especially with the Scribner log rules) causing bias.

The process of developing conversions is also complicated by attempting to convert logs with parameters which do not exist where the scaling methods originate. Scaling methods develop to suit the needs and fit the applications that exist within a region. For example, the Doyle formula is commonly criticized as grossly understating the volume of small logs, while it is true that in principle small logs (4–7″ d) are grossly understated relative to bigger logs, in practice the Doyle formula is virtually never used to measure logs less than 7.5″ in diameter on the small-end, making a comparison which includes logs under 7.5″ misleading (other methods such as weight or stacked measure are utilized for these smaller logs).

2.5.1 Modelling conversion factors

Despite the difficulties of developing generic conversion factors for the various log scaling methods, it is useful to develop a 'regionally neutral' model with closely categorized diameter and length classes to allow proper weighting of size information. In this case, the author chose to utilize a model based on 175 spruce logs that were thoroughly measured in regard to diameters, length, taper and defects. These logs are not unlike conifers elsewhere (and not unlike many hardwoods), in terms of the important parameters that can affect conversion ratios (taper, ovality and defect). The distribution of log lengths and percentage of butt cut logs in the model are based on a typical mix that would be common in western North America. Obviously there will be populations of logs that do not fit well with this model, but the listed conversions are a place to start, in lieu of developing population specific conversions.

The following is a summary of the attributes of the logs used for the model:

Range of small-end diameters .... 11.4–90.2 cm (4.5″–35.5″)
Range of large-end diameters .... 13.7–116.6 cm (5.4″–37.3″)
Range of lengths .... 2.5–12.5 m (8.3′–41′)

Average taper for all logs.. 1.11 cm/m (0.119″/ft)
Average taper for butt cut logs,
DBH to small-end... 0.97 cm/m (0.109″/ft)
Av. log dimensions 11.43–19.05 cm
group (4.5–7.49″)......................... 7.9 m–14.7 cm–22.4 cm (25.9′–5.8″–8.8″)
Av. log dimensions 19.06–29.19 cm
group (7.5–11.49″)..................... 7.2 m–23.7 cm–30.4 cm (23.6′–9.3″–12.0″)
Av. log dimensions 29.2–39.35 cm
group (11.5–15.49″)................. 8.5 m–34.9 cm–42.7 cm (27.9′–13.8″–16.8″)
Av. log dimensions >39.35 cm (>15.5″)
group .... 7.6 m–55.6 cm–62.2 cm (24.9′–21.9″–24.5″)
Av. log dimensions total group of
logs in the model .... 7.8 m–32.5 cm–39.7 cm (25.5′–12.8″–15.6″)

The log-scaling procedures from each of the 15 methods (16 including the non-revised version of the Scribner Short Log Rule) listed were applied to the key dimensions and defects of the 175 logs. For those rules that do not have procedures for scaling long logs as one piece, or which stipulate the need to segment when taper exceeds a certain level (BC Firmwood), long logs were segmented and the actual midpoint diameters were used in the absence of taper distribution rules. In other words, the tree stems were theoretically cut into separate logs, the volume was ascertained, and in those situations where comparisons were made with long logs scaled by other scaling methods, the segment volumes were added together so that a comparison could be made. In the case of the New Zealand 3-D method, 4 cm was added to the projected large-end diameter of all butt cut logs to account for the convention of including butt flare in the large-end diameter measurement. This number was chosen as a simple average by analysing stem profiles of an assortment of various sized Radiata pine.

The log volumes were summarized into 16 categories (four diameter and four length categories), shown in Table A.1.M and from these categorized volumes, an index was created (Table 2.9), which is also categorized into four length classes and four diameter classes.

Table 2.9 and Fig. 2.41 indexes the volume from the listed log scaling methods in relation to one cubic metre as measured by the BC Firmwood Scale. The BC Firmwood Scale was chosen as the benchmark as it uses an unbiased rounding logic, well-documented and consistent procedures, and accounts for lengths, diameters and defects in a consistent, tangible way. The product output rules, all of which are based on the board foot, are represented in units of 1000 bf (mbf) in relation to 1 m^3 BC Firmwood. In the case of Swedish cubic, Russian standard, Cubage au Réel, Brereton, Hoppus and JAS scale, the lack of definitive documentation (and actual knowledge on the part of the author) on application of defect deductions resulted in comparisons made for these log scales based solely on gross volumes. As New Zealand 3-D is only used on plantation wood that is considered free of defect, it is also only compared on a gross scale basis. It is important to note that defect deductions can affect conversion ratios for those log scales that either use grades to address defects, only deduct for

Table 2.9. Volume index by length and small-end diameter class (1.00 m^3 BC Firmwood = 1.000).

	2.5–4.6 m 8–15 ft		4.7–6.4 m 16–21 ft		6.5–9.5 m 22–31 ft		9.6–12.5 m 32–41 ft		Total all lengths	
	Gross	Net	Gross	Net	Gross	Net	Gross	Net	Gross	Net
	Small-end diameter 11.43–19.05 cm (4.5–7.49″)									
USFS Cubic	0.941	0.901	0.925	0.871	0.966	0.904	0.920	0.897	0.930	0.897
BC Firmwood	1.000	1.000	1.000	1.000	1.000	1.000	1.000	1.000	1.000	1.000
Alberta Cubic	0.979	0.929	0.962	0.912	0.980	0.928	0.979	0.956	0.978	0.946
Ontario Cubic	0.952	0.956	0.989	0.989	0.963	0.963	0.955	0.961	0.959	0.963
Swedish Cubic*	0.840	–	0.700	–	0.817	–	0.787	–	0.789	–
Russian Standard*	1.009	–	0.979	–	0.978	–	0.988	–	0.987	–
Cubage au Réel*	0.936	–	0.929	–	0.943	–	0.941	–	0.940	–
New Zealand 3-D*	1.140	–	1.040	–	1.019	–	0.980	–	1.000	–
Brereton (PNG)	0.944	–	0.896	–	0.902	–	0.881	–	0.890	–
Hoppus	0.789	–	0.768	–	0.766	–	0.772	–	0.771	–
JAS Scale*	0.908	–	0.723	–	0.835	–	0.820	–	0.821	–
Scribner Short R**	0.160	0.162	0.169	0.155	0.153	0.137	0.154	0.148	0.155	0.147
Scribner Short NR**	0.160	0.162	0.169	0.126	0.142	0.132	0.148	0.139	0.149	0.138
Scribner Long Log**	0.160	0.144	0.126	0.112	0.116	0.105	0.105	0.106	0.112	0.108
Doyle**	0.038	0.035	0.023	0.022	0.080	0.075	0.061	0.059	0.060	0.058
International ¼″**	0.133	0.126	0.126	0.112	0.179	0.171	0.170	0.164	0.166	0.159

Continued

Table 2.9. *continued*

	2.5–4.6 m 8–15 ft		4.7–6.4 m 16–21 ft		6.5–9.5 m 22–31 ft		9.6–12.5 m 32–41 ft		Total all lengths	
	Gross	Net	Gross	Net	Gross	Net	Gross	Net	Gross	Net
Small-end diameter 19.06–29.19 cm (7.5–11.49″)										
USFS Cubic	0.905	0.876	0.935	0.882	0.935	0.911	0.930	0.892	0.930	0.893
BC Firmwood	1.000	1.000	1.000	1.000	1.000	1.000	1.000	1.000	1.000	1.000
Alberta Cubic	0.973	0.941	0.963	0.909	0.974	0.948	0.970	0.932	0.970	0.933
Ontario Cubic	0.977	0.977	0.972	0.971	0.967	0.944	0.966	0.964	0.968	0.962
Swedish Cubic*	0.858	–	0.850	–	0.861	–	0.838	–	0.847	–
Russian Standard*	0.957	–	0.988	–	0.985	–	0.978	–	0.979	–
Cubage au Réel*	0.948	–	0.969	–	0.952	–	0.981	–	0.969	–
New Zealand 3-D*	1.021	–	1.027	–	0.977	–	0.947	–	0.974	–
Brereton PNG	0.928	–	0.945	–	0.924	–	0.918	–	0.925	–
Hoppus	0.761	–	0.784	–	0.772	–	0.781	–	0.777	–
JAS Scale*	0.932	–	0.890	–	0.963	–	0.905	–	0.919	–
Scribner Short R**	0.161	0.156	0.155	0.138	0.170	0.159	0.168	0.155	0.166	0.153
Scribner Short NR**	0.156	0.151	0.146	0.131	0.166	0.159	0.168	0.156	0.163	0.153
Scribner Long Log**	0.127	0.127	0.140	0.128	0.134	0.132	0.123	0.119	0.129	0.124
Doyle**	0.125	0.122	0.111	0.105	0.141	0.137	0.131	0.125	0.130	0.125
International $^1/_4$″**	0.193	0.185	0.200	0.186	0.205	0.196	0.206	0.196	0.204	0.193
Small-end diameter 29.2–39.35 cm (11.5–15.49″)										
USFS Cubic	0.943	0.930	0.954	0.914	0.968	0.905	0.941	0.924	0.946	0.921
BC Firmwood	1.000	1.000	1.000	1.000	1.000	1.000	1.000	1.000	1.000	1.000
Alberta Cubic	0.977	0.958	0.965	0.933	0.989	0.945	0.971	0.955	0.973	0.952
Ontario Cubic	0.997	0.997	0.965	0.959	0.993	0.993	0.982	0.970	0.982	0.974
Swedish Cubic*	0.916	–	0.877	–	0.921	–	0.889	–	0.893	–

Russian Standard*	0.989	–	0.985	–	0.996	–	0.987	–	0.988	–
Cubage au Réel*	0.955	–	0.957	–	0.982	–	0.969	–	0.969	–
New Zealand 3-D*	1.013	–	0.993	–	1.034	–	0.961	–	0.976	–
Brereton	0.961	–	0.944	–	0.963	–	0.947	–	0.949	–
Hoppus	0.779	–	0.775	–	0.786	–	0.784	–	0.783	–
JAS Scale*	0.993	–	0.975	–	1.082	–	1.000	–	1.007	–
Scribner Short R**	0.202	0.198	0.190	0.182	0.209	0.197	0.203	0.195	0.202	0.194
Scribner Short NR**	0.202	0.198	0.190	0.182	0.209	0.197	0.203	0.195	0.202	0.194
Scribner Long Log**	0.187	0.183	0.188	0.180	0.160	0.155	0.153	0.144	0.159	0.151
Doyle**	0.188	0.185	0.174	0.165	0.181	0.169	0.191	0.183	0.188	0.180
International ¼″**	0.231	0.227	0.225	0.214	0.240	0.224	0.235	0.227	0.234	0.225
Small-end diameter > 39.36 cm (15.5″)										
USFS Cubic	0.929	0.867	0.964	0.952	0.974	0.953	0.969	0.956	0.966	0.947
BC Firmwood	1.000	1.000	1.000	1.000	1.000	1.000	1.000	1.000	1.000	1.000
Alberta Cubic	0.978	0.915	0.963	0.951	0.994	0.981	0.972	0.956	0.978	0.960
Ontario Cubic	1.003	0.994	0.984	0.970	0.991	0.986	1.002	0.999	0.995	0.989
Swedish Cubic*	0.965	–	0.941	–	0.964	–	0.932	–	0.947	–
Russian Standard*	0.982	–	0.978	–	1.000	–	0.985	–	0.988	–
Cubage au Réel*	0.962	–	0.960	–	0.980	–	0.987	–	0.977	–
New Zealand 3-D*	1.060	–	1.055	–	1.061	–	1.015	–	1.041	–
Brereton	0.974	–	0.964	–	0.982	–	0.971	–	0.974	–
Hoppus	0.773	–	0.770	–	0.789	–	0.789	–	0.784	–
JAS Scale*	1.086	–	1.063	–	1.095	–	1.090	–	1.086	–
Scribner Short R**	0.250	0.227	0.255	0.248	0.265	0.255	0.247	0.242	0.255	0.246
Scribner Short NR**	0.250	0.227	0.255	0.248	0.264	0.254	0.247	0.242	0.255	0.246
Scribner Long Log**	0.245	0.220	0.241	0.233	0.233	0.225	0.201	0.197	0.223	0.215
Doyle**	0.264	0.241	0.257	0.251	0.267	0.258	0.242	0.235	0.255	0.246
International ¼″**	0.259	0.236	0.271	0.263	0.271	0.261	0.266	0.259	0.268	0.259

Continued

Table 2.9. *continued*

	2.5–4.6 m 8–15 ft		4.7–6.4 m 16–21 ft		6.5–9.5 m 22–31 ft		9.6–12.5 m 32–41 ft		Total all lengths	
	Gross	Net	Gross	Net	Gross	Net	Gross	Net	Gross	Net
					Total all diameters					
USFS Cubic	0.928	0.883	0.958	0.935	0.969	0.940	0.950	0.932	0.955	0.931
BC Firmwood	1.000	1.000	1.000	1.000	1.000	1.000	1.000	1.000	1.000	1.000
Alberta Cubic	0.977	0.928	0.964	0.942	0.991	0.970	0.972	0.953	0.976	0.954
Ontario Cubic	0.995	0.990	0.980	0.969	0.988	0.981	0.987	0.981	0.987	0.980
Swedish Cubic*	0.933	–	0.914	–	0.941	–	0.893	–	0.912	–
Russian Standard*	0.980	–	0.980	–	0.997	–	0.985	–	0.987	–
Cubage au Réel*	0.957	–	0.960	–	0.975	–	0.976	–	0.972	–
New Zealand 3-D*	1.048	–	1.041	–	1.047	–	0.985	–	1.014	–
Brereton	0.963	–	0.957	–	0.970	–	0.949	–	0.957	–
Hoppus	0.773	–	0.772	–	0.786	–	0.785	–	0.782	–
JAS Scale*	1.036	–	1.020	–	1.068	–	1.014	–	1.030	–
Scribner Short R**	0.223	0.207	0.231	0.222	0.243	0.232	0.214	0.207	0.225	0.216
Scribner Short NR**	0.222	0.206	0.230	0.220	0.242	0.232	0.213	0.206	0.224	0.215
Scribner Long Log**	0.212	0.195	0.217	0.208	0.208	0.201	0.167	0.161	0.189	0.181
Doyle**	0.218	0.202	0.221	0.213	0.235	0.226	0.196	0.188	0.212	0.203
International ¼″**	0.238	0.221	0.252	0.242	0.256	0.245	0.240	0.232	0.246	0.236

Note: The indexes for 'Total all diameters' are not necessarily representative of any specific population.
*Swedish Cubic, Russian Standard, Cubage au Réel, New Zealand 3-D, Brereton, Hoppus and JAS Scale are indexed only to gross scale.
**Product output rules are reflected in units of 1000 board feet (mbf), indexed against 1.00 m^3 BC Firmwood. Scribner Short R = revised Scribner used in California, Oregon, Washington and Alaska, Scribner NR = non-revised Scribner used elsewhere.

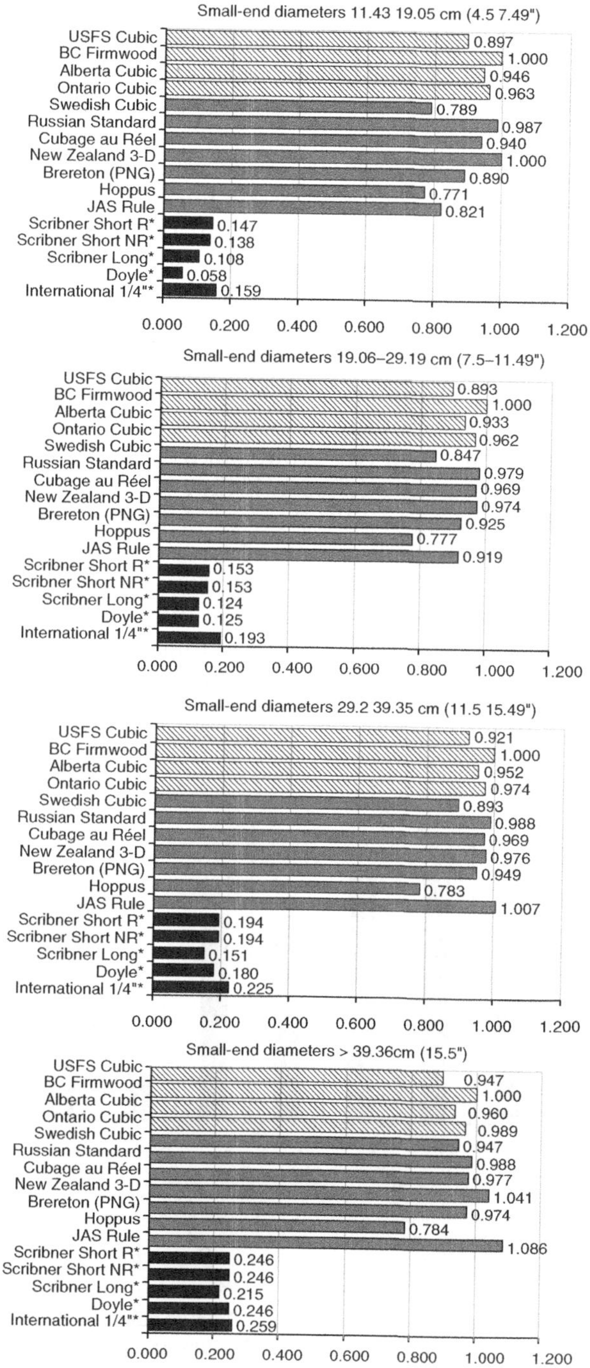

Note: Swedish Cubic, Russian Standard, Cubage au Réel, New Zealand 3-D, Brereton, Hoppus, and JAS Scale are indexed only to gross scale. Scribner Short R = revised Scribner used in California, Oregon, Washington and Alaska, Scribner NR = non-revised Scribner used elsewhere. *Product output rules are reflected in units of 1000 board feet (mbf), indexed against 1.00 m^3 BC Firmwood.

Source: Calculated by author.

Fig. 2.41. Net volume index by small-end diameter class (1.000 m^3 BC Firmwood = 1.000).

void, soft-rot and char, or utilize some or both of these conventions. If a comparison is needed with any of the cubic log scaling methods using cubic feet, one needs to multiply the m^3 by 35.315.

Even within each of the 16 L and diameter groupings listed, there can be substantial variation of conversion ratios, especially amongst JAS and the product output rules. For example, a log which is 20′–6″–9″ has 20 bf when scaled via Scribner Short Log and $0.182\,m^3$ ($6.4\,ft^3$) using USFS Cubic. This gives a ratio of 0.110 mbf per m^3; a log which is 16′–5″–6″ also scales 20 bf under the Scribner Short Log Rule, but only has $0.0765\,m^3$ ($2.7\,ft^3$) using USFS Cubic, which gives a ratio of 0.262 mbf per m^3 (a 138% increase over the first ratio). Variation is more pronounced in the smaller diameter classes, primarily due to the effects of rounding diameters and lengths, and in the case of the two Scribner based rules, rounding the volumes to 10 bf volume classes.

Figure 2.42 is based on two groups of hypothetical logs measured via some selected scaling rules (chosen if they showed substantial variation due to taper) indexed against BC Firmwood. It is included to illustrate the effects of taper on conversion ratios (it is not necessarily a statistically correct representation of any particular group of logs). The two groups of logs have identical lengths and small end diameters, are 5 m and 10 m in length (16.5′ and 33′) and have small end diameters ranging from 15 cm to 36 cm (6″–14″). The low taper group averages 0.6 cm/m (0.07″/ft) taper

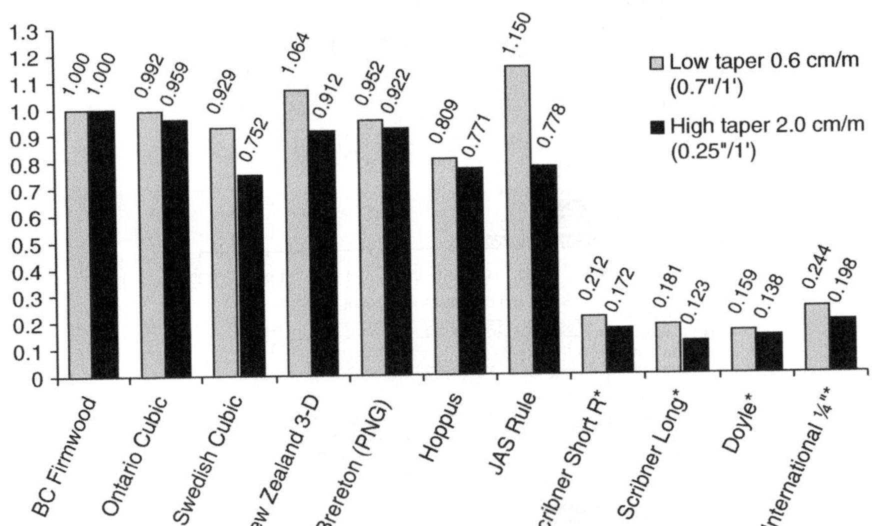

Note: All scale rules are indexed only to gross scale. *Product output rules are reflected in units of 1000 board feet (mbf), indexed against 1.00 m^3 BC Firmwood.

Source: Calculated by author.

Fig. 2.42. Effects of log taper on converting between selected log scales (1.000 m^3 BC Firmwood = 1.000).

and the high taper group averages 2.0 cm/m (0.25″/ft). As shown in Fig. 2.42, taper will affect the conversion ratios significantly in the product output rules, Swedish cubic and JAS which view a log segment as a form of a cylinder (thus ignoring volume outside of the scaling cylinder). Increased taper can also reduce the volume of the Huber based methods (Ontario Cubic, Brereton) relative to the volume of Smalian based measure. This is because the area of both log-end areas ÷ 2 will always exceed the area as determined by both diameters ÷ 2. It should be noted that neither Huber nor Smalian is necessarily the more accurate; the actual volume could go either way, dependent on the form characteristics of the log, and methods of determining diameters and length.

New Zealand 3-D also appears (Fig. 2.42) to have a higher relative volume for low tapered logs vs. high tapered logs, but this is a result of the formula, which makes assumptions as to stem-form (neiloid, conic, paraboloid) based on taper, and thus is sensitive to hypothetical scenarios such as this one, which classifies taper without butt flare.

Figure 2.43 shows the product output rules indexed against Scribner Short Log R (revised) which is used in California, Oregon, Washington and Alaska. Scribner Short Log NR (non-revised), which is used elsewhere in the USA, is included to show that the difference between it and revised Scribner in the smaller diameter logs is significant, but relatively unimportant in larger diameters. Table A.1.N shows the Scribner Long Log Rule indexed against Scribner Short Log Scale and is included in the annex as a guide for individual log dimensions. Table A.1.O shows the relationship of Scribner Long Log and Scribner Short Log to BC Firmwood

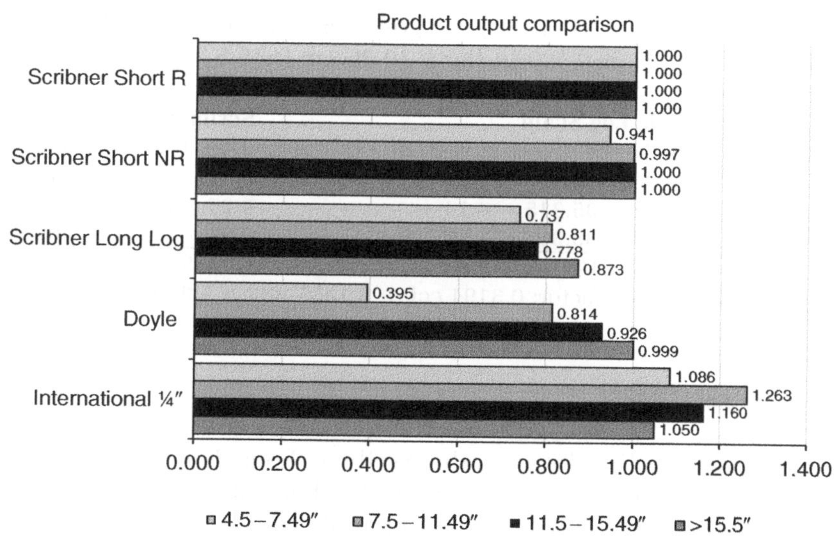

Fig. 2.43. Net volume index by small-end diameter class for product output (bf) rules (1.000 mbf Scribner Short Log Rule = 1.000).

Scale m^3, based on standard Oregon and Washington Department of Natural Resource size classifications.

2.5.2 Examples of using conversions

The following examples of converting log volumes are based on the information in Table 2.9.

Example 2.2

Assume that one would like to estimate the approximate conversion ratio to BC Firmwood, for logs that are scaled via Scribner Long Log and come from a region that harvests approximately 20% in the 11.43–19.05 cm (4.5–7.49″) category, 30% in the 19.06–29.19 cm (7.5–11.49″) category, 30% in the 29.2–39.35 cm (11.5–15.49″) category, and 20% in the >39.36 cm (>15.5″) category.

$$\text{Conversion factor} = (0.2 \times 0.108) + (0.3 \times 0.124) + (0.3 \times 0.151) + (0.2 \times 0.215) = 0.1471\,\text{mbf per m}^3 \text{ or } 6.8\,\text{m}^3 \text{ per mbf}$$

Example 2.3

Assume that a conversion is needed to convert Scribner Short Log Scale to USFS Cubic (reflected in cubic feet or cunits) for timber made up of 40% in the 11.43–19.05 cm (4.5–7.49″) category, 30% in the 19.06–29.19 cm (7.5–11.49″) category, 25% in the 29.2–39.35 cm (11.5–15.49″) category, and 5% in the > 39.36 cm (>15.5″) category.

USFS Cubic Log Scale	**Scribner Short Log**
$0.4 \times 0.897 \times 35.315 = 12.67$	$0.4 \times 0.147 = 0.0588$
$0.3 \times 0.893 \times 35.315 = 9.46$	$0.3 \times 0.153 = 0.0459$
$0.25 \times 0.921 \times 35.315 = 8.13$	$0.25 \times 0.194 = 0.0485$
$0.05 \times 0.947 \times 35.315 = 1.67$	$0.05 \times 0.246 = 0.0123$
31.94 ft^3 (or 0.3194 ccf)	0.1655 mbf

Conversion factor: 0.3194 ccf $\div$ 0.1655 mbf = **1.93 ccf per mbf**

Example 2.4

A company selling logs on the Scribner Long Log Rule for $400/mbf would like to obtain a conversion to Scribner Short Log Rule in order to find what the equivalent price per mbf would be. The mix of logs is typically 55% 7.5–11.49″, 30% 11.5–15.49″, and 15% > 15.5″. Using the information from Fig. 2.43:

Conversion factor $= (0.55 \times 0.811) + (0.30 \times 0.778) + (0.15 \times 0.873) =$
0.81 mbf Scribner Long Log per 1 mbf Scribner Short Log;
$0.81 \times \$400 =$ an equivalent price of **$324/mbf**

2.6 Sample Scaling

Sample scaling is a method for estimating the total volume (and often other values) of a group of logs, without having to physically measure all the logs. Generally, this entails measuring the volume and attributes of a quantifiable sample portion of a log population, and assigning the known attributes from the sample population to the non-sampled population via the quantifier. Normally, in dealing with roundwood, the sample is quantified to weight, by the truckload or by piece count. Log volume, especially cubic measured and weight correlate well, and there is also good correlation between log volume and space occupied. Thus weight, or truck load when there is good consistency of load size, are the most common methods of quantifying volume. Of course, volume is not always the only attribute that gets extrapolated from the sample loads, often species, log diameter and length data, grade and manufacturing quality data are also recorded from the sample loads and expanded into the non-sampled population.

The aim of sample scaling is to give acceptably accurate estimates for roundwood volume without the cost and effort of 100% scaling. The realization of reduced effort and cost can be especially relevant given: fee wood, where absolute accuracy may not be needed; low valued wood, where scaling and log handling costs can be prohibitive; larger log sales (large total populations) which have lower standard errors, thus more reliable estimates; and situations where the extra handling of logs needed for scaling causes value losses from additional damage, breakage, and dirt/rock contamination of logs.

In the case of any sampling system, the sample selection should be based on random sample selection given the number of samples desired. There are many systems, both computerized and manual that will give a randomly selected sample at a given frequency.

Besides assuring the randomness of sample selection, it is also very important to eliminate other biases from the sample system, such as an incorrectly defined stratum, changing a stratum after a load was selected as a sample, and changing sample frequencies without closing a sample base and restarting a new one.

While it is not the author's intentions to delve too deeply into the conventions and logic behind statistical sampling (there are whole publications on this subject), we will cover the fundamentals of determining delivered log volumes using sampling techniques.

2.6.1 Determining sample size

Before starting the process of sample scaling, it is important to determine the intensity of sampling needed to obtain the desired level of accuracy. Higher accuracy will require more intensive sampling. Obviously, the intensity of the sample ratio is not the only variable that drives accuracy. The variability of the value per unit measured (can be volume or other attributes such as monetary value) and the size of the total population have effects that have to be quantified in order to determine an acceptable sampling intensity.

The procedures for determining sample size have been adapted from the procedures listed in the *Alberta Scaling Manual* (Alberta Land and Forest Service, 1992). To calculate the number of samples needed to obtain an estimate that fits within a predetermined accuracy standard, one needs to determine the critical value to be estimated. This critical value will likely be volume (such as m^3, ccf, mbf), or monetary value. It may be that a sale is delivering logs of different species or grades with very similar volume-to-weight ratios, but with large differences in monetary values. In this scenario, the sampling intensity required to estimate the volume of the sale will likely be insufficient to measure the monetary value of the sale (assuming significant variation of load to load species or grade content). Under this scenario, monetary value per unit would likely be the preferred choice of the critical value to be used in the formula, unless it is feasible to stratify the sale (stratification will be discussed in Section 2.6.3).

Once the value critical to determining sampling intensity is decided, the allowable sampling error and probability interval need to be determined. The allowable sampling error (SE) is a measure of the precision of the estimate, and is generally set between 5% and 1%, with exponential increases in the samples needed as the allowable sampling error is decreased. Probability is reflected as the *t* value and is often set at 2, which statistically means that there is a 95.3% (or roughly 19 out of 20 times) likelihood that the number of samples taken from the sample size formula will give results within the allowable sampling error specified in the formula.

The final variables needed to determine sample size are an estimate of the number of loads (statistically referred to as the 'total population') that will be delivered from the timber sale, and a control group of loads of logs, which would ideally be at least 10 representative loads from the sale in question.

Listed below is a methodology of the statistical procedures for determining sample frequency from a measured control group of loads:

1. Determine average value per unit (e.g. tons per m^3, ccf per load, $ per ton, mbf per load, etc.): average value per unit = total value of control group ÷ total units of control group.
2. Determine average value per load: total value of control group ÷ total loads of control group.

3. Apply the average value per unit (from step 1) to the units of each load in the control group, e.g. net tonnes from each load ÷ the average tonnes per m^3 from the control group = theoretical extrapolated volume for each load.
4. Calculate the difference between the actual value of each load, and the extrapolated value from above, e.g. if the actual m^3 of a load is 24.212 and the theoretical extrapolated volume is 25.003, the difference is 0.791 m^3.
5. Square the difference of each load, and sum all of the squared differences from each load in the control group.
6. Calculate the Standard deviation of the loads in the control group by the following formula: standard deviation = √(the sum of the squared differences ÷ (number of samples −1)).
7. Determine the coefficient of variation (CV): CV = standard deviation ÷ mean load.
8. Calculate the number of samples needed: samples needed = (total number of loads × t^2 × CV^2) ÷ ((total numbers of loads × allowable sampling $error^2$) + (t^2 × CV^2)). If less than 30 samples go to next step.
9. Calculate the number of samples needed using revised t value (rt) from Table 2.10: samples needed = (total number of loads × rt^2 × CV^2) ÷ ((total numbers of loads × allowable sampling $error^2$) + (rt^2 × CV^2)).

Note: The t value is a constant based on the level of probability and the size of the sample. The t value generally is 2, which represents a 95% confidence level.

Example 2.5 is a hypothetical control group of loads of logs and calculations used to determine the recommended sample size.

Step 1. Average tons/ccf: 270.9 ÷ 83.718 = **3.236**
Step 2. Average volume per load: 83.718 ÷ 10 = **8.3718**

Example 2.5. Examples of determining sample size.

Step	Net weight	Actual	Extrapolated ccf based on average *ton* ccf	Absolute difference	Absolute difference^2
	Ton	Scaled ccf			
1	27.1	7.795	8.375	−0.580	0.336
2	25.7	8.327	7.942	0.385	0.148
3	26.7	8.440	8.251	0.189	0.036
4	29.1	8.148	8.993	−0.845	0.714
5	26.4	7.991	8.159	−0.168	0.028
6	27.5	8.607	8.499	0.108	0.012
7	28.3	8.702	8.746	−0.044	0.002
8	27.4	8.603	8.468	0.135	0.018
9	26.7	8.477	8.251	0.226	0.051
10	26.0	8.628	8.035	0.593	0.352
	270.90	**83.718**	**83.718**		**1.697**

Step 3. Apply the average value per unit (from step 1) to the units of each load in the control group, e.g. 27.1 ÷ 3.236 = 8.375, etc.

Step 4. Calculate the difference between the actual value of each load, and the extrapolated value from above, e.g. 8.375 − 7.795 = 0.580, etc.

Step 5. Square the difference between the actual and extrapolated volume (or value) from each load, e.g. $0.580^2 = 0.336$, etc.

Step 6. Calculate the standard deviation (SD): $\sqrt{}(1.697 \div (10 \text{ sample loads} - 1)) =$ **0.43423**

Step 7. Determine the CV: standard deviation of 0.43423 ÷ mean load of 8.3718 = **5.187%**

Step 8. Number of samples needed: $(57 \text{ loads expected} \times 2^2 \times 5.187^2) \div ((57 \text{ loads expected} \times 5^2) + (2^2 \times 5.187^2)) =$ **4.003**

As this is less than 30 samples, look up new revised *t* value (*rt*) in Table 2.10 and go to step 9 (in this case the revised *t* value is 3.18).

Step 9. Number of samples needed: $(57 \text{ loads expected} \times 3.18^2 \times 5.187^2) \div ((57 \text{ loads expected} \times 5^2) + (3.18^2 \times 5.187^2)) =$ **9.14 or a 16% sample.**

Note: $t = 2$, which means a 95% confidence level (unless less than 30 sample loads; in which case go to Table 2.10, revised *t* value (*rt*)).

SE = 5, which is the allowable sampling error in this example.

Remaining total number of loads expected = sale volume estimated to be 561 ccf: (561 ccf – 83.7 ccf already delivered) ÷ 8.37 = 57.02 loads.

Table 2.10. Revised *t* value.

Calculated (*n* – 1)	Revised *t*	Calculated (*n* – 1)	Revised *t*
1	12.7	16	2.12
2	4.3	17	2.11
3	3.18	18	2.1
4	2.78	19	2.09
5	2.57	20	2.09
6	2.45	21	2.08
7	2.36	22	2.07
8	2.31	23	2.07
9	2.26	24	2.06
10	2.23	25	2.06
11	2.2	26	2.06
12	2.18	27	2.05
13	2.16	28	2.05
14	2.14	29	2.05
15	2.13		

Note: Use this table for *t* when the number of samples is less than 30.
Source: Alberta Land and Forest Service, 1992.

Based on the above statistics, 9 sample loads taken on the remaining 57 loads will give an answer within ± 5% of the true mean 95% of the time.

It should be noted that the procedures outlined above are generally geared toward large sale volumes, and while it is true that sample scaling works better given large populations, it can also be used for smaller populations when the variation is low. Given smaller sale sizes, it is feasible to estimate the CV based on historical evidence for the timber type in question, and calculate a sample frequency without a control group of loads of logs, e.g. if one is about to receive a relatively small sale comprised of homogeneous plantation timber, of approximately 30 loads in total, the information which would normally be gathered from the control group sample can be estimated based on historical norms for the timber type in question, and what is known about the sale. In this case, if we assume that for this timber type, the CV will be about 4.5% (based on knowledge of similar timber harvested in the same area), we would need 10 loads in total to meet a 5% allowable sampling error at the 95% confidence level. Setting the sample frequency at a 1 in 3 ratio should suffice to obtain the sampling objectives.

Opposite the above scenario is a common situation where the timber in question is very homogeneous and well suited for weight sample scaling, but the timber owner is sceptical of statistical estimates and wants all of the logs scaled. In this case, being able to give clear explanations of the statistical processes of sample scaling to the timber owner and a willingness to oversample the timber sale, are often better options than having to 100% scale the sale, or not getting the sale at all.

2.6.2 Types of sample scaling

Timber sale volume is generally expanded based on one of three different types of sample systems: simple sample, weight and 3-P (which is short for probability proportional to prediction). In the case of weight sampling versus simple sampling, weight is generally a better way to quantify and expand sample and non-sampled volume, but requires a weight-scale, which may not be available. Three-P is a method of sampling, which is generally best suited for sales with a high coefficient of variation that would normally require a high sampling intensity.

2.6.2.1 Simple sample (count)

Simple sampling utilizes count as the quantifying factor for extrapolation, e.g. volume per load, volume per log, etc. rather than volume per weight unit. Simple sampling has the potential to have more variation than weight sample because truck loads or log size can vary substantially. It should be noted though, that in many regions, truck configurations are uniform, and thus often the CVs per truck load are not so different from the CVs of weight-to-volume ratios.

Sample scaling based on volume per stem or log, works best in uniform stands of wood, especially where trees are manufactured into tree-length logs. This system works exactly like an inventory on standing trees, excepting that the level of accuracy is potentially better as the tree can be more accurately measured when on the ground. In Canada, the forest services of both Alberta and Ontario use a similar system for scaling tree-length logs, based on taking a sample of tree-length stems, scaling the stem (in segments) with callipers or by cutting the stem into mill-length logs, and developing a volume per stem by species ratio (Alberta Land and Forest Service, 1992; Ontario Ministry of Natural Resources, 2000). The non-scaled component is extrapolated by measuring the butt diameter class and species, and the volume expanded from the sample stem averages for a particular butt diameter class/species stratum.

2.6.2.2 Weight sample

Weight sample scaling is probably the most common method of sample scaling. Timber has a good correlation between weight and volume, and drive-on weight-scales are often located along the transportation routes as well as on the plant site of many roundwood conversion facilities. Load cells (pads which can register weight), and devices that record weight via static hydraulic pressure are also being installed on many log hauling trucks, and cable operated or hydraulic operated machines for handling logs.

The principle behind weight scaling is simple; the weight-to-volume relationship from the sample population is applied to the non-sampled population to extrapolate volume (and often other attributes of the non-sampled population).

2.6.2.3 3-P (corrected estimate)

Three-P scaling is a system where logs are sampled rather than loads, and the non-scaled logs are given an estimated value. The scaling process is generally accomplished via a handheld data recorder that also generates random numbers, which are used to select sample logs at a predetermined frequency. The scaler makes a quick visual estimate of the key values of the log (generally species and gross volume in ft^3, m^3, or bf). After the log volume is estimated and entered into the data recorder, a random number chosen from between the lowest and highest expected volume estimates signifies whether a log has to be physically scaled. If the random number is equal to or less than the estimate, the log will be physically scaled, if the random number is greater than the estimated value, the scaler will go to the next log without taking any measurements. The larger a log is, and thus its estimated volume, the higher the probability that the log will be selected as a sample log. A log that is estimated as having a value of 40 is four times more likely to be sampled than a log with an estimated value of 10. Volume is expanded based on the factor generated by dividing the actual volume of the scaled logs by the estimated volume from the same logs. The coefficient of variation used to determine the sample frequency is taken from the difference between the estimate and actual volume.

The advantage of 3-P is that variation is controlled by the scaler's ability to estimate consistently, and not so much from the variability of the logs themselves, or the inability to stratify the sale. This means that scaling costs can be reduced on timber sales that contain highly variable wood, by estimating the values of all the logs, but only physically scaling (measuring) a small percentage. Weight or simple sample scaling can be combined with 3-P (so that one is taking a sample of a sample), but this reduces the advantages gained from visually accounting for all the variables of the timber sale.

2.6.3 Population and subpopulations to be expanded (stratum)

When devising a sample plan, thought has to be given to at what level the population to be expanded will be defined. Some of the typical levels are: total sale area, delivery point (mill sort-yard, etc.), harvest contract (when the sale is divided into separate areas and there are more than one contractor harvesting the timber), timber type, etc. Dividing a timber sale into smaller more homogeneous units is called stratification and is a very important aspect of sample scaling.

Example 2.6 simulates a sampling plan for a timber stumpage sale that has been sold to a purchaser that has two mills (a large log plant and a small log plant). The purchaser has hired two logging contractors to harvest the timber from this sale. Each of the logging contractors is segregating the logs into loads based on size (which determines if the logs go to a big log mill or a small log mill) and species. In this example, the buyer, seller, the harvest contractor and more than likely, some government agencies (for taxation and reporting purposes), will each have their own data needs. Volumes for all the above key variables (two mills, two contractors and two species; eight different combinations) will be needed by one or more of the parties involved. Under this scenario, the timber sale should be stratified into eight different strata.

Volume and load counts (sampled and total) for each of the eight strata is listed in Example 2.6. The eight areas shaded are the expansion levels (strata) used in this scenario, e.g. big mill pine from contractor 'A', big mill pine from contactor 'B', big mill spruce from contractor 'A', big mill spruce from contractor 'B', small mill pine from contractor 'A', small mill pine from contractor 'B', small mill spruce from contractor 'A', small mill spruce from contractor 'B'. While the sale in the example was not small at 1962 cunits (230 loads), the populations in each stratum were not large, thus one might assume that this would increase the number of samples needed. In fact, given the need for good data on

Example 2.6. Breaking a timber sale into strata.

Estimated sale volume from (ccf)

	Big log mill			Small log mill			Total sale		
	Pine	Spruce	Total	Pine	Spruce	Total	Pine	Spruce	Total
Harvest contractor 'A'	132	254	386	760	29	789	892	283	1175
Harvest contractor 'B'	603	112	715	46	26	72	649	138	787
Total timber sale	735	366	1101	806	55	861	1541	421	1962

Estimated number of all loads

	Big log mill			Small log mill			Total sale		
	Pine	Spruce	Total	Pine	Spruce	Total	Pine	Spruce	Total
Harvest contractor 'A'	16	26	42	95	3	98	111	29	140
Harvest contractor 'B'	70	11	81	6	3	9	76	14	90
Total timber sale	86	37	123	101	6	107	187	43	230

Estimated number of sample loads needed

	Big log mill			Small log mill			Total sale		
	Pine	Spruce	Total	Pine	Spruce	Total	Pine	Spruce	Total
Harvest contractor 'A'	6	6	12	3	3	6	9	9	18
Harvest contractor 'B'	10	4	14	1	3	4	11	7	18
Total timber sale	16	10	26	4	6	10	20	16	36

eight different variables, it would require far more loads to obtain the desired accuracy levels if the sale was not stratified.

The estimate of sample loads needed with an allowable sampling error of 5% at a 95% confidence level is also listed in Example 2.6. The reason for 100% scaling of the small log spruce strata (from contractor 'A' and contractor 'B'), was that there is a risk of not getting a sample load at all due to the randomness of sample selection. It should also be noted that due to the small size of each stratum, it is not practical in this case to get a control group of 10 loads, which is considered the minimum to check the CV. In a situation such as this, the administrator of the sale would likely use historical CVs to calculate frequency and then monitor the CV as sample loads accumulate (the estimated sample loads needed in example 2.6 were calculated using estimated CVs of 8% for big pine, 7% for big spruce, 4% for small pine, and 6% for small spruce). The total estimate of needed sample loads is 36.

When making a sampling plan and determining strata, it is important to keep in mind that a stratum needs to be definable and distinguishable to all parties involved, e.g. if the person determining the stratum of a load misidentifies species on the load and wrongly classifies the stratum, it can have a very bad influence on the accuracy of the statistics. At worst, predicted volumes can be made very inaccurate, and at best, the stratum may not improve the accuracy of the predicted volumes.

Once the sale is under way, it is good policy to review sample loads to determine that the sale is properly stratified, and that the party responsible for determining stratum is doing it properly, e.g. check to see if the sample loads are classified correctly (are there sample loads of pine in the spruce stratum?). While reviewing sample loads, if it is discovered that loads have been put in the wrong stratum, one needs to avoid the temptation of moving sample loads from one stratum to another, as it needs to be assumed that non-sampled loads were misidentified as well.

In summary, stratification is a very useful tool in improving the accuracy level of components of a population. One of the rules of statistical sampling is that the larger the population, the smaller the sample size needed to obtain a level of accuracy. Thus, breaking a large population into smaller populations (strata) would appear to increase the sample size needed; while this is true when just looking at an overall value such as sale volume, it is not likely the case when looking at components of the sale, e.g. the volume of small spruce logs.

2.6.4 Level of expansion

The data collected in the scaling process often contains much more data than just volume. Logically the drivers of the transaction processes are volume, species and grade, but there is often much more information that is noted during the scaling process which can be extrapolated. Length and diameter information, defect deductions, and piece-count, can be expanded. Other attributes such as whether or not the log is a butt-cut (indicating stem removals and DBH), and log manufacturing quality information, which gives a tool for feedback to the timber harvester, are also commonly expanded from the sample loads. The statistical reliability of this data may not be as good as the primary values expanded (the CV will likely be higher than for volume), but the data is valuable none the less, especially when combined with the statistics from other strata (making the overall statistics much more accurate). Expanding volume at this detailed level is referred to as 'expanding at the log level', because the log-by-log detail from the sample loads is quantified and expanded into the non-sampled population by weight. The weight of the non-sampled volume is divided by the weight of the sampled volume to determine the expansion factor, and each individual log record including the desired level of detail is multiplied by this ratio.

Example 2.7 demonstrates expanding sample loads at a detailed level; in this case the log level. In this simplified example (for ease of understanding), the weight of the non-sampled loads are divided by the weight of the sample loads to obtain the expansion factor (6.7102). The factor is then multiplied by the accumulative attributes (volume, log count) in the sampled population, in order to expand individual logs in the non-sampled population. The non-accumulative attributes (species, grade, diameter, length, etc.), are then given to each of the components (in this case logs) which have been expanded, e.g. a log from a sample load that was a spruce, 6 m long with a 94 cm d and a 104 cm D, becomes 6.71 logs with these dimensions in the non-sampled population. In total there are 7.71 spruce logs 6 m – 94 cm – 104 cm from the sale so far (1 from the sample load + 6.71 from the non-sampled load = 7.71).

While this very detailed level of data expansion can be useful for quantifying the drivers of product recovery, log manufacturing quality control, product size and quality, it can also create a great deal of data and may be reserved for those with the resources and desire to process, store and analyse large amounts of data.

2.6.5 Expansion time window

Many factors dictate the amount of time a stratum accumulates and expands volume. In some cases, it is advantageous or even necessary to close the sample base and expansion process of a stratum, and restart the process all over again. The reasons for doing this may include: an unplanned change in the log population (log manufacturing specification change, changing the log sorts going into a stratum), a need to change the sample frequency, significant seasonal variation in weight-to-volume ratios, a change in the harvest contractor or the need for fixed delivery numbers for use in log billing and logger payment. While fixed delivery numbers for interim periods can be calculated for sales expanded over the life of the sale, there is a risk of non-representative volatility in interim period volume calculations or even negative numbers.

To illustrate the possibility of interim volatility, assume that the data presented in Example 2.7 is the first month of deliveries from a timber sale. The purchaser is paying the timber owner $50/m^3$, which totalled $8998.60 for the 179.972 m^3 delivered. Only three loads were delivered in the following month. There were two non-scaled loads with a total net weight of 24,000 kg; one sample load with a net weight of 13,280 kg, a gross volume of 11.362 m^3, and a net volume of 7.150 m^3. The total volume from the sale is re-expanded at the end of the second month of deliveries, and because the sample load that was taken during the second month of the sale is very defective, the weight-to-volume ratio now stands at 1179 kg per m^3 (vs. the 971 kg for the previous month). As shown in Example 2.8, only three loads in total were delivered, and a volume was already calculated with a payment of $8998.60 made for the previous month, the changes to the entire volume has to be absorbed by the second month. Given that the total volume for the sale at the end of the first month was calculated at 179.972 m^3, and the total at the end of the second month is calculated at 179.873 m^3; the timber owner gets a debit of $4.95 for three loads of logs.

The above scenario is a situation that is mathematically correct, but is difficult to explain to a log seller, who may not be well versed in statistical sampling. Expansion volatility, such as negative numbers, are more likely

Example 2.7. Expanding sample volume at a detailed level.

	Sample logs						Non-sample loads	
	Species	Ln	d	D	Gross m^3	Net m^3		Net kg
Sample load 1	Spruce	6	94	104	4.630	3.667	Non-sample load 1	12,100
11,390 kg	Pine	4.8	54	58	1.184	1.122	Non-sample load 2	11,210
	Spruce	4.8	76	80	2.295	2.062	Non-sample load 3	14,040
	Pine	4.8	40	46	0.700	0.700	Non-sample load 4	13,180
	Pine	4.8	106	112	4.482	3.922	Non-sample load 5	10,960
					13.292	11.474	Non-sample load 6	12,300
							Non-sample load 7	12,020
Sample load 2	Spruce	8.2	42	54	1.509	1.509	Non-sample load 8	11,440
11,280 kg	Pine	9.2	44	56	1.829	1.740	Non-sample load 9	11,020
	Pine	9.2	102	114	8.619	8.619	Non-sample load 10	10,680
					11.957	11.868	Non-sample load 11	9,830
							Non-sample load 12	11,960
Total sample base					25.249	23.342	Non-sample load 13	11,380
22,670 kg							**Total**	**152,120**

Expansion factor: 152,120 ÷ 22,670 = 6.7102

	Expanded totals						
	No. logs	Species	Ln	d	D	Gross m^3	Net m^3
Log detail	6.7	Spruce	6	94	104	31.071	24.607
	6.7	Pine	4.8	54	58	7.943	7.531
	6.7	Spruce	4.8	76	80	15.401	13.835
	6.7	Pine	4.8	40	46	4.700	4.700
	6.7	Pine	4.8	106	112	30.078	26.318
	6.7	Spruce	8.2	42	54	10.126	10.126
	6.7	Pine	9.2	44	56	12.272	11.678
	6.7	Pine	9.2	102	114	57.835	57.835
		Average					
	No. logs	Species	Ln	d	D	Gross m^3	Net m^3
Expanded pine totals	33.6	Pine	6.6	69	77	112.828	108.062
Expanded spruce totals	20.1	Spruce	6.3	71	79	56.597	48.568
Total expanded	53.7		6.5	70	78	169.425	156.630
Sample pine totals	5	Pine	6.6	69	77	16.814	16.104
Sample spruce totals	3	Spruce	6.3	71	79	8.435	7.238
	8		6.5	70	78	25.249	23.342
Total volume pine	38.6	Pine	6.6	69	77	129.642	124.166
Total volume spruce	23.1	Spruce	6.3	71	79	65.032	55.806
Total volume	**61.7**		**6.5**	**70**	**78**	**194.674**	**179.972**

to occur at the beginning of a timber sale when the sample base is small, and are often associated with a high coefficient of variation resulting from a poor sampling plan. Because of the risk of this type of situation, some timber purchasers choose to use an expansion period that is the same as

Example 2.8. Expansion volatility.

	Net kg	Gross m^3	Net m^3	kg/m^3		Net kg
Month 1, Sample load 1	11,390	13.292	11.474	993	Month 1 non-sample total	152,120
Month 1, Sample load 2	11,280	11.957	11.868	950	Month 2 non-sample total	24,000
Month 2, Sample load 3	13,280	11.362	7.150	1,857	Total sale to date	176,120
Total Sample	35,950	36.611	30.492	1,179		

Expansion factor : 176,120 ÷ 35,950 = **4.899**

	Net kg	Gross m^3	Net m^3	kg/m^3	@ \$50/m^3
Totals at end of month 2	212,070	215.970	179.873	1,179	\$8,993.66
Totals at end of month 1	174,790	194.674	179.972	971	\$8,998.60
Deliveries month 2	37,280	21.296	−0.099		−\$4.94

their pay period for timber procurement. Short-term expansion periods can also have a level of volatility, because there will never be many samples in the sample base, and thus deviation from one period to the next is more pronounced in comparison to the long-term sale which can have very big swings initially but which will smooth out after the sample base becomes more populated.

Typical time frames for expansion include: life of the sale, yearly, seasonally (winter/spring and summer/autumn), quarterly, monthly, bimonthly, or sample load to sample load also known as prior load expansion (PLE). PLE is a method that uses a very short expansion term, from one sample load to the next. When a new load is sampled, the sample base is reset and the process continues. Some of the advantages of this system are that one can change sampling frequencies without having to close and restart the expansion process, it cannot generate negative numbers, and it is simple to calculate and thus explain to people involved in the transaction process. In most applications, it is a system that expands forward (the attributes are taken from the sample load and repeated in the non-sampled loads until the next sample load is selected), and thus hard removal/delivery volumes can be determined on an ongoing basis, quickly. Example 2.9 illustrates PLE. In the example, the net weight of the load to be expanded is divided by the net weight of the prior load to obtain the expansion factor (12,100 ÷ 11,390 = the factor 1.0623), this factor is then multiplied by the volumes of the prior load to obtain the volume for the expanded load, e.g. 1.0623 × 6.367 gives a gross m^3 of pine for non-sample load 1 of 6.764.

The loads used in Example 2.9 are the same loads as were used to illustrate methods of expanding loads of logs at a detailed level in

Example 2.9. Prior load method of expansion (PLE).

	Net kg	Expansion factor	Pine Gross m³	Pine Net m³	Spruce Gross m³	Spruce Net m³	Total Gross m³	Total Net m³
Sample load 1	**11,390**		**6.367**	**5.745**	**6.926**	**5.729**	**13.292**	**11.474**
Non-sample load 1	12,100	1.0623	6.764	6.103	7.357	6.086	14.121	12.189
Non-sample load 2	11,210	0.9264	6.266	5.654	6.816	5.638	13.082	11.293
Non-sample load 3	14,040	1.2525	7.848	7.082	8.537	7.062	16.385	14.143
Non-sample load 4	13,180	0.9387	7.367	6.648	8.014	6.629	15.381	13.277
Non-sample load 5	10,960	0.8316	6.126	5.528	6.664	5.513	12.790	11.041
Non-sample load 6	12,300	1.1223	6.875	6.204	7.479	6.187	14.354	12.391
Non-sample load 7	12,020	0.9772	6.719	6.063	7.309	6.046	14.027	12.108
Non-sample load 8	11,440	0.9517	6.395	5.770	6.956	5.754	13.351	11.524
Sample load 2	**11,280**		**10.448**	**10.359**	**1.509**	**1.509**	**11.957**	**11.868**
Non-sample load 9	11,020	0.9770	10.207	10.120	1.474	1.474	11.681	11.595
Non-sample load 10	10,680	0.9691	9.892	9.808	1.429	1.429	11.321	11.237
Non-sample load 11	9,830	0.9204	9.105	9.028	1.315	1.315	10.420	10.343
Non-sample load 12	11,960	1.2167	11.078	10.984	1.600	1.600	12.678	12.584
Non-sample load 13	11,380	0.9515	10.540	10.451	1.522	1.522	12.063	11.973
	174,790		**121.997**	**115.547**	**74.907**	**63.493**	**196.903**	**179.040**

Example 2.7. While the net volume in both examples is extremely close (0.51%), the volumes at a species level are not so close. This is a result of the oversimplified example used, and the fact that two sample loads would not supply a statistically accurate prediction given the very high CV that exists at the species level. If volume at a species level is an important attribute, the sale should be stratified as shown in Example 2.6. If it is not feasible to sort the loads into pure species strata, consideration should be given to other stratum classifications such as estimated species percentages (e.g. > 67% pine, > 67% spruce, < 67% pure), 3-P scaling or using the percentage of each load at a species level to calculate the recommended sample frequency (often this results in a high percentage to be scaled).

Sample scaling is a very useful method of determining the values and volumes of roundwood. However, it requires a good deal of managing to be accomplished successfully. Fortunately, most computerized log accounting systems can perform the calculations on an automatic and ongoing basis, but someone still needs to manage the sampling plan by applying lessons learned from the past, knowledge of the current sample population, and monitoring results as the sale progresses.

3 Measuring Log Yard Inventories and Mill Usage Volume

Logs are often inventoried in log decks in order to provide the needed flow of raw materials for times when log deliveries are insufficient to supply usage. Payments are often made on decked inventory volumes. Log deck inventory volumes also often generate the usage numbers, which are a key parameter in allocating raw material costs to the mill, and thus are a major component of periodic profit and loss statements, and mill efficiency evaluations.

The focus of the information on measuring decked log inventories will be on how to apply measurement techniques in the log yard of a timber conversion facility, although these procedures can be applied elsewhere also.

3.1 Basic Concepts

Techniques for measuring decked log inventories range from 100% scaling into a deck, 'backing into' the inventory volume when log deliveries and usage are known, determining volume from weight (as discussed in Section 2.4.1), stacked wood scale techniques (Section 2.4.2), and sample scaling methods (Section 2.6). In a typical large mill, more than one of the above-mentioned techniques is often used for accounting for log volumes, and generally decks are operated under two types of accounting schemes: closed decks and open decks.

Closed decks are closely accounted for: logs are placed in the deck during the building phase, and no removals are allowed until the deck is opened for depletion. Before the deck is opened for depletion the volume is determined, and when the deck is depleted, the volume is charged to usage. With closed decks it is not permissible to put volume back into the deck during the depletion phase. The advantage of closed decks is that if there is bias or inaccuracy in the methods used to calculate volume or the expectations of conversion efficiency on the part of the mill, the inaccuracies tend to show up early without compiling very large volume errors. Closed decks are also needed for the most accurate methods of measuring log inventories (scaling into inventory, weight, sample scale).

Open decks (occasionally called 'hot decks') are taken from, and added to, as needed, thus limiting volume-accounting procedures to stacked measure. Most mills that operate with closed decks also use open decks to reduce handling costs and facilitate supplying the infeed of the mill with logs during very busy times (when it is not cost-effective or practical to use closed decks). Open decks allow a great deal of flexibility, but volume determination is not generally as accurate using stacked measure as by other means, and log rotation is often not the best as the deck is often only partly depleted before placing more logs back in the deck (isolating older logs behind new ones).

Periodic (annual, quarterly, etc.) 'cut-outs' are a good accounting process with log decks, especially if one is utilizing a good deal of open decks. A cut-out is when deliveries after a cut-off point are kept separate from the prior period's deliveries. This allows the older log inventory to be depleted and accounted for, with any biases or inaccuracies in the inventory finally reconciled when the old inventory is gone and checked against output.

3.2 Methods of Measuring Log Yard Inventory

3.2.1. Stacked measure (deck factors)

A common method of measuring decked log volume is stacked measure. It allows the greatest flexibility for building and depleting decks; all of the decks can be 'open decks' and thus one can take from, and put back without restrictions, as the method of determining inventory is not dependent on accounting for the weight or count of what is being put into or removed from inventory. Another advantage of stacked measure is cost; it is likely the least costly method of determining log deck volume, but it also is prone to generating different volumes from what is actually in the deck as determined by scaling.

Many of the procedures discussed in Section 2.4.2 on stacked wood scale apply to measuring log decks. The standard procedure for determining the volume of log decks with stacked measure is to measure the height, length and width of the deck to determine the cubic area including voids (air space between logs), bark, log defects and any differences between nominal and actual sizes. The cubic area of the deck is generally multiplied by a factor(s) that accounts for the voids, bark, defects, and nominal sizing, and if applicable, converts it to a unit of measure other than cubic, such as bf (Fig. 3.1).

If one wants to convert the cubic volume to a product output rule, locally developed conversions are the best, followed by conversions specific to length and diameter categories, such as those from Section 2.5. Product output rules are notoriously poor at correlating with cubic area and are extremely sensitive to diameter, length and defects.

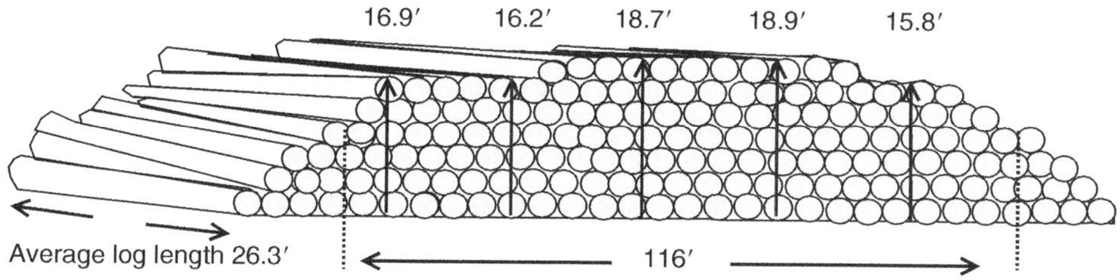

Cubic area of log deck: 116 × 26.3 × ((16.9 + 16.2 + 18.7 + 18.9 + 15.8) ÷ 5) = **52,778.8** ft^3 (1494.52 m^3).

Gross volume solid wood factor: 0.73 × 52,778.8 = 38,528.5 ft^3 (1091.0 m^3); net volume assuming an average 4.7% defect 38,528.5 ft^3 × (1 – 0.047) = **36,717.7** ft^3 (1039.72 m^3).

Fig. 3.1. Determining log deck volume using stacked measure.

Assuming that one would want to convert the cubic volumes as determined in Fig. 3.1 to Scribner Short Log using a ratio of 1.93 ccf per mbf (taken from Example 2.3 on p. 92), one would get a volume of 190.25 mbf. Note that product output rule (bf) to cubic volume ratios are very variable (see Section 2.5). The accuracy of stacked measure for estimating the volume of log decks is driven by the following.

1. **The physical measurements.** The most sensitive aspect of physically measuring a deck is the height measurements. There need to be enough measurements to obtain a representative height of the deck, with measurements taken on both sides of the deck, as often one side of the deck has more large-ends than the other (generally due to the fact that skidding and loading equipment favour handling logs from the large-end). Deck length while not so sensitive as height, should be measured from points, which are representative of the length measured longitudinally and equidistant between the top and bottom of the deck (as shown with the dashed lines in Fig. 3.1).
2. **The solid wood factor.** This should be specific to species, sort and equipment used to build the decks (there can be big differences between the tightness of decked logs given the equipment used). Factors need to be rechecked periodically and often enough to gain confidence in the factors used.
3. **The estimated average log length and estimated defect percentage.** It is best to obtain average log length and defect percentages from log delivery reports for the species sort in the deck. In lieu of actual scale data, one can measure a representative sample in the deck, but be cautioned that logs may be shorter than average at the tops of the decks (where samples are easiest to obtain), as loader operators often put long logs at the bottom and shorter logs on top.
4. **The factor used to convert to a product output rule** (if applicable). Product output rules are extremely sensitive to different length and diameter combinations, so it is very important to use the most applicable conversion factor for the logs being measured. As derived from the mbf/m^3 ratios in Table 2.9, conversions for the Scribner Short Log Rule to the USFS

Cubic Log Scale ranged from 2.3 ccf/mbf (16–21′ L, 4.5–7.5″ d) to 1.27 ccf/mbf (22–31′ L, 15.5″+ d). This huge difference (81%) in ratios shows the difficulty of obtaining accurate factors for the product output rules.

5. The quality and consistency of the log stacking in the deck. It is very difficult to accurately calculate the volume of a deck that is not evenly stacked with a consistent density throughout. Heights and lengths are difficult to measure under these circumstances and factors may not be applicable.

3.2.2 Scaled inventory

The most accurate method of determining log deck inventory is to scale each log. For operations that already physically scale each log from all, or the majority of, the log deliveries, this is the best method. It is also useful for mills that utilize sample scale to build decks from scaled logs for developing solid wood factors, and conducting recovery tests. The procedure is to assign a deck number to each log during the scaling process. Once the load is scaled, the volume going into the deck(s) is totalled from each scale load and compiled with the totals from other loads. It is extremely important that the scaler is in good communication with the loader operators who are decking the logs. The scaler needs to know the current deck number where each log sort is being decked and if there is going to be any changes in deck assignments for log sorts. It may also be necessary for the scaler to mark logs that are near the boundaries of each log sort or that are not easily identifiable, to ensure that the log is put into the deck it was assigned. For example, if logs are being sorted into two groups – 30 cm d and smaller, and 31 cm d and larger – it may be necessary for the scaler to identify the logs of 24–37 cm d, which are close enough to the boundaries of the log sorts to be misidentified (normally with a spot of paint or a coloured tag) to ensure that the logs are not put into the wrong deck by the loader operator.

Once the deck is filled and the total volume tabulated, the cubic area (stacked measure) or length (lineal measure) of the deck is measured so that a deck factor (m^3 log volume/m^3 deck area volume, ft^3 log volume/lineal foot, etc.) can be calculated for use when the deck is partially depleted. This factor can also be utilized for determining the volume of ‘open decks’.

Volume can also be measured during the depletion stage of a deck by accounting for the volume removed from the deck. This is sometimes accomplished by scaling the logs removed (a duplication of labour and a possible area of problems when trying to reconcile volumes). In other cases (when dealing with big and valuable logs) logs are individually tagged with a bar-coded label that corresponds to the scale data, the tag numbers of the logs removed from the deck are recorded, and thus the volume left in the deck can be calculated (expensive and labour-intensive, but accurate).

3.2.3 Sample scaled inventory

Sample scaling is a commonly utilized and viable method of determining deck volumes. There are many variations of sample scaling such as weight-based or count-based. The sample scale system can be part of the normal sample log scale system as described in Section 2.6. Log deck is an attribute that can be assigned to logs in a sample load and expanded at a detailed level into the non-sampled population. This system works best where loads are delivered already sorted into the stratum that they will eventually be decked in, in other words, where the loads can go from the truck to the log deck without being sorted. Care needs to be taken that sample loads, which may sit on the ground for several days waiting to be scaled, are put in the deck they were assigned.

Example 3.1 assumes that a sale is being delivered in two strata (pine and spruce). The loads are expanded (in this case using prior load expansion), and a deck number is assigned to the load. The deck number is normally assigned to the load by the equipment operator, who puts the load into the log deck. The pine stratum was going into deck-23, but starting at load six, the pine went to deck-29.

Example 3.1. Determining deck volume from sample scaling deliveries.

	Net kg	Gross m^3	Net m^3	Deck #	
		Pine stratum			
Sample load 1	**11,390**	**13.292**	**11.474**	23	
Non-sample load 1	12,100	14.121	12.189	23	
Non-sample load 2	11,210	13.082	11.293	23	
Non-sample load 3	14,040	16.385	14.143	23	
Non-sample load 4	13,180	15.381	13.277	23	
Non-sample load 5	10,960	12.790	11.041	23	
Non-sample load 6	12,300	14.354	12.391	29	**Total volume #23 = 73.417 m^3**
Non-sample load 7	12,020	14.027	12.108	29	
Non-sample load 8	11,440	13.351	11.524	29	
					Total volume #29 = 36.023 m^3
		Spruce stratum			
Sample load 1	**11,280**	**11.957**	**11.868**	11	
Non-sample load 1	11,020	11.681	11.595	11	
Non-sample load 2	10,680	11.321	11.237	11	
Non-sample load 3	9,830	10.420	10.343	11	
Non-sample load 4	11,960	12.678	12.584	12	**Total volume #11 = 45.043 m^3**
Non-sample load 5	11,380	12.063	11.973	12	
					Total volume #12 = 24.557 m^3

3.2.4 Weight expanded inventory

Often loads are not delivered sorted into the same strata as they are put into inventory. In this scenario, the loads have to be sorted, and thus the integrity of the net weight of the load is lost. One can still use a weight sample system to determine deck volumes, albeit a system that is separate from the delivery system and dependent on reweighing the logs after they are sorted.

Many of the procedures discussed in Section 2.4.1 on weight scaling and in Section 2.6 on sample scaling also apply to measuring log deck volume by weight. Generally, the logs are weighed on scales mounted on the log yard equipment; by load cells mounted on the log bunks that are used for sorting the logs; or via drive on weight-scale platform. In applications following the first method, the log yard equipment is also generally equipped with data recorders, which record the data on each load put into inventory (net weight, sort, deck number, time, date, operator, etc.). The data recorders can be uploaded easily into a PC for quick and easy data processing.

Sample loads and expansion can be handled in any number of ways, but generally the number of samples needed is very few as the weight-to-volume relationship is usually very good within a particular species and size classification, and the total population is large. In many cases, the weight conversion used to extrapolate volume is based on a rolling average (e.g. the last five samples).

Depleting deck volume using weight is tricky. Log weight changes while in the deck; moisture evaporates (and at different rates during the year depending on temperature and humidity); and a certain amount of bark and even broken log ends are often left as log yard waste, further reducing the weight. In some cases (during particularly wet weather or when decks are sprinkled with water to prevent blue stain and weather checks) the logs can become even heavier. Rather than using weight to get interim volumes when depleting decks, it is generally easier to use stacked measure, or apply a ratio of volume per length (taken when the deck was full), e.g. a deck that is 35.6 m long and has 1500 m^3 when full (giving it a ratio of 42.135 m^3/m length) would be inventoried as having 922.75 m^3 during the depletion phase when only 21.9 m long.

3.2.5 Count (load or log)

Sample methods based on count are commonly used for measuring deck volume where weighing may not be possible. The typical procedure is to develop an average volume per load or log, and count the number of logs or loads in the deck. In the case of small logs, it is very difficult and time-consuming to count the logs in a deck, and in some cases it is impossible.

The most common count-based unit of sample measure for log decks are loads, both truck and sort bunk. As previously mentioned, log trucks tend to be uniformly configured and generally hold similar-sized loads. Sorting bunks are used in a log yard as a container for holding logs of the same sort until the log yard equipment can pick up the load and put it into inventory. In general, it is best to keep the bunk loads separate from truck loads (two separate sample and expansion strata), as there will likely be some differences in the size of the loads. Individual loads can be identified with a brightly painted log end or a bright tag (possibly one colour for truck loads, and another colour for bunk loads), so that a deck volume can be obtained (if needed) during the depletion stage by counting the number of loads remaining in the deck.

3.2.6 Book estimated inventory

Most mills have an expected recovery factor; in other words, there is an expectation as to the output in finished products, based on the input volume of raw materials. This 'book estimated inventory' works well when used in conjunction with a physical inventory as a corroborative check on the overall volume, or when recovery factors are extremely accurate and reliable. Even when the overall inventory volume is accurately estimated, the volume at a species or sort level may not be very accurate when using the book estimated inventory. The basic formula is:

Log yard inventory volume = beginning inventory volume + delivery volume − (production volume ÷ recovery factor)

Note: (production volume ÷ recovery factor) can be replaced by usage volume if it is known by other means.

Example 3.2 shows a hypothetical book estimated inventory volume calculation. The calculated usage was derived by dividing the monthly production volume by the historical recovery factor of 0.307 mbf lumber produced per m^3 of log used. The example assumes that starting on 1 May new deliveries were kept separate from old deliveries in order to get a cut-out, which occurred on 23 May (thus lumber production and log usage are only shown from 23 May).

Normally, the book estimated inventory is tracked simultaneously with the physical inventory, and a reconciliation made if the two numbers drift apart beyond an allowable tolerance level; e.g. if the physical inventory at the end of October is 73,421 m^3 vs. a book inventory of 76,935 m^3, the book inventory may be reset to 73,421 m^3 with a change to a historical recovery factor of 0.300 for the start of November. Obviously, such a scenario would result in a write-off of 3514 m^3, which could have substantial financial consequences.

Example 3.2. Book estimated inventory volume.

	Beginning inventory (m^3)	Deliveries (m^3)	Production (mbf lumber)	Historical recovery (mbf/m^3)	Calculated usage (m^3)	Ending inventory (m^3)
May	0	24,934	1,991.3	0.307	6,486	18,448
June	18,448	39,167	9,053.0	0.307	29,489	28,126
July	28,126	43,185	9,344.4	0.307	30,438	40,873
August	40,873	44,032	9,387.2	0.307	30,577	54,328
September	54,328	40,420	9,202.9	0.307	29,977	64,771
October	64,771	41,771	9,089.3	0.307	29,607	76,935
November	76,935	38,610	8,832.3	0.307	28,770	86,776
December	86,776	24,299	8,654.2	0.307	28,190	82,885
January	82,885	31,773	9,052.7	0.307	29,488	85,170
February	85,170	20,431	8,568.2	0.307	27,909	77,692
March	77,692	9,328	9,207.1	0.307	29,991	57,029
April	57,029	1,013	9,001.5	0.307	29,321	28,722
		358,963	101,384.1		330,241	

3.3 Calculating Mill Log Usage Volume

The usage number is extremely important to log conversion facilities in order to calculate monthly profit and loss, and to determine the efficiency of the mill at converting logs into products. There are three approaches for measuring usage:

1. Measure the log inventory and deliveries, and solve for the usage.
2. Measure the production and solve for the usage by utilizing a recovery factor (see Example 3.2).
3. Measure the usage directly.

3.3.1 Measure the inventory and deliveries, and solve for the usage

The formula to solve for usage is a variation of the formula mentioned in Section 3.2.6 (book estimated inventory):

$$\text{usage} = \text{beginning inventory} + \text{deliveries} - \text{ending inventory}$$

An example of this would be: beginning inventory is 11,325.64 mbf; deliveries are 4,213.81 mbf; ending inventory is 10,966.94 mbf; usage is 4572.51 mbf. This approach to determining usage is likely one of the most common methods in use.

3.3.2 Measure the production and solve for the usage by utilizing a recovery factor

As discussed in Section 3.2.6 production can be used to estimate usage if one has a recovery factor (input volume of logs to output volume of product). The formula is:

usage = production volume ÷ recovery factor

This method is typically used to support figures derived from other methods of determining usage, or where there is great confidence in the accuracy of the recovery factor. The scenario in Example 3.2 also utilizes the production and a recovery factor to estimate usage, e.g. June production of 9053 mbf ÷ historical LRF of 307 = usage of 29,489 m^3.

3.3.3 Measure usage directly

The direct measurement of usage can be accomplished by manually measuring each piece that goes into the mill for processing (expensive and redundant if the logs have already been scaled for payment). In some log yards, logs have a tag with a number and/or bar code, which can be cross-referenced with the log data and volume information. These tags can be recorded or scanned on the log conveyer leading into the mill, which gives a very accurate reconciliation between delivered volume and volume charged to usage (but also expensive, and possibly troublesome if the tag is missing or unreadable).

Scanning technology (see Section 2.4.3) is already a proven method of measuring volume of logs going into the mill. Many mills now have optimized bucking stations and/or optimized 'primary log breakdowns'. Most of these optimizers are equipped with three-dimensional log scanning, which can measure and record volume. The volume may differ to some degree from the volume on the same logs as measured by scaling, but the biases (if any) are typically consistent and easy to factor (especially when measuring log volume in cubic). The product output rules and possibly JAS may be somewhat more difficult to factor, especially if the scanner is seeing logs after they are bucked into shorter lengths. Figure 3.2 is a copy of a production report printed out from an optimizer.

PAGE 4

SUMMARY

Total logs processed	351		
Total PP_LPP logs	351	= 100%	
Average log length	13.0'	= 3.96 m	
Average log top diameter	10.6"	= 26.9 cm	
Average Smalian volume/log	9.53 ft^3	= 0.270 m^3	
Total Smalian log volume	3346.3 ft^3	= 94.756 m^3	
Total board volume	1938.89 ft^3	= 54.903 m^3	= 57.9%
Total Chip volume	1064.5 ft^3	= 30.143 m^3	= 31.8%
Total sawdust volume	342.94 ft^3	= 9.711 m^3	= 10.2%
Total mbf lumber	25.125 mbf		
Average bf lumber/log	71.581 bf/log		
Projected LRF	7.508		
Projected sawmill recovery	57.9%		
Total number of boards	3849		
Total center cant boards	3227	= 83.8%	
Total center cant edger boards	861	= 22.4%	
Total side board flitches	621	= 16.1%	
Total side boards	622	= 1 board(s)/flitch	
Total edger split side boards	1	= 0.16%	
Pieces routed to edger	1482	= 38.5%	
Total lumber value	$11,934.29	= $34.00/log	
Total chip value	$958.05		
Total sawdust value	$102.88		
Total manufacturing costs	$3,571.84		
Net total product value	$9,423.38		
Material under 4.0" diameter	0	= 0.0 lin ft	= 0.000 ft^3
Material over 14.0" diameter	47	= 74.0 lin ft	= 82.34 ft^3
Total downtime	00:33:23	(HH:MM:SS)	
Manual overrides in auto	24		
Logs processed in manual	0		
EDLF productivity based on target of 2800 logs		= 12.50%	

Fig. 3.2. Optimizer production report.

4 Measuring Log Quality

An important aspect of measuring roundwood is measuring its value (both as presented and as opportunities lost). The present value, referred to as log grade, is linked to quality characteristics that affect the value and profitability of the products manufactured from the log. Another important aspect of measuring log quality is to record its manufacturing quality (often referred to as 'log quality control'). A tree stem can lose substantial value through improper log manufacturing practices. Normally these errors in log manufacture are taken into account via the scaling method (through defect deductions) or via log grade (by downgrading the log), but these types of errors cause a preventable loss in value and thus should be tracked to ensure performance accountability and future improvement if problems are detected.

4.1 Log Grading

There are too many different log grading rules to cover individually. On top of regional grading rules are rules specific to particular species or types of logs, manufacturing processes and proprietary to individual private enterprises. There are also diverse views as to what is an issue of grade and what is an issue of volume reduction (which is handled by the log scale). Many log scales have provisions for deducting defects, at least in principle, proportional to which the available volume from a log is reduced, and thus most grading rules would be defined as determining the quality, not the quantity, of product. This loss of volume, however, is linked to the product being manufactured; a log may be completely unsuitable for making lumber, but be entirely suitable for making chips. Without looking at the particular specifications of grading rules, there are universal commonalities to log grading in regard to the quality of the products being produced, manufacturing process of these products and physical structure of a tree stem.

4.1.1 Grading sawlogs and peelers

Listed below are the commonly accounted for attributes and/or degrading factors when grading sawlogs and veneer logs:

- **Diameter** is likely the most important driver per unit volume of log value and, as such, is a key parameter of log grade. In general, bigger logs are worth more than smaller logs. There are many reasons for this, including the fact that bigger diameter logs have a higher relative product recovery ratio and allow the manufacture of wide lumber, which usually carries a premium in price. Big diameter logs have lower manufacturing and handling costs, from harvesting through processing into finished products. There is also a very strong correlation between bigger diameters and knot-free area, finer grain and heartwood area. Mills are often limited as to a maximum diameter that can be processed, and thus occasionally, logs that are too large can be downgraded. Elliptical diameters are also a grade problem for rotary peeled veneer logs as volume recovery is reduced, and the quality of veneer produced can be compromised.
- **Length** of logs is tied very closely to the products that can be produced. Some products require fixed lengths, such as studs, peelers, railroad ties (sleepers), etc., and so receiving logs in multiples other than these lengths can cause waste, which will reduce values. In addition, some products have higher values when produced in particular lengths (generally longer lengths), and thus logs that facilitate producing these preferred lengths yield more value. Longer logs also have more volume, which reduces handling costs and allows manufacturers more options to optimize value, when merchandising the stem, e.g. bucking out crook or sweep, and bucking at diameter and length breaks that maximize value. Finally, longer logs have larger diameter portions attached for a given small-end diameter vs. the same small-end diameter log, which is shorter, again increasing value.
- **Knot-free area** facilitates the manufacture of clear knot-free lumber and knot-free sliced veneer, which commands a premium in price. Because of the nature of rotary peeling, clear veneer from rotary peeled logs requires that the entire circumference for the length of the peeler block must be clear. Distance between knots is also very important as many products are produced from the clear cuttings that occur between knots. While the presence of knots on the log provides the most obvious evidence as to clear faces or clear cuttings in a log, one also has to take into account knot indicators, which are swirls in the bark indicating the past presence of limbs on the log bole. Knot-free face area is normally specified in grading rules as percentage of surface area of log free of knots, number of clear faces, or by a prediction of product output in clear product or products with clear cuttings.
- **Knot size and number of knots** have an impact on the strength and appearance of lumber and veneer. Because of the cross-grained nature of the area within and surrounding a knot, the strength characteristics of a

board can be lessened. Even when strength is not the primary consideration, such as with appearance of boards, higher value grade recovery is obtained with fewer and smaller knots. In the case of veneer value, it is generally better to have many small knots (which can be plugged and tend to be affixed) than to have just one knot that is too large to plug.

- **Green/dead knots** are important in that green knots have a much better chance of surviving the manufacturing, drying and surfacing process with the knot still affixed to the board or veneer. Dead knots come from branches that are dead, and these will generally be encased or loose in the log until the point in the stem's growth at which the branch died. Dead knots often become a hole in the board and affect boards or veneer sheets more severely because the entire branch diameter (including the sapwood area of the branch) is considered a degrading factor as opposed to a green knot where only the heartwood of the knot is generally counted as being a degrading factor.
- **Stain** is a degrading factor in lumber with regard to its appearance, but not normally as it relates to strength characteristics (unless it is associated with rot). Blue stain is a common problem for species with pitch-saturated sapwood such as the pines, and to a lesser degree, the spruces. Normally blue stain occurs in felled trees and logs, on areas of the log missing bark, or under the bark from spores carried onto the sapwood by bark-boring beetles. Once present, the stain spreads quickly over the sapwood area. Heartwood stains can occur due to mineral deposits, but also because of wetwood and decay. When accounting for stain in grading logs, one needs to consider that once even a slight amount of stain shows on the log surface, it can quickly spread to the entire sapwood portion unless the weather is cool (temperature < 0°C or 32°F), the log is quickly milled and dried, or the log is saturated with water. Heartwood stain needs to be closely examined to ensure that it is not decay that will affect strength characteristics as well as appearance.
- **Grain characteristics** normally refer to the characteristics of the grain of the log, both radially (growth ring characteristics) and longitudinally (slope of grain etc.). The density of growth rings are commonly accounted for in log grades, but the ratio of latewood to earlywood within the growth rings also has a bearing on the quality of the lumber. Tight growth rings and a greater ratio of latewood to earlywood not only increase the density of the wood (improving strength characteristics) but also tend to provide a smoother finish when surfaced through a planer, thus improving appearance. Compression and tension wood (also referred to as off-centre heart) is also considered a degrading factor in lumber and veneer quality. Longitudinal grain characteristics such as slope of grain (also called twist), and curly grain (also called horse mane) affect both structural and appearance characteristics. Grain that is very sloped yields lumber with very poor strength, but also with a propensity toward warping and twisting in the drying process and roughness despite surfacing (due to 'end grain' characteristics).
- **Deductible defects**, even when accounted for via volume reduction, can further reduce the value of a log, and thus are often accounted for in log

grades. An example of this is centre rot or breakage in logs, which may only cause a small loss in scale volume, but which render the log unchuckable on a lathe for rotary peeling into veneer. Grain separations such as checks and shake, centre rot, as well as sectional defective areas such as basal scars can have serious consequences on logs utilized for rotary peeling, which may otherwise be very good for making lumber. Weather checks (radial splits from the perimeter of the log into the heart), worm and insect holes are linked primarily to logs cut from dead trees (but also occasionally occur in green logs), and are important degrading factors. Sweep and crook, even when accounted for via deductions, can seriously reduce usability, value and volume recovery, and thus are commonly addressed by grading rules (especially when using log scales that do not deduct for sweep or crook). Defects that are accounted for via defect deductions can also create added expense in manufacturing costs due to increased processing time incurred by manufacturing defective portions of a log that yields no products. For those scaling systems that do not account for defects which reduce the output of lumber or veneer, defect attributes are extremely important to account for in the grading process.

- **Taper** has an effect on recovery, and depending on which methods are utilized to determine log volume, taper can either reduce or increase product recovery. When using cubic measure (excepting Swedish Cubic and JAS Scale), higher taper equals lower product recovery of lumber or veneer, and less taper equals higher recovery. This situation is the opposite with the board foot rules and Swedish Cubic and JAS Scale, where the log is viewed as a cylinder or is tapered at a fixed rate, and thus higher taper increases the actual volume of a log segment without increasing the recorded volume. High taper is also indicative of trees that are growing in open stands, which are often predisposed to more and bigger branches and less distance between knot whorls – all issues reducing quality. In general it can be said that high taper decreases recovery for cubically scaled logs, increases recovery for product output scaled logs and often indicates a log that has grade-reducing characteristics (numerous big knots) relative to low tapered logs.
- **Sapwood-to-heartwood** ratio is important for certain products where there is a preference for one or the other. Some products specify a preference for heartwood, as it may be more resistant to decay, and in some species, the heartwood has a colour that is preferred for certain products. High ratios of heartwood to sapwood indicate older logs – generally bigger, which typically have beneficial grade attributes (as listed under diameter and grain characteristics).

4.1.2 Grading chip logs

Chip logs, often referred to as pulp logs, have certain attributes that contribute to their value for making their intended end products, e.g.

paper and cardboard, OSB, fibreboard, etc. While many of the important attributes for making lumber or veneer are not so critical for chip logs, the issues that affect manufacturing costs such as size of log and defects are. Listed below are some important characteristics when grading chip logs:

- **Diameter and length** are primarily a driver of value in regard to handling costs and suitability for manufacture (given the manufacturing process); e.g. if one is using a single stem debarking and chipping process, bigger logs mean more production, or if one is using a drum debarker and can chip multiple stems simultaneously, the size of the log is not such an issue.
- **Grain characteristics** are important, in that strength and density of fibre are key issues for all products produced from woodchips. Chips are often purchased, not by volume, but by BD content (the weight of the chips with all moisture removed). There is a wide variation of wood density between species, and even within species due to age class and environmental factors.
- **Defects** cause reduced productivity even if accounted for by defect deductions, and can cause problems in that rotten areas can become fines (small particles) that can create difficulties in some of the manufacturing processes, especially for producing paper. Char is particularly troublesome for all products made from chips, and is generally forbidden entirely from chip logs intended for the production of paper. Any defects that prevent the removal of bark from the log or impede the manufacturing process, such as forks, excessive crookedness, bark seams or rotten logs that will collapse in the debarking process, are often rejected or downgraded.
- **Freshness** is an issue for chip logs as logs that have been down for some time are often more difficult to debark and may have stains, such as blue stain, which may have effects on the appearance of some products produced. Logs cut from dead trees are often utilized for chipping, but can give somewhat lower recoveries in chips, as a result of increased generation of fines from rot, and the fact that dry wood is more brittle and breaks more in the chipping and conveying processes.

4.2 Log Manufacturing Quality

Log manufacturing quality is commonly tracked during the scaling process. A typical procedure is to account for log manufacturing errors in a load of logs, and provide immediate feedback, when serious problems (such as incorrect lengths) are encountered. Long-term statistics on log manufacturing errors are often used as a ‘report card’ for the contract logger and the timber sale administrator. Some important log manufacturing quality items to watch for are listed below:

- **Lengths** can normally be controlled by the logger, and facilitate products that require a fixed length such as studs, ties (sleepers) or peeler blocks. There are also product lengths that command a premium price and thus log length manufacture should favour these lengths when possible. In some regions, such as North America, lumber lengths are traditionally utilized in certain length multiples, e.g. 2′ multiples (0.6096 m), and are normally sold in lengths from 6′ to 20′ (1.83–6.1 m). Most manufacturers need to have a certain amount of trim allowance (additional length) to ensure that lengths can be re-squared, end-checking (which occurs in storage) can be trimmed away and allow for shrinkage, which will occur when the wood is dried. Nevertheless, too much additional trim is wasteful. Common practice is to list preferred lengths, allowable lengths (because it is not always possible to cut the preferred lengths) and an acceptable tolerance on length, e.g. −1″, + 2″(− 2.5 cm, + 5 cm).

- **Diameters** are normally specified as to the minimum and maximum allowed. Substandard top diameters can severely reduce recovery and productivity (depending on products produced and plant configuration). Maximum diameters are normally set at the limit of diameter that can be processed by the breakdown equipment. Regardless of effects on recovery, productivity or whether the log is even too big to be processed, the specifications as to minimum and maximum are controlled by the purchase agreement or by utilization standards.
- **Limbing** (the removal of branches) is important for conveying logs in the manufacturing process, and preventing torn-out limbs, which are caused by the debarker, from reducing the quality and quantity of products produced.
- **Squareness of bucked ends** are important for allowing proper manufacturing processes of sawlogs and veneer logs. Un-square (also called bevelled) ends often cause log blocks to be cut short at mill chopsaws as the protruding end of the bevel will come into contact with the log stop, causing the chopsaw to cut a length that is too short on the short end of the bevel. There may also be problems with chucking a bevelled end on a lathe for rotary peeling.
- **Long-butting** is a term used to describe the removal of cull (worthless) material through bucking. Cull material creates unnecessary manufacturing, handling and transportation costs, and can cause downtime in the mill. Like for diameters, a purchase agreement or contractual utilization specifications should give guidance as to what should be long-butted off from the log and what should not.
- **Mis-sort** is normally defined as a log that does not meet the specifications for a mill, because it was manufactured with the intention of delivering it to another destination. An example of this would be delivering a properly manufactured pine log to a mill that cuts only hardwoods, or delivering a log to a sawmill that is a cull for the manufacture of lumber due to rot or crook, but which is quite suitable as a chip log delivered to the pulp mill.
- **Unnecessary log damage** is usually used to classify damage that occurs in the harvesting process, which should be preventable, such as breakage, split butts from feller-bunchers or cut to length log processors (CTLs), damage from improper chainsaw felling, split tops from improper

bucking or not supporting the top when using a CTL, spike incisions or 'cat tracks' caused by using CTLs with long spiked rolls on trees with thin bark. All these damages can add up to substantial losses of value.

- **Crook or sweep allowance** is usually specified to encourage the timber harvester to cut log lengths at, or near, crooks in order to minimize losses in product recovery from sweep or crook, and to eliminate possible problems in the manufacturing process from trying to handle and convey crooked logs. Normally maximum crook or sweep specifications are presented as the maximum deflection allowed with the 'bowstring method' (measured at the point of maximum deflection with a tape line or string held from end to end on the concave side of the log).
- **Excessive bark removal** is a problem if logs are stored for periods of dry hot weather. The bark forms a protective sheath around the log, which holds in moisture and slows the process of fungal stains and weather checks. Some equipment used in harvesting timber can inadvertently strip the bark off the stem. In many cases, the excessive removal of bark can be prevented by using special feed-rolls and well-maintained delimbing knives.
- **Excessive delay in delivery** can contribute to stained wood, weather checks and difficult-to-debark logs. Weather conditions have a big influence on what is excessive and what is not; obviously, when the weather is cold and damp, damage from delay is minimized. In general, if one is finding the beginning stages of stain and weather checks on delivered logs cut from green trees, the time lag between felling and delivery is too long.

5 Roundwood Weight and General Physical Properties

As discussed in Section 2.4.1, weight is useful for measuring roundwood. For most commercial woodland and wood manufacturing operations, it would be impossible to conduct business without utilizing weight metrics in some form. In order to obtain the highest level of usefulness from weight and its relationship with volume, it is important to understand the drivers that determine weight.

5.1 The Variables that Determine Weight-to-Volume Ratios

5.1.1 Moisture content

A living tree is made up of large quantities of moisture, and the level of moisture retained within a log can have a big impact on its weight. Water weighs 62.4 lb/ft^3 (1000 kg/m^3). Normally the moisture in a log is measured as a percentage of the bone-dry (BD) (0% moisture) weight. For example, if a BD m^3 of wood weighs 340 kg (750 lb) while it weighed 880 kg (1940 lb) when green, it is considered to have a 159% mc: $(880 - 340) \div 340 = 1.59$. The moisture content of wood typically varies from around 30% to more than 200% (USDA, 1974). Aside from the variance attributable to the norms within a given species, the drivers of moisture content are as follows.

Heartwood-to-sapwood ratio can have a large effect on weight-to-volume ratios. This is because the heartwood generally has less moisture (especially with coniferous species). The heartwood is in essence dead and only serves a structural purpose. The sapwood is the living part of the stem with a primary purpose of storing moisture and nutrients. Table 5.1 shows some examples of average moisture contents of heartwood and sapwood by species. It is also important to note that young trees, which tend to be the smaller trees, typically have a higher percentage of sapwood than older trees; this gives many species a decreasing weight-to-volume ratio as log diameter increases. This characteristic can be seen in Table A.2.A for most of the species that have size class information.

Table 5.1. Average physical characteristics of logs from some common North American tree species.

Common name	Latin name	% moisture heartwood	% moisture sapwood	*kg/m^3	*lbs/ft^3	Wood specific gravity**	Bark % whole log weight (bark & wood)	Bark % log wood volume
Conifers								
Balsam fir	*Abies balsamea*	88	173	864.1	54.0	0.33	10.7	15.0
Grand fir	*Abies grandis*	91	136	884.6	56.0	0.35	11.7	14.3
Western larch	*Larix occidentalis*	54	119	921.4	57.2	0.48	10.1	19.5
Engelmann spruce	*Picea engelmannii*	51	173	858.5	53.6	0.33	11.0	13.0
Black spruce	*Picea mariana*	52	113	848.8	53.0	0.38	7.2	14.0
Lodgepole pine	*Pinus contorta*	41	120	905.2	56.7	0.38	5.4	6.0
Shortleaf pine	*Pinus echinata*	32	122	1025.0	64.0	0.47	9.4	15.0
Western white pine	*Pinus monticola*	62	148	842.5	52.6	0.35	12.4	15.0
Longleaf pine	*Pinus palustris*	31	106	1140.3	71.2	0.54	8.4	15.0
Ponderosa pine	*Pinus ponderosa*	40	148	1015.6	62.3	0.38	11.7	20.3
Red pine	*Pinus resinosa*	32	134	859.1	53.6	0.41	8.7	16.0
Eastern white pine	*Pinus strobus*	50	175	839.9	52.4	0.34	18.0	16.0
Loblolly pine	*Pinus taeda*	33	110	1026.6	64.1	0.47	9.4	15.0
Douglas fir	*Pseudotsuga menziesii*	30	112	950.8	58.8	0.45	13.7	20.1
Northern white cedar	*Thuja occidentalis*	32	240	649.6	40.6	0.29	11.2	12.0
Western red cedar	*Thuja plicata*	58	249	621.1	38.8	0.31	11.8	12.6
Eastern hemlock	*Tsuga canadensis*	97	119	979.1	61.1	0.38	18.2	21.0
Western hemlock	*Tsuga heterophylla*	85	170	942.8	58.9	0.42	9.2	10.3

Hardwoods								
Soft maple	*Acer saccharinum*	58	97	993.0	62.0	0.49	11.2	12.0
Hard maple	*Acer saccharum*	65	72	1121.1	70.0	0.56	10.5	12.0
Yellow birch	*Betula alleghaniensis*	74	72	1089.1	68.0	0.55	11.1	12.0
Paper birch	*Betula papyrifera*	89	72	1009.0	63.0	0.48	14.7	16.0
Pecan hickory	*Carya illinoensis*	71	49	1057.0	66.0	0.60	11.6	13.0
Hickory	*Carya* spp.	97	62	1201.2	75.0	0.64	17.0	17.0
Hackberry	*Celtis* spp.	61	65	961.0	60.0	0.49	13.3	15.0
American beech	*Fagus grandifolia*	55	72	1025.0	64.0	0.56	6.6	7.0
White ash	*Fraxinus americana*	46	44	961.0	60.0	0.54	9.1	16.0
Sweetgum	*Liquidambar styraciflua*	79	137	1121.1	70.0	0.48	12.0	11.0
Yellow poplar	*Liriodendron tulipifera*	83	106	1025.0	64.0	0.44	12.4	15.0
Water tupelo	*Nyssa aquatica*	150	116	944.9	59.0	0.45	11.0	11.0
Black gum	*Nyssa sylvatica*	87	115	1041.0	65.0	0.48	11.0	12.0
Sycamore	*Platanus occidentalis*	114	130	1089.1	68.0	0.45	4.0	4.6
Cottonwood	*Populus* spp.	162	146	944.9	59.0	0.37	15.0	15.0
Aspen	*Populus tremuloides*	95	113	944.9	59.0	0.39	18.9	18.0
White oak	*Quercus alba*	64	78	1185.2	74.0	0.64	10.0	11.0
Water oak	*Quercus nigra*	81	81	1249.2	78.0	0.57	12.8	13.0
Southern red oak	*Quercus falcata*	83	75	1249.2	78.0	0.57	17.0	20.0
American basswood	*Tilia americana*	81	133	784.8	49.0	0.32	18.3	16.0
Elm	*Ulmus* spp.	95	92	1089.1	68.0	0.46	10.1	14.0

Note: The data presented in this table are based on empirically determined averages, individual logs or populations of logs may differ substantially from these data.
*Weight-to-volume ratios is for green logs, includes weight of bark, and volume of wood only.
**Specific gravity based on oven-dry weight (0% moisture content), volume measured when green.
Source: Moisture contents and specific gravity from USDA Forest Products Laboratory, 1974. Sources for weight-to-volume ratios for wood and bark content of logs listed at end of Table A.2.A on p. 227.

Environmental factors and seasonality have a big effect on moisture content. Because of seasonal biological processes, many evergreens and most deciduous trees are lighter in the bole of the tree during summer and autumn in relation to winter and spring. While conducting a 2-year-long study of weight-to-volume ratios on pure species loads of softwood logs in the inland northwest and the southeast USA (loblolly, shortleaf and longleaf pines only), the author noted that the log weight-to-volume ratios of some species were significantly higher during the heavy season (November–June) vs. light season (July–October): ponderosa pine +4.0%, Douglas fir +4.6%, western larch +5.4%, and grand fir +7.8%; while no significant changes were observed for subalpine fir, Engelmann spruce, lodgepole pine, red cedar or any of the Southern pines. These trends may vary significantly for the same species in different climates and regions.

There are also non-biological events such as drought and heat or particularly cool wet weather that affect the amount of moisture retained in the bole of the tree. Other non-biological factors that can impact weight significantly in felled trees include the time between felling and weighing (reducing weight-to-volume ratios), and snow and ice sticking to the log (increasing weight-to-volume ratios). According to published data (Hamilton, 1975), a delay of 20 days of delivery will reduce the weight-to-volume ratio by:

Species	April–September	October–March
Spruce, Douglas fir	−6.5%	−1.5%
Pine, larch	−5%	−1%

The rates shown are for timber and conditions in the UK, and one would assume that the trees are primarily second growth (i.e. small) trees. One would also assume that these rates of drying would vary according to temperature, humidity, precipitation, wind, etc. and may slow down beyond 20 days.

5.1.2 Wood density

The density of wood can vary significantly from one species to another, within a given species, and even within a single tree stem. The density of wood is normally measured by its BD weight in relation to a given volume that was displaced when the wood was green (negating the effects of shrinkage). Because units of volume are variable, wood density is commonly reflected as an index, called 'specific gravity' (SG). SG uses the weight of water displacing the same volume as the indexed material, e.g. wood (or bark, or both) that measures 1 m^3 when green (before shrinkage occurs), and having a weight of 340 kg when BD is said to have an SG of 0.34 (1 m^3 of water weighs 1000 kg $\div$ 340 kg BD weight = 0.34). This calculation will yield the same SG when using imperial measure (35.315 ft^3 of water weighs 2205 lb $\div$ 750 lb BD weight = 0.34), and thus gives a universal measure of density. In metric countries, BD measure is commonly referred to as 'basic density', and is usually reflected in kg/m^3 (m^3 being measured in the green state, before shrinkage occurs). This measure equals SG $\times$ 1000; thus, an SG of 0.34 equals a basic density of 340 kg/m^3. Table 5.1 gives the SG of the wood of some common tree species (Table A.2.A provides a more comprehensive list).

5.1.3 Bark volume and weight

Bark also has an impact on weight-to-volume ratios as it is generally not accounted for volumetrically, but is usually present when the logs are weighed. The weight of bark is typically around 12% of the total weight of the stem for most species (bark and wood), but can vary from less than 6% to more than 25% dependent on species, age, log diameter, environmental factors, location relative to the entire tree stem and man-made factors such as bark inadvertently stripped off during the logging and handling processes. In general, bark has a similar weight-to-volume ratio as wood, but this can vary significantly from one species to another. Bark weight and volume percentages for some common tree species are shown in Table 5.1. (Table A.2.A provides a more comprehensive list.)

5.1.4 Deducted defect volume

Most deducted defects carry weight but not volume, and thus weight factors will increase as the percentage of defect deductions increase. As there is often tremendous variation in defects depending on differing log populations, utilization standards in practice and the scale in use, it is best to use gross volume for conversions to and from weight, and then factor defect later, if needed, by using stratum-specific knowledge of defect percentage.

5.2 Conversions to and from Weight

While it is true that there is a certain level of variability in weight-to-volume ratios of logs (there are many contrasting variables to account for), which can make use of conversions tricky, thoughtful application of conversions and accounting of the variables mentioned in Sections 5.1.1–5.1.4 can reduce surprises. Essentially this means, where possible, stratifying logs to be converted into homogeneous groups and utilizing locally developed stratum-specific conversions to the greatest extent practicable. It can also be said that the level of accuracy of weight conversions is linked to the unit of measure being converted. Weight correlates more favourably with volume when the unit of measure in use relates linearly to the actual displaced volume of the log, e.g. it is easier to convert weight to BC Firmwood (which closely measures actual displaced volume) than it is to convert to Scribner Long Log (which measures a perceived recoverable volume in product rounded in large steps while ignoring taper). The weight-to-volume ratios listed in Tables 5.1 and A.2.A should be considered to be based on actual displaced cubic volume, and thus Table 2.9 should be used to convert to other units of measure. The ratios listed in Tables 5.1 and A.2.A are best used for general estimates and should not be used as a substitute for locally developed stratum-specific averages.

Example 5.1.

Assume that an estimate of tons per gross mbf is needed for Eastern hemlock measured with the International ¼-Inch Log Rule in log lengths that will be primarily 16′ long (5 m). The log mix is estimated to be 20% 4.5–7.49″ (11.43–19.05 cm), 40% 7.5–11.49″ (19.06–29.19 cm), 30% 11.5–15.49″ (29.2–39.35 cm) and 10% 15.5″ + (> 39.36 cm).

The conversion factor from actual gross cubic (BC Firmwood) to gross mbf International ¼″ $= (0.2 \times 0.126) + (0.4 \times 0.200) + (0.3 \times 0.225) + (0.1 \times 0.271) = 0.1998$ gross mbf/m^3 or 5.005 m^3/mbf. Eastern hemlock is listed as having a typical weight of 979 kg/m^3 (2158 lb/m^3); $2158 \times 5.005 = 10,801$ lb/mbf (5.40 t/mbf gross). Assuming a defect percentage of approximately 10%, $5.40 \div (1 - 0.10) = 6.0$ t/mbf net.

6 Metrics of Lumber Recovery

Lumber recovery, defined here as the amount and type of lumber recovered out of a given quantity of logs, is often misunderstood and difficult to gain insight into. This is due to the dynamics of many variables that often have opposing and inconsistent effects on recovery. Many sawmill managers have been confounded by trying to predict and plan for recovery ratios.

While lumber recovery is an important issue to focus on, trimming off some lumber length (at the expense of lumber volume recovery) and converting it to chip volume is good business if one can improve lumber value enough to offset the volume loss from lumber, which has instead become chips. Besides lumber, mills produce chips, sawdust, shavings and even products from bark, such as boiler fuel, cogeneration fuel and decorative landscape mulch. Most modern sawmills achieve recoveries that approach 100% of the delivered wood fibre when accounting for all the products produced. Table 6.1 shows a typical allocation of recovered wood fibre from logs manufactured in four different types of sawmills. Note that Table 6.1 (as do all the lumber recoveries shown in this book) assumes the lumber will be surfaced. For rough lumber recovery, add the percentage loss in shavings to the lumber percentage, e.g. studs equal 56%.

Recoveries are generally calculated as a ratio of output of product volume to the input of log volume, e.g. if a mill produces 0.45 m^3 of lumber from 1 m^3 of log volume, it is said to have 0.45 or 45% recovery. This percentage method of reflecting recovery is common in regions that use the same unit of measurement for output as for input, e.g. m^3 log volume in – m^3 lumber volume out.

Where cubic log scale is used, recovery is either reported as percentage of input (as shown above) or as is the case in North America, where the bf measure is still the unit of measure used for lumber, lumber recovery factor (LRF) is often used. LRF is the volume of lumber in bf divided by the log volume, either in cubic feet (USA) or cubic metres (Canada). For example, if the lumber produced is 400,000 bf, and usage is 50,000 ft^3 (1415.8 m^3) of logs, LRF is said to be 8.0 (283 if using m^3 for log scale).

In regions where product output (bf) log scale is used, recoveries are determined by dividing the output of lumber in bf, by the input of logs in

Table 6.1. Typical distribution of fibre in the lumber manufacturing process.

	Lumber (%)	Chips (%)	Sawdust (%)	Shavings (%)	Shrinkage (%)
Board	41	31	12	13	3
Stud	48	34	7	8	3
Dimension	48	33	7	9	3
Hardwood	44	25	14	10	7

Note: Based on recoveries on the following average small-end diameter logs 10″ – boards, 7″ – studs, 10″ – dimension, 13″ – hardwood.
Source: Calculated by the author from Tables 6.6–6.9.

bf. If the output volume of lumber exceeds the inputted log volume, the ratio is called overrun (OR); if the output volume is less than the inputted log volume, the ratio is called an underrun (UR): e.g. if lumber produced is 400 mbf and usage is 300 mbf, recovery is said to be 1.33 or a 33.3% OR ($400 \div 300 = 1.333$); if instead the usage were 420 mbf, the recovery would be 0.95 or a 5% UR ($400 \div 420 = 0.952$). Often a mill will show an OR as the product output rules tend to understate the actual recovered volume, but this can vary by log rule, log dimensions and products produced. Example 6.1 shows the same inputted log volume, outputted lumber volume (at exactly the same conversion efficiency) reflected via different scales, different methods of calculating lumber volume (bf or m^3), and the traditional units of recovery. Note that it is easy to assume significant differences in efficiency of recovery.

Example 6.1. Same log volume, product volume and mill efficiency reflected in different units of measure.

Scale method	Log usage	Lumber produced	Recovery
USFS Cubic	44,198 ft^3	400 mbf	LRF = 9.05
BC Firmwood	1374.6 m^3	400 mbf	LRF = 291
Alberta Cubic	1304.7 m^3	400 mbf	LRF = 307
Ontario Cubic	1339.9 m^3	400 mbf	LRF = 299
Swedish Cubic	1191.9 m^3	640 m^3	54%
Russian Standard	1382.8 m^3	640 m^3	46%
Cubage au Réel	1281.7 m^3	640 m^3	50%
New Zealand 3-D	1358.8 m^3	640 m^3	47%
Brereton (PNG)	1284.3 m^3	640 m^3	50%
Hoppus	1070.7 m^3	640 m^3	60%
JAS Scale	1260.3 m^3	640 m^3	51%
Scribner Short Log	241.8 mbf	400 mbf	1.65 or 65% OR
Scribner Long Log	188.4 mbf	400 mbf	2.12 or 112% OR
Doyle	206.7 mbf	400 mbf	1.93 or 93% OR
International ¼″	270.4 mbf	400 mbf	1.48 or 48% OR

6.1 Measuring Lumber Volume

Lumber volume is often measured in different stages of the manufacturing process, which can drastically change the volume reported. Lumber production can be measured at the rough green stage (the initial milling process); after drying, but still in the rough form (normally creating a volume loss of 3–12% from shrinkage); or after the lumber is reprocessed and surfaced (reducing volume from surfacing off volume, re-trimming lumber lengths, and even culling some lumber). Table 6.2 shows typical lumber volume at various stages of manufacture for four categories of lumber (boards, dimension, studs, hardwood) in actual cubic volume as well as in the bf measure. Note that while the actual loss of lumber cubic volume between the rough-green tally and finished-dry tally is 23–32%, the bf tally only dropped by 3–8% (mostly from grade trimming in the planer) because the bf tally does not change as a result of shrinkage or planing.

The method of measuring lumber is also important to the overall understanding of lumber recovery. It is critical to know whether lumber is measured in actual measure or nominal measure, and if measured nominally, to what degree. In North America the bf measure is utilized as the standard unit of measure for lumber, while in most of the world, the cubic-metric system is utilized. In both the metric and bf systems, nominal sizes are normally the basis for production and sales volumes, but with the board foot system, volume determination conventions can be somewhat inconsistent, misleading and complicated in comparison with the simpler and more accurate cubic-metric measure.

6.1.1 Lumber board foot measure

The bf method of measuring lumber is an old system of measure, which, despite its complexity and inaccuracies, manages to maintain use for measuring domestically consumed lumber products in North America.

Table 6.2. Comparison of sawmill lumber volume as measured when rough-green, rough-dry, and finished-dry.

	Rough-green			Rough-dry			Finished-dry lumber			% change rough-green to finished-dry	
	mbf	ft^3	m^3	mbf	ft^3	m^3	mbf	ft^3	m^3	mbf	ft^3, m^3
Boards	1.087	83.6	2.37	1.087	80.5	2.28	1	56.6	1.60	−8	−32
Studs	1.03	72.5	2.05	1.03	70.1	1.98	1	55.6	1.57	−3	−23
Dimension	1.045	74.8	2.12	1.045	72.3	2.05	1	56.6	1.60	−4	−24
Hardwood	1.064	100.4	2.84	1.064	90.1	2.55	1	69.2	1.96	−6	−31

Source: Calculated by the author.

A bf of lumber is defined as representing a board that is 1 ft^2 (12″ × 12″), and 1″ thick. The formula for determining the bf volume is ostensibly:[1]

$$\text{bf} = (\text{nominal thickness in inches} \times \text{nominal width in inches} \times \text{actual length in feet}) \div 12$$

In principle, this would indicate that a bf displaces 1⁄12 of a cubic foot (0.0833 ft^3). At the time of its inception, a bf likely did represent approximately 1⁄12 of a cubic foot and was tallied while still green and before surfacing. Later when lumber manufacturing became more sophisticated, it was discovered that the wood could be dried and surfaced to a smooth finish. Obviously, however, this piece of wood that started out with a 0.0833 ft^3 would end up with something less volumetrical once dried down to a stable moisture content (8–19%), and surfaced. Standards were established as to the actual final size specifications of different lumber products (generally based on traditional sizes and use). Because lumber size standards were based on the dimensions finally produced, it was quickly discovered that the original green target sizes (GTS) could occasionally be narrowed while still producing the desired nominal size; conversely, it was discovered that with some products, due to high shrinkage or mill inefficiency, GTS actually had to be increased over the stated nominal size in order to produce a saleable product at the size standard. Unfortunately, there is no consistent actual-to-nominal size convention or ratio that applies to all lumber types or dimensions.

Table 6.3 shows a typical range of lumber widths by log small-end diameter, given a sawing strategy toward maximizing wide lumber widths. It should be noted that beyond 17″ there were no clear trends in width recovery. Table 6.4 shows the nominal sizes, GTS (which are generalizations, actual target sizes can vary) and finished sizes for selected board, dimension and stud lumber, softwood shop, and hardwood lumber products as well as the actual cubic volume per bf. Nominal sizing is generally individual to the lumber product lines listed in Table 6.4. As an example of the difference in nominal sizing between different products measured by the bf, consider a hypothetical situation of a mill producing all 6⁄4″ softwood shop lumber vs. a mill producing all 2″ × 4″ studs. It would appear that the mill producing 6⁄4″ softwood shop is roughly 43% less efficient than a mill producing all 2″ × 4″ studs, as it takes (in principle) 43% more fibre to make a bf of 6⁄4″ softwood shop lumber than it does to make a bf of 2″ × 4″ stud lumber (all other things being equal). In practice, the actual fibre needed to make a bf will vary depending on saw kerf, surfacing allowance, lumber dimension, edging and trimming practices, etc., but likely this number is not so far from the actual.

[1] *Note*: Some stud lumber lengths are based on nominal measure (notably 92⅝″ long studs, which get tallied as being 96″); actual widths are used in the case of softwood shop that is 5⁄4″ and thicker, as well as for hardwood lumber.

Table 6.3. Example of lumber width % yield in North American sizes by small-end diameter of log.

Lumber width inches = nominal cm = actual	Small-end diameter class of log in inches (cm) 5″ (13 cm)	6″ (15 cm)	7″ (18 cm)	8″ (20 cm)	9″ (23 cm)	10″ (25 cm)	11″ (28 cm)	12″ (30 cm)	13″ (33 cm)	14″ (36 cm)	15″ (38 cm)	16″ (41 cm)	17″ (43 cm)
4″ (8.9 cm)	100	99	36	18	11	6	3	3	2	2	2	2	2
6″ (14.0 cm)	0	1	64	82	51	21	14	10	7	7	6	5	4
8″ (18.4 cm)	0	0	0	0	38	69	47	18	8	8	8	8	5
10″ (23.5 cm)	0	0	0	0	0	4	36	69	21	17	17	16	16
12″ (28.6 cm)	0	0	0	0	0	0	0	0	62	66	67	69	73

Note: The above percentages reflect recoveries observed by the author in several tests on mill processed log lengths (< 20′ long (6.1 m)). Actual width recovery will vary depending on log taper, log lengths, products produced and cutting strategies.
Source: Studies conducted by the author.

Table 6.4. Board foot actual to nominal sizes and volumes by lumber product type.

	Nominal			Green target size*			Finished		
	Thick	Width	bf/ft^3	Thick	Width	bf/ft^3	Thick	Width	bf/ft^3
Softwood boards									
1 × 4	1.00	4.00	12.00	0.94	3.80	13.44	0.75	3.50	18.29
1 × 6	1.00	6.00	12.00	0.94	5.88	13.04	0.75	5.50	17.45
1 × 8	1.00	8.00	12.00	0.94	7.88	12.97	0.75	7.25	17.66
1 × 10	1.00	10.00	12.00	0.94	9.88	12.93	0.75	9.25	17.30
1 × 12	1.00	12.00	12.00	0.94	11.88	12.90	0.75	11.25	17.07
Dimension and studs									
2 × 4	2.00	4.00	12.00	1.75	3.80	14.44	1.50	3.50	18.29
2 × 6	2.00	6.00	12.00	1.75	5.88	14.01	1.50	5.50	17.45
2 × 8	2.00	8.00	12.00	1.75	7.88	13.93	1.50	7.25	17.66
2 × 10	2.00	10.00	12.00	1.75	9.88	13.89	1.50	9.25	17.30
2 × 12	2.00	12.00	12.00	1.75	11.88	13.86	1.50	11.25	17.07
Softwood shop									
5/4	1.25	rw	12.00	1.417	rw	10.58	1.156	rw	12.97
6/4	1.50	rw	12.00	1.680	rw	10.71	1.406	rw	12.80
7/4	1.75	rw	12.00	1.878	rw	11.18	1.594	rw	13.18
8/4	2.00	rw	12.00	2.108	rw	11.39	1.813	rw	13.24
9/4	2.25	rw	12.00	2.404	rw	11.23	2.094	rw	12.90
10/4	2.50	rw	12.00	2.700	rw	11.11	2.375	rw	12.63
Hardwood lumber									
2/4**	0.50	rw	24.00	0.605	rw	19.84	0.313	rw	38.40
3/4**	0.75	rw	16.00	0.874	rw	13.74	0.563	rw	21.33
4/4	1.00	rw	12.00	1.142	rw	10.50	0.813	rw	14.77
5/4	1.25	rw	12.00	1.411	rw	10.63	1.063	rw	14.12
6/4	1.50	rw	12.00	1.680	rw	10.71	1.313	rw	13.71
7/4	1.75	rw	12.00	1.882	rw	11.16	1.500	rw	14.00
8/4	2.00	rw	12.00	2.151	rw	11.16	1.750	rw	13.71
9/4	2.25	rw	12.00	2.487	rw	10.86	2.063	rw	13.09
10/4	2.50	rw	12.00	2.688	rw	11.16	2.250	rw	13.33
11/4	2.75	rw	12.00	2.957	rw	11.16	2.500	rw	13.20
12/4	3.00	rw	12.00	3.226	rw	11.16	2.750	rw	13.09
13/4	3.25	rw	12.00	3.495	rw	11.16	3.000	rw	13.00
14/4	3.50	rw	12.00	3.763	rw	11.16	3.250	rw	12.92
15/4	3.75	rw	12.00	4.032	rw	11.16	3.500	rw	12.86
16/4	4.00	rw	12.00	4.301	rw	11.16	3.750	rw	12.80

* Green target sizes are set by the manufacturer based on shrinkage, size control, etc., thus these are just examples of what may be 'typical'.

**North American lumber product types tallied as 1″ in nominal thickness when the nominal thickness is 1″ or less.

Sources: Western Wood Products Association, 1998; National Hardwood Lumber Association, 1994.

6.1.1.1 *Softwood dimension lumber and studs*

This type of lumber is generally used for applications where appearance is secondary to serviceability and structural strength. Dimension lumber tends to be manufactured in longer lengths of 10–20′, with a nominal thickness of 2″, and widths produced in 2″ nominal increments from 4″ up to 12″. A stud is also a structural type of lumber, but is generally produced in shorter lengths (6–9′, with the majority being 92⅝″), is primarily used for vertical applications such was wall framing (thus requiring less strength and even less appearance characteristics) and, unlike dimension lumber, can be produced in nominal lengths, notably the 92⅝″ (7.719′) stud, which is tallied as having an 8′ length. Studs are normally produced in nominal widths of 4″ and 6″ (with the majority being 4″). As appearance is not a primary focus when grading this category of lumber, surfacing imperfections and relatively large amounts of wane are tolerated. To market this category of lumber as being dried, it must have no more than 19% mc.

An example of calculating the bf volume of a nominal 2″ × 4″ × 8′ is (actual dimensions are 1.5″ × 3.5″ × 92⅝″): $2 \times 4 \times 8 \div 12 = 5.333$ bf. As this particular board has an actual cubic content of 0.2814 ft^3, there are 18.95 bf of lumber per cubic foot of actual wood fibre (669 bf/m^3).

6.1.1.2 *Softwood boards (appearance type lumber)*

Boards (nomenclature for softwood lumber that is manufactured for applications where appearance is of primary concern) are generally manufactured to a nominal thickness of 1″ and in nominal widths on the even inch (generally from 4″ to 12″). Lengths are normally produced in 2′ multiples, and are based on actual length. Because this type of lumber sees much of its service in visible applications (furniture, fascia, trim, etc.), it is desirable to have a high finish quality, and small amounts, if any, of wane are allowed. To market this category of lumber as being dried, it must have no more than 19% moisture content (mc).

An example of calculating the bf volume of a nominal 1″ × 8″ × 16′ is (actual dimensions are 0.75″ × 7.25″ × 16′): $1 \times 8 \times 16 \div 12 = 10.667$ bf. As this particular board has an actual cubic content of 0.6042 ft^3, there are 17.66 bf of lumber per cubic foot of actual wood fibre (623 bf/m^3).

6.1.1.3 *Softwood factory lumber (shop)*

Factory lumber or shop is often used in the construction of very specific products such as solid wooden doors, window frames, etc. where appearance is very important. As shop lumber is intended to be remanufactured into components (clear cuttings between imperfections in the board), it is only surfaced on the wide faces (S2S). Shop can be manufactured in many different sizes (generally in 0.25″ nominal increments), but two sizes $^5/_4$″ and $^6/_4$″ shop (nominal thickness of 1.25″ and 1.5″ respectively) are the most popular. Shop lumber is measured based on actual random width (but rounded to the nearest inch, e.g. 8.7 rounds to 9 and 8.4 rounds to 8), and lengths can be in 1′ multiples. To market this category of lumber as being dried, it must have no more than 12% mc. Moulding and $^4\!/\!_4$″ shop also fall into the category of factory lumber, but it should be noted that these two products are measured in the same manner as softwood boards (typically nominal widths finished on all four sides, 1″ nominal thickness). Shop lumber is normally produced in tandem with boards.

An example of calculating the bf volume of a $^6\!/\!_4$″ shop board that is nominally 1.5″ thick (actual thickness is 1.406″), with an actual width of 12.3″ and an actual length of 13′, is: [(12 × 13) ÷ 12 rounded to nearest whole number] × 1.5 = 19.5 bf. As this board has an actual cubic content of 1.561 ft^3, there are 12.5 bf of lumber per cubic foot of actual wood fibre (441 bf/m^3).

6.1.1.4 Hardwood lumber

Hardwood lumber is similar to softwood shop in that much of it is intended for remanufacture. It is used heavily in the manufacture of furniture and flooring amongst other products. Like softwood shop, width and length are measured in actual measure. Hardwoods tend to shrink quite a bit during drying, and as hardwoods are utilized heavily in applications with a priority toward appearance, the finishing requirements are high. The target moisture content for hardwood lumber is typically around 8%, which works best in the indoor and heated environments where hardwoods see most of their use, but which contributes to the already high levels of shrinkage that commonly occur with hardwoods.

An example of calculating the bf volume of a $^6\!/\!_4$″ shop board that is nominally 1.5″ thick (actual thickness is 1.313″), with an actual width of 12.5″ and an actual length of 13′, is: [(12.5 × 13) ÷ 12 rounded to nearest whole number] × 1.5 = 21 bf. As this board has an actual cubic content of 1.482 ft^3, there are 14.17 bf of lumber per cubic foot of actual wood fibre (500.4 bf/m^3).

6.1.2 Lumber cubic volume

In most of the world, outside of North America, lumber volume is measured and reflected via cubic-metric measure, and in general, the volume calculated should closely reflect the actual volume. The formula for determining lumber cubic volume is:

$$m^3 = \text{thickness in millimetres} \times \text{width in millimetres} \times \text{length in metres} \div 1{,}000{,}000$$

$$ft^3 = \text{thickness in inches} \times \text{width in inches} \times \text{length in feet} \div 144$$

A tolerance level is given for size variation, e.g. + 6 mm, − 2 mm. It is also often common in the case of green lumber measured before drying to give an additional allowance (normally around 5%) for shrinkage, based on the predicted size once the lumber reaches its desired dried moisture content. For example, green lumber that measures 55 mm × 210 mm may be tallied based on 50 × 200 (shrinkage allowance + size tolerance) (Alberta Land and Forest Service, 1997).

An example of calculating the lumber volume of a dried board with an actual size of 52 mm × 199 mm × 5.02 m is: (50 × 200 × 5.0) ÷ 1,000,000 = 0.0500 m^3. The actual volume of this board is 0.0519 m^3.

6.2 Factors Affecting Lumber Recovery

6.2.1 Milling efficiency

6.2.1.1 Saw kerf

Saw kerf varies by machine centre and type of saw, with primary and secondary breakdown saws commonly ranging from 0.09″ to 0.32″ thick (2–8 mm). In general, band saws are thinner than circular saws, and sawing configurations for large logs or cants, hardwoods, frozen logs and high feed speeds tend to require thicker saws. According to published data, for each decrease of 0.01″ (0.254 mm) in saw kerf, LRF will increase by 0.026 bf of lumber per cubic foot (roughly 0.4%) of log when using circular saws, and 0.052 bf per cubic foot (roughly 0.8%) when using band saws (Wade *et al.*, 1992). As can be seen from the examples in Table 6.1, the results of saw kerf (sawdust) take up a large percentage of fibre used in a sawmill.

6.2.1.2 Target sizing and size control

A mill's ability to produce uniformly sized products is a key component in setting the target size. The target size needs to be large enough to account for shrinkage and still leave enough fibre for planing in order to fit the finish requirements for the size and type of product being produced. Size variation, shrinkage and surfacing allowance are all key components of the calculation on setting the GTS in order to produce finished lumber of an acceptable size and finish quality. Obviously, if the finished thickness of a board needs to be 0.75″ (19 mm) after planing and shrinkage, and sawing variation is ±0.065″ (1.7 mm), the target size will need to be developed primarily with the minus variation in mind (thus wasting a significant volume of the lumber that falls into the plus side of the variation). Size variation can be caused by saws that are too thin for the feed speeds, dull saws, worn saw spacers or arbor bearings, misalignment on double arbor saws, poor control of the log, cant or boards during the sawing process, frozen wood, slope of grain, etc.

6.2.1.3 Shape sawing optimization

There are optimal ways of bucking log stems into sawmill lengths and cutting lumber of a particular size and orientation given the shape of a log in order to capture the maximal volume and value from a log. In the past, these decisions were made by the chop-saw operator and the sawyer through human judgement. With the advent of true shape scanning and computer processors, which can rapidly process information and instruct equipment as to the best sawing solution and even adjust sawing lines to cut parallel to the curvature of a log (curve sawing), optimization equipment is now incorporated into many mills, including secondary processing equipment such as edgers and trim saws, and this has the potential to significantly improve product volume recovery.

6.2.1.4 Products manufactured

As discussed in Section 6.1 and shown in Table 6.4, different lumber products require differing fibre contents to produce as a result of nominal sizing. Additionally, variability in the fibre needed to produce a given volume of lumber can be the result of a relatively high percentage of the wood fibre being converted into sawdust (necessary for manufacturing less thickness), and producing appearance-type lumber, which does not allow much wane and requires plenty of planing allowance to prevent skip.

Some facilities also capture short lumber and trim-ends that were formerly converted into chips, as stock for finger-jointing or other products where these short pieces can be utilized. Tallying these short lumber lengths can significantly improve volume recovery, but may not have the same relative effect on value recovery.

6.2.2 Log characteristics

Log characteristics have a dramatic influence on lumber recovery. There are many geometric influences of manufacturing products with a square or rectangular cross section of given sizes and multiples from a raw material with a round cross section which may not be shaped or sized in a manner that is conducive to gaining maximal or consistent utilization.

It is very important to note that trends that make complete sense when looking at log volume in a cubic volumetric manner will often not make sense when looking at the same trends from a product output scale rule standpoint. This is because the architects of product output scaling rules tried to account for many of the influences of log characteristics within the rule (when the scale rule was designed, long ago), and often there are overcompensations, biases and inaccuracies in these assumptions when applied to current milling technology, wood utilization practices and products made.

6.2.2.1 Diameter

CUBIC SCALED LOGS. Because of limitations in lumber sizes resulting from the need to produce marketable and profitable lumber sizes of minimum and standardized thickness and widths, lumber recovery tends to trend upward steeply from 5″ d (13 cm) up to about 16″ d (41 cm) where it levels off significantly. Slab loss is obviously much higher for small logs given the needed opening face size to produce profitable products.

Using dimension lumber as an example, lumber recovery starts at a 5″ d (11.4–14.0 cm) with roughly 33% lumber recovery, and reaches a peak at 24″ d (60–62 cm) with approximately 63% lumber recovery.

PRODUCT OUTPUT SCALED LOGS. Excepting for the Scribner log rules in some diameter classes, recovery tends to trend downward as diameters increase. This is primarily the result of slab loss allowances assumptions made when the rules were designed, which may have been appropriate for the large logs, crude technology and wane-free products of that time, but which no longer apply to the smaller logs, technology and products of today. In other words, the lumber volume predicted from product output rules tends to be understated for small logs relative to large logs.

6.2.2.2 Length

When discussing the effects of length on recovery, it is very important to identify the length being discussed. The scaling length of a log can be 32′ (9.8 m), made up of two segments of 16′ (4.9 m), and the mill-processed length of the same log can be 8′ (2.45 m) when cut into four equal mill-processing lengths.

CUBIC SCALED LOGS. Shorter log lengths processed in the mill generally increase recovery. This is because longer lengths will incur more slab loss due to taper, and related to this, shorter log lengths afford the opportunity to reassess sawing patterns more frequently given a changed log cross section. In addition, longer lengths tend to lose recovery because of crook and sweep, while shorter lengths can minimize these effects.

Longer scaling lengths, especially in small-diameter wood where the recovery by diameter trend is the steepest, can also increase the recovery of cubic scaled wood relative to the same small-end diameter logs that are shorter. This is because the large-end of the log is bigger relative to a shorter log with the same small-end diameter. In the case of the Swedish National Log Scale and JAS, this trend is magnified by the fact that taper is not accounted for until log lengths reach 5.9 and 6 m respectively.

PRODUCT OUTPUT SCALED LOGS. For the same reasons listed for the cubic scaled logs, short mill-processed lengths will increase the recovery of product output scaled logs. The opposite is true for scaling segment length, as the longer the unadjusted scaling cylinder is in a log, the better the recovery will be. This is because the amount of log volume outside the scaling cylinder increases due to taper, and lumber volume can be recovered from this unaccounted for volume (see Fig. 2.26 on p. 49). With all of the rules except the International ¼-Inch Rule, the scaling cylinder is the same length as the log segment length; the International ¼-Inch Rule scaling cylinder increases by ½″ every 4′ (see Fig. 2.34 on p. 68). Thus, excepting the International ¼-Inch Rule, longer segment lengths generally translate into greater recovery. This trend is rather erratic and less pronounced with the Scribner log rules due to the step function (see Figs 6.6 and 6.8). Note that this trend is tied to segment length and not necessarily to scaling length, e.g. a log that is scaled by the Scribner Short Log Rule can have a scaling length of 20′ with a 20′-long scaling cylinder, but a log with a 22′ scaling length has two segments: 10′ for the top segment and 12′ for the bottom segment. Thus, the scaling cylinders are shorter and the second scaling cylinder readjusts to a new and larger diameter, thereby accounting for more of the log volume, which will likely reduce the recovery relative to the same small-end diameter log that has a 20′ scaling length. Also crook and sweep deductions will be minimized in the 22′ log relative to the 20′ log because the log is in theory two log segments, which allows the scaling cylinder to assume a new direction at the segment break.

6.2.2.3 Taper

CUBIC SCALED LOGS. Taper has a negative impact on lumber recovery when accounting for the actual volume of a log. Figure 6.1 indexes the effects of taper on four size classes of 16.5′ (5 m) logs, with taper ranging from 1″ (2.5 cm) to 4″ (10.2 cm). Note that for 6″ d (15 cm) logs with 2″ of taper in 16.5′ (5.1 cm in 5 m), recovery is 5.9% less than what it is at 1″ in 16.5′ (2.5 cm in 5 m), and at 4″ in 16.5′ (10.2 cm in 5 m) recovery falls off by almost 17%. A notable exception to this trend is in regard to taper assumptions for cubic measure via standard taper factors. Obviously, if the actual taper ratio is higher than what is assumed, recovery will be improved, not because taper truly helps recovery but because the actual volume of the log was higher than what was assumed. In the case of JAS, other things being equal, recovery is better for logs that are 5.8 m long versus logs that are 6.0 m long, as the 5.8 m log is scaled without any taper allowance while logs that are 6.0 m long and longer are given some allowances for taper (thus increasing the relative log volume).

PRODUCT OUTPUT SCALED LOGS. Taper is also a key component of recovery ratios for product output scaled logs, and like cubic measure where taper is assumed, increased amounts of taper improve the recovery ratio (Fig. 6.1).

6.2.2.4 Log defects

When only looking at a log's gross volume, log defects will certainly reduce the volume of lumber recovered. However, since many scaling methods account for defects due to deductions of volume from the gross scale, defects can even increase the recovery ratio. This is because lumber can often be made from portions of the deducted log volume. Often the lumber from the defective areas is of very low quality containing some unsound fibre, fibre separations, oversize knots, etc., but with a value higher than chips. However, if the scaling method used only accounts for void, soft-rot and char, or does not account for defect at all, any defects whatsoever are likely to reduce recovery.

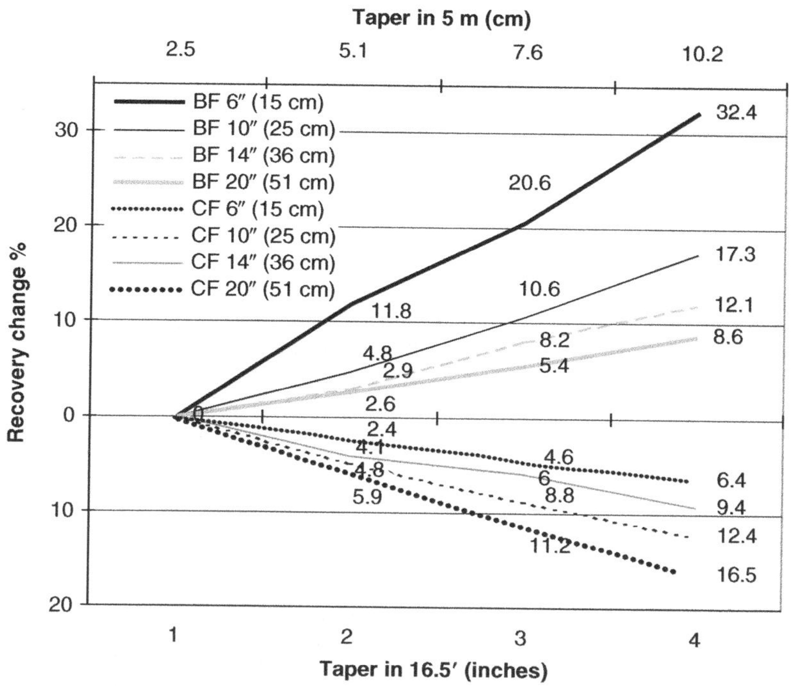

Source: Hallock *et al.*, 1979
Note: BF is Scribner Short Log; CF is cubic foot. Logs are assumed to be 16.5′ (5 m) long, with a small-end diameter of 6″, 10″, 14″, 20″ (15, 25, 36, 51 cm).

Fig. 6.1. Effects on lumber recovery from taper in cubic and board foot scaled logs (% change in recovery for taper exceeding 1″ in 16.5″ (2.5 cm in 5 m)).

It is also common for defect deductions to be made for crook and sweep, with assumptions made as to how a log will be processed and segmented. If a mill has the option of bucking the log into shorter lengths than were assumed for calculating defect deductions, much of the crook and sweep can be eliminated by cutting the log into short lengths, thus nullifying some of the assumed loss. Curve sawing technology also allows mills to produce lumber from sweep or crook that may be deducted during the scaling procedures.

6.2.2.5 Wood property issues

An important wood property factor affecting recovery is moisture content. Water exists in the wood cell cavities and within the cell walls. Shrinkage normally begins to occur once the water from the cell cavities is absent, and the water in the cell walls begins to evaporate. Normally, for most species this occurs at about 30% mc (USDA Forest Products Laboratory, 1974). From the 30% point down to the target dry moisture content for the lumber product, lumber will shrink. Shrinkage can vary a great deal based on wood grain orientation and species.

Table 6.5. Shrinkage rates of common North American trees (green to oven-dry).

Conifers		Shrinkage (%)		Hardwoods		Shrinkage (%)	
Common name	Latin name	Radial	Tangential	Common name	Latin name	Radial	Tangential
Balsam fir	*Abies balsamea*	2.9	6.9	Hard maple	*Acer saccharum*	4.8	9.9
Grand fir	*Abies grandis*	3.4	7.5	Yellow birch	*Betula alleghaniensis*	7.3	9.5
Western larch	*Larix occidentalis*	4.5	9.1	Paper birch	*Betula papyrifera*	6.3	8.6
Engelmann spruce	*Picea engelmannii*	3.8	7.1	Hickory	*Carya* spp.	7.0	10.5
Black spruce	*Picea mariana*	4.1	6.8	Hackberry	*Celtis* spp.	4.8	8.9
Lodgepole pine	*Pinus contorta*	4.3	6.7	American beech	*Fagus grandifolia*	5.5	11.9
Shortleaf pine	*Pinus echinata*	4.6	7.7	White ash	*Fraxinus americana*	4.9	7.8
Western white pine	*Pinus monticola*	4.1	7.4	Sweetgum	*Liquidambar styraciflua*	5.3	10.2
Longleaf pine	*Pinus palustris*	5.1	7.5	Yellow poplar	*Liriodendron tulipifera*	4.6	8.2
Ponderosa pine	*Pinus ponderosa*	3.9	6.2	Water tupelo	*Nyssa aquatica*	4.2	7.6
Red pine	*Pinus resinosa*	3.8	7.2	Black gum	*Nyssa sylvatica*	5.1	8.7
Eastern white pine	*Pinus strobus*	2.1	6.1	Sycamore	*Platanus occidentalis*	5.0	8.4
Loblolly pine	*Pinus taeda*	4.8	7.4	Cottonwood	*Populus* spp.	3.6	8.6
Douglas fir	*Pseudotsuga menziesii*	4.8	7.6	Aspen	*Populus tremuloides*	3.5	6.7
Northern white cedar	*Thuja occidentalis*	2.2	4.9	White oak	*Quercus alba*	5.3	10.8
Western red cedar	*Thuja plicata*	2.4	5.0	Southern red oak	*Quercus falcata*	4.7	11.3
Eastern hemlock	*Tsuga canadensis*	3.0	6.8	American basswood	*Tilia americana*	6.6	9.3
Western hemlock	*Tsuga heterophylla*	4.2	7.8	American Elm	*Ulmus* spp.	4.2	7.2

Source: USDA Forest Products Laboratory, 1974.

Table 6.5 shows typical shrinkage percentages of common woods in North America by grain orientation (radial or tangential). Wood normally shrinks about half as much radially (across the growth rings) as it does tangentially along the growth rings, and it shrinks very little longitudinally (generally between 0.1 and 0.2%).

Table 6.5 shows average total shrinkage from green to 0% mc, but shrinkage to the target moisture content of dried wood can be calculated with the following formula:

$$\text{Shrinkage} = \text{total shrinkage to 0\% mc} \times (30 - \text{target mc}) \div 30$$

For example, to calculate the shrinkage of Douglas fir if dried to 19% mc:

$4.8 \times (30 - 19) \div 30 = 1.76\%$ radial;
$7.6 \times (30 - 19) \div 30 = 2.79\%$ tangential

To determine the volume loss to shrinkage the formula is:

$$\text{\% volume loss due to shrinkage} = 100 - \{100 - (\text{\% radial shrinkage}) \times [1 - (\text{\% tangential shrinkage} \div 100)]\}$$

For example, to calculate the percentage of volume loss from drying Douglas fir lumber from green to 19% mc:

$$100 - (100 - 1.76) \times [1 - (2.79 \div 100)] = 4.5\%$$

Green lumber produced from conifers will lose roughly 4% volume during the drying process, while green lumber produced from hardwoods will lose roughly 8% volume (USDA Forest Products Laboratory, 1974). There is a great deal of variation, however, within different tree species, and even within the same tree stem.

Aside from shrinkage, many species are prone to cupping, warping, splitting and twisting during the drying process, which will further reduce recovery. These tendencies are further aggravated by wood property issues such as tension/compression of wood, and excessive slope of grain.

6.3 Recovery Trends by Log Size and Lumber Products Produced

With the exception of hardwood, the preceding graphs and tables were constructed from recovery studies conducted by the author. The softwood recovery data were taken from numerous 'return to log' studies that were compiled by product type (board, stud, dimension and hardwood). The studies were conducted in facilities that, in the author's opinion, have excellent equipment and practices (curve sawing, cut-in-two solutions for re-edging, etc.), and thus recovery is somewhat better in comparison with the absolute recoveries shown in *Regional Softwood Sawmill Processing Variables as Influenced by Productive Capacity* (Steele *et al.*, 1991). The hardwood recovery graphs and tables were constructed from data taken from Hanks *et al.* (1980).

The data presented in Tables 6.6–6.9 and Figs 6.2–6.4 represent typical recovery ratios and trends. As with many ratios involving roundwood,

Table 6.6. Board mill, lumber and residual product recovery by small-end diameter of cubically scaled logs.

SED		USFS National CF Lumber recovery		BC Firmwood Lumber recovery			Green		Planer	Dry
inches	cm	bf/ft^3	%	bf/m^3	%	Sawdust (%)	chips (%)	Shrinkage (%)	shavings (%)	chips (%)
5	**13**	6.4	34.8	207	32.0	10.4	42.3	2.1	10.4	2.8
6	**15**	6.7	36.4	216	33.4	10.8	39.7	2.2	10.9	2.9
7	**18**	7.0	39.8	228	36.6	11.1	35.5	2.4	11.3	3.2
8	**20**	7.3	41.6	240	38.6	11.5	32.0	2.5	12.0	3.4
9	**23**	7.5	42.5	246	39.5	11.6	30.1	2.6	12.7	3.4
10	**25**	7.7	43.6	252	40.5	11.9	28.4	2.7	13.1	3.5
11	**28**	7.7	44.0	252	40.9	11.7	28.4	2.7	12.8	3.6
12	**30**	7.7	44.2	257	41.8	11.9	27.1	2.7	12.8	3.6
13	**33**	7.8	45.1	260	42.6	11.9	26.2	2.8	12.8	3.7
14	**36**	7.9	45.9	263	43.4	12.0	25.3	2.8	12.7	3.8
15	**38**	8.0	46.5	266	43.9	12.1	24.4	2.9	12.8	3.8
16	**41**	8.2	47.6	278	45.9	12.7	21.0	3.0	13.4	4.0
17	**43**	8.2	47.6	278	45.9	12.7	21.0	3.0	13.4	4.0
18	**46**	8.2	47.6	278	45.9	12.7	21.0	3.0	13.4	4.0
19	**48**	8.2	47.6	278	45.9	12.7	21.0	3.0	13.4	4.0
20	**51**	8.2	47.6	278	45.9	12.7	21.0	3.0	13.4	4.0
21	**53**	8.2	47.6	278	45.9	12.7	21.0	3.0	13.4	4.0
22	**56**	8.2	47.6	278	45.9	12.7	21.0	3.0	13.4	4.0
23	**58**	8.2	47.6	278	45.9	12.7	21.0	3.0	13.4	4.0
24	**61**	8.2	47.6	278	45.9	12.7	21.0	3.0	13.4	4.0
25	**64**	8.2	47.6	278	45.9	12.7	21.0	3.0	13.4	4.0
26	**66**	8.2	47.6	278	45.9	12.7	21.0	3.0	13.4	4.0
27	**69**	8.2	47.6	278	45.9	12.7	21.0	3.0	13.4	4.0
28	**71**	8.2	47.6	278	45.9	12.7	21.0	3.0	13.4	4.0
29	**74**	8.2	47.6	278	45.9	12.7	21.0	3.0	13.4	4.0
30	**76**	8.2	47.6	278	45.9	12.7	21.0	3.0	13.4	4.0

Table 6.7. Stud mill, lumber and residual product recovery by small-end diameter of cubically scaled logs.

SED		USFS National CF Lumber recovery		BC Firmwood Lumber recovery						
inches	cm	bf/ft^3	%	bf/m^3	%	Sawdust (%)	Green chips (%)	Shrinkage (%)	Planer shavings (%)	Dry chips (%)
5	**13**	8.0	42.9	259	39.4	6.7	41.9	2.3	7.3	2.4
6	**15**	8.7	47.0	284	43.2	7.3	36.3	2.5	8.0	2.6
7	**18**	9.4	52.3	305	48.1	7.4	30.6	2.8	8.2	2.9
8	**20**	9.7	53.9	318	50.1	7.5	27.6	2.9	8.6	3.3
9	**23**	9.7	53.9	318	50.1	7.3	27.5	2.9	8.6	3.5
10	**25**	9.8	54.4	321	50.6	7.3	26.8	3.0	8.7	3.6
11	**28**	10.0	55.5	327	51.5	7.3	25.4	3.0	8.9	3.8
12	**30**	10.2	56.5	339	53.4	7.5	22.7	3.1	9.3	4.0
13	**33**	10.3	57.1	342	53.9	7.4	21.9	3.2	9.4	4.2
14	**36**	10.3	57.1	342	53.9	7.3	21.9	3.2	9.4	4.2
15	**38**	10.3	57.1	342	53.9	7.3	21.9	3.2	9.4	4.2
16	**41**	10.3	57.1	350	55.1	7.5	20.3	3.3	9.6	4.3
17	**43**	10.3	57.1	350	55.1	7.5	20.3	3.3	9.6	4.3
18	**46**	10.3	57.1	350	55.1	7.5	20.3	3.3	9.6	4.3
19	**48**	10.4	57.6	353	55.6	7.5	19.6	3.3	9.7	4.4
20	**51**	10.5	58.1	356	56.1	7.6	18.8	3.3	9.7	4.4
21	**53**	10.9	60.8	372	58.6	8.0	15.1	3.5	10.2	4.6
22	**56**	11.4	63.4	389	61.2	8.3	11.4	3.6	10.6	4.8
23	**58**	11.4	63.4	389	61.2	8.3	11.4	3.6	10.6	4.8
24	**61**	11.4	63.4	389	61.2	8.3	11.4	3.6	10.6	4.8
25	**64**	11.4	63.4	389	61.2	8.3	11.4	3.6	10.6	4.8

Table 6.8. Dimension mill, lumber and residual product recovery by small-end diameter of cubically scaled logs.

SED		USFS National CF Lumber recovery		BC Firmwood Lumber recovery			Green		Planer	Dry
inches	cm	bf/ft^3	%	bf/m^3	%	Sawdust (%)	chips (%)	Shrinkage (%)	shavings (%)	chips (%)
5	**13**	6.7	36.4	216	33.4	5.6	50.1	2.0	6.3	2.5
6	**15**	8.1	44.2	262	40.6	6.9	39.4	2.4	7.6	3.1
7	**18**	8.9	50.5	290	46.4	7.1	32.1	2.7	8.0	3.5
8	**20**	8.9	50.8	293	47.1	7.0	31.2	2.8	8.2	3.6
9	**23**	9.0	51.1	296	47.5	6.9	30.5	2.8	8.7	3.6
10	**25**	9.0	51.1	296	47.5	6.8	30.5	2.8	8.9	3.6
11	**28**	9.3	53.2	305	49.5	6.9	28.3	2.9	8.7	3.7
12	**30**	9.5	54.6	317	51.6	7.1	25.5	3.0	8.9	3.9
13	**33**	9.5	55.0	317	52.0	7.0	25.5	3.0	8.6	3.9
14	**36**	9.5	55.4	317	52.3	6.9	25.5	3.0	8.3	4.0
15	**38**	9.5	55.4	317	52.3	6.9	25.5	3.0	8.3	4.0
16	**41**	9.5	55.4	324	53.4	7.0	23.9	3.1	8.5	4.0
17	**43**	10.0	58.1	340	56.1	7.4	20.1	3.3	8.9	4.2
18	**46**	10.0	58.1	340	56.1	7.4	20.1	3.3	8.9	4.2
19	**48**	10.0	58.1	340	56.1	7.4	20.1	3.3	8.9	4.2
20	**51**	10.3	59.8	350	57.7	7.6	17.9	3.4	9.1	4.4
21	**53**	10.5	60.9	356	58.8	7.7	16.3	3.4	9.3	4.5
22	**56**	10.9	63.7	372	61.4	8.1	12.5	3.6	9.7	4.7
23	**58**	10.9	63.7	372	61.4	8.1	12.5	3.6	9.7	4.7
24	**61**	11.2	65.3	382	63.0	8.3	10.3	3.7	10.0	4.8
25	**64**	11.0	64.2	376	62.0	8.2	11.8	3.6	9.8	4.7
26	**66**	10.8	63.1	369	60.9	8.0	13.3	3.5	9.6	4.6
27	**69**	10.6	62.0	363	59.8	7.9	14.8	3.5	9.5	4.5
28	**71**	10.5	60.9	356	58.8	7.7	16.3	3.4	9.3	4.5
29	**74**	10.3	59.8	350	57.7	7.6	17.9	3.4	9.1	4.4
30	**76**	10.1	58.7	343	56.6	7.4	19.4	3.3	9.0	4.3

Table 6.9. Hardwood mill, lumber and residual product recovery by small-end diameter of cubically scaled logs.

SED		USFS National CF Lumber recovery		BC Firmwood Lumber recovery						
inches	cm	bf/ft^3	%	bf/m^3	%	Sawdust (%)	Green chips (%)	Shrinkage (%)	Planer shavings (%)	Dry chips (%)
8	**20**	5.62	38.9	184	36.1	11.3	36.3	5.3	8.1	2.8
9	**23**	6.00	41.5	197	38.6	12.0	32.0	5.7	8.6	3.0
10	**25**	6.10	42.2	200	39.2	12.2	30.9	5.8	8.8	3.1
11	**28**	6.25	43.2	205	40.2	12.5	29.2	5.9	9.0	3.1
12	**30**	6.35	44.0	212	41.5	13.0	26.8	6.1	9.3	3.2
13	**33**	6.80	47.1	227	44.5	13.9	21.6	6.6	10.0	3.5
14	**36**	6.83	47.3	228	44.7	13.9	21.2	6.6	10.0	3.5
15	**38**	6.87	47.6	229	44.9	14.0	20.8	6.6	10.1	3.5
16	**41**	6.99	48.4	238	46.7	14.6	17.7	6.9	10.5	3.6
17	**43**	7.34	50.8	250	49.0	15.3	13.6	7.3	11.0	3.8
18	**46**	7.38	51.1	251	49.3	15.4	13.2	7.3	11.0	3.8
19	**48**	7.45	51.6	254	49.8	15.5	12.3	7.4	11.2	3.9
20	**51**	7.52	52.1	256	50.2	15.7	11.5	7.4	11.3	3.9
21	**53**	7.44	51.5	253	49.7	15.5	12.4	7.3	11.1	3.9
22	**56**	7.47	51.7	255	49.9	15.6	12.1	7.4	11.2	3.9
23	**58**	7.55	52.3	257	50.5	15.7	11.1	7.5	11.3	3.9
24	**61**	7.86	54.4	268	52.5	16.4	7.4	7.8	11.8	4.1
25	**64**	7.39	51.2	252	49.4	15.4	13.0	7.3	11.1	3.9
26	**66**	7.71	53.4	263	51.5	16.1	9.2	7.6	11.5	4.0
27	**69**	7.66	53.0	261	51.2	16.0	9.9	7.6	11.5	4.0
28	**71**	7.47	51.7	255	49.9	15.6	12.0	7.4	11.2	3.9
29	**74**	7.52	52.1	256	50.2	15.7	11.5	7.4	11.3	3.9
30	**76**	7.56	52.3	258	50.5	15.8	11.0	7.5	11.3	3.9

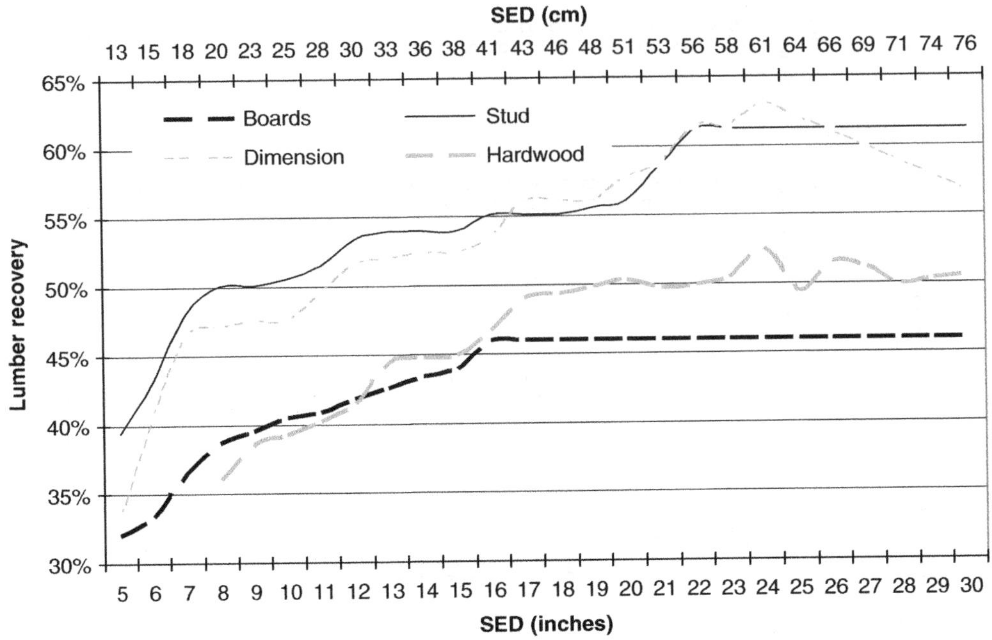

Fig. 6.2. Lumber recovery by small-end diameter (SED) of log and mill type, actual log volume (BC Firmwood). (% of actual displaced log volume).

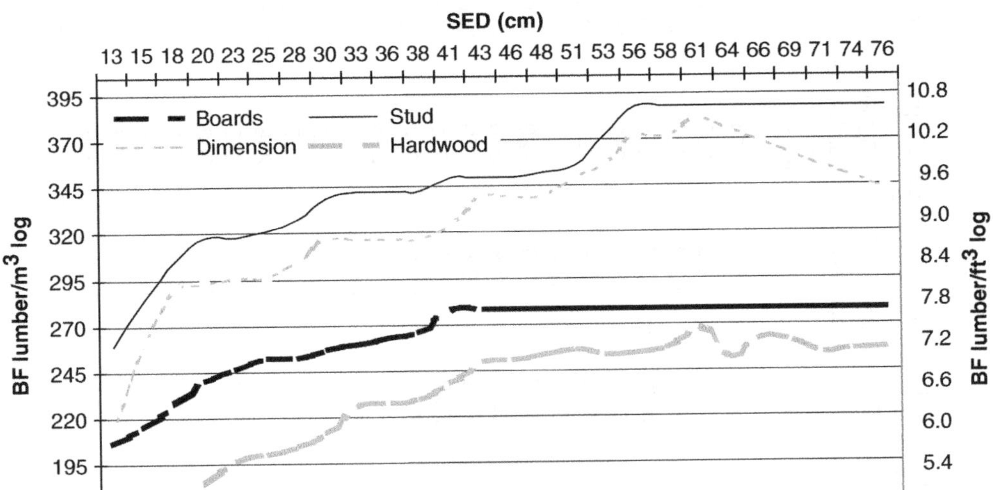

Fig. 6.3. Lumber recovery by small-end diameter (SED) of log and mill type, actual log volume (BC Firmwood). (LRF).

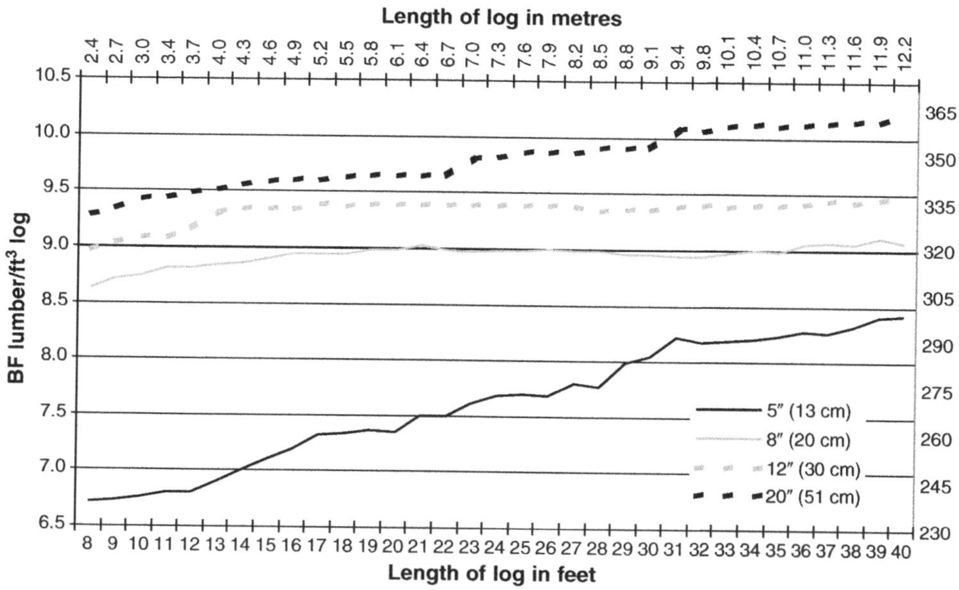

Note: Product recovery tables and graphs based on mill lengths of 20′ and less in nominal lengths (< 6.1m). Lengths from 21′ to 40′ in nominal length are included in this figure for comparison with Fig. 6.8.

Fig. 6.4. Cubic lumber recovery trends by log length (LRF).

these data should not be construed as universal ratios that are applicable to all log populations, sawmill configurations or cutting strategies. Despite every effort to show results and trends that are typical and universal, the data may have attributes unique to the recovery tests on which these data were based, and thus, there may be certain data points or outcomes unique to the source data. It is more important to focus on the relative recovery rates by log size and lumber type than on the absolute recovery numbers, which can vary substantially as a result of the variables discussed in Section 6.2.

All recovery data are based on log segments 8.3–20.5′ (2.5–6.2 m) long. Data and ratios were applied from stud blocks into the longer segment lengths used for comparison. Comparisons assume the original range of diameters covered in the log recovery tests from which the data were extracted. Recovery ratios for the different scaling methods were calculated using the conversion ratios from Section 2.5. All lumber except hardwood is assumed to be finished S4S (surfaced on all four sides). Hardwood is assumed to be S2S (only surfaced on the wide faces).

6.3.1 Cubic scaled logs

Recovery for cubic scaled logs have similar trends amongst the different log scales, and so recovery data is displayed for BC Firmwood (representing the assumed actual displaced volume), although in the case of Tables 6.6–6.9, the USFS Cubic Log Scale is also shown, as it was the original measure used for the tests (excepting the hardwood scenario). If one wants to estimate cubic recoveries for the other cubic scales, use the index figures from Table 2.9 on pp. 85–88 to adjust estimates. Note that because BC Firmwood theoretically measures the actual volume, the total product recoveries (% lumber + % sawdust + % green chips + % planer shavings + % dry chips + shrinkage) will add up to 100%; this will likely not occur when using recovery ratios of other log scales, e.g. the USFS Cubic Log Scale. Lumber recovery factors (LRFs) are shown in three different units of measure: bf lumber as it relates to m^3 log, bf lumber as it relates to the ft^3 log, and actual volume of lumber recovered as a percentage of the total volume of log.

Note that the mills producing structural lumber (studs and dimension) tend to have higher recovery than the mills producing appearance-type lumber (boards and hardwood). It is also interesting to note, in Fig. 6.2 based on recovery as a percentage, that stud mills and dimension mills share similar recoveries, and their graph lines even cross in several places. The same holds true of board mills and hardwood mills, with the board mill showing better recovery for small logs and the hardwood mill showing better recovery for larger logs. In Fig. 6.3, which is based on lumber recovery of bf lumber to cubic volume of the logs, it appears that the dimension mill recovery is consistently less than the stud mill's, but it also appears in a more pronounced manner that hardwood recovery is significantly less than what a board mill attains. In terms of actually measuring efficiency (especially across product lines), Fig. 6.2 gives the more accurate measure. Figure 6.4 shows lumber recovery trends by diameter and log length for cubic scaled logs based on BC Firmwood but they are applicable to all of the cubic methods excepting Swedish Cubic and JAS because taper is ignored for lengths less than 6 m.

6.3.2 Product output scaled logs

Recovery trends for product output scaled logs can be very diverse, and so recovery ratios and trends for each of the four product output rules are discussed at length in this book. These ratios were extrapolated from the log segment volumes taken from the logs shown in Table A.1.M.

In the case of Scribner Long Log, which utilizes segments up to 40′ in length, volume was allocated from these long segments (over 20′) to the shorter mill-processed lengths via the cubic volume contained in the mill-processed lengths, e.g. a log that is 32′ × 5″ × 9″ (with a 7″ midpoint) has a volume of 30 bf using the Scribner Long Log Rule. Since comparisons were made against mill-processed lengths, the 30 bf was allocated to the top 16′ × 5″ × 7″ segment and bottom 16′ × 7″ × 9″ segment via the actual cubic volume contained in each segment. In this case, the top segment was allocated 36% of the 30 bf (11 bf) and the bottom segment was allocated 64% of the 30 bf (19 bf). In addition, the small-end diameter used for comparison is based on conventional rounding rules – not the truncated rounding rules of Scribner Long Log. While this methodology could be construed as being somewhat arbitrary, it is a workable approach to making the various scale rules comparable on an equal basis with conventionally rounded diameter classifications and log lengths that are likely used in a mill.

When reviewing the data in Tables 6.10–6.13 and Figs 6.5–6.9, it is interesting to note the erratic trend lines that occur in the two Scribner diagram rules versus the more predictable trends in the formula-based rules (Doyle and International ¼-Inch) and cubic. As Doyle is rarely used for diameters under 8″, recovery is only shown for 8″ and larger logs. All of the product output rules display more linear trends as the diameters get larger, and the International ¼-Inch does a particularly good job of predicting lumber output when making hardwood lumber. It is also important to note that the recovery ratios shown for the two Scribner rules for the smaller-end diameter logs (5″ and 8″ d) are particularly volatile to changes in log lengths (Figs 6.6 and 6.8) in comparison to the same trends shown for cubic scaled logs (Fig. 6.4). Doyle and the International ¼-Inch Rule were not graphed by length as they will have linear trend lines.

Table 6.10. Board mill, lumber recovery by small-end diameter (SED) of product output scaled logs (bf lumber/bf log).

SED					
inches	cm	Scribner SL	Scribner LL	Doyle	International ¼-Inch
5	13	1.19	1.86		1.95
6	15	1.55	1.83		1.44
7	18	1.43	2.21		1.32
8	20	1.76	2.14	2.97	1.31
9	23	1.66	1.99	2.28	1.26
10	25	1.51	2.08	1.99	1.26
11	28	1.48	2.18	1.79	1.18
12	30	1.45	1.79	1.66	1.20
13	33	1.37	1.64	1.53	1.16
14	36	1.36	1.69	1.47	1.13
15	38	1.25	1.57	1.33	1.12
16	41	1.31	1.63	1.37	1.16
17	43	1.25	1.60	1.33	1.10
18	46	1.18	1.46	1.25	1.09
19	48	1.17	1.47	1.23	1.07
20	51	1.14	1.44	1.23	1.08
21	53	1.11	1.26	1.16	1.04
22	56	1.14	1.32	1.18	1.07
23	58	1.07	1.24	1.09	1.04
24	61	1.08	1.27	1.08	1.03
25	64	1.03	1.23	1.07	1.03
26	66	1.04	1.08	1.05	1.04
27	69	1.00	1.17	1.03	1.02
28	71	0.99	1.13	1.01	1.01
29	74	1.06	1.10	1.03	1.00
30	76	1.04	1.13	1.00	1.02

Table 6.11. Stud mill, lumber recovery by small-end diameter (SED) of product output scaled logs (bf lumber/bf log).

SED inches	SED cm	Scribner SL	Scribner LL	Doyle	International ¼-Inch
5	**13**	1.49	2.33		2.45
6	**15**	2.04	2.41		1.89
7	**18**	1.91	2.96		1.77
8	**20**	2.34	2.83	3.93	1.74
9	**23**	2.14	2.57	2.94	1.63
10	**25**	1.92	2.64	2.53	1.60
11	**28**	1.92	2.82	2.32	1.53
12	**30**	1.92	2.37	2.19	1.59
13	**33**	1.81	2.17	2.02	1.52
14	**36**	1.77	2.19	1.91	1.47
15	**38**	1.61	2.02	1.71	1.44
16	**41**	1.64	2.04	1.72	1.45
17	**43**	1.57	2.01	1.67	1.38
18	**46**	1.49	1.83	1.57	1.37
19	**48**	1.48	1.87	1.55	1.35
20	**51**	1.46	1.84	1.58	1.38
21	**53**	1.48	1.68	1.56	1.39
22	**56**	1.60	1.84	1.65	1.50
23	**58**	1.49	1.72	1.52	1.45
24	**61**	1.51	1.77	1.51	1.44
25	**64**	1.44	1.71	1.50	1.43

Table 6.12. Dimension mill, lumber recovery by small-end diameter (SED) of product output scaled logs (bf lumber/bf log).

SED inches	SED cm	Scribner SL	Scribner LL	Doyle	International ¼-Inch
5	**13**	1.24	1.94		2.04
6	**15**	1.88	2.23		1.74
7	**18**	1.81	2.81		1.68
8	**20**	2.15	2.61	3.63	1.60
9	**23**	2.00	2.39	2.74	1.51
10	**25**	1.77	2.44	2.34	1.48
11	**28**	1.79	2.63	2.17	1.43
12	**30**	1.79	2.21	2.05	1.49
13	**33**	1.68	2.01	1.87	1.41
14	**36**	1.64	2.03	1.77	1.36
15	**38**	1.49	1.87	1.59	1.34
16	**41**	1.52	1.89	1.59	1.34
17	**43**	1.52	1.96	1.62	1.34

continued

Table 6.12. *continued*

SED					
inches	cm	Scribner SL	Scribner LL	Doyle	International ¼-Inch
18	**46**	1.44	1.78	1.53	1.33
19	**48**	1.43	1.80	1.50	1.30
20	**51**	1.43	1.80	1.55	1.35
21	**53**	1.41	1.61	1.49	1.33
22	**56**	1.53	1.76	1.58	1.44
23	**58**	1.42	1.65	1.46	1.39
24	**61**	1.49	1.74	1.49	1.41
25	**64**	1.39	1.65	1.45	1.39
26	**66**	1.38	1.44	1.39	1.38
27	**69**	1.30	1.53	1.35	1.32
28	**71**	1.26	1.45	1.29	1.29
29	**74**	1.33	1.38	1.29	1.26
30	**76**	1.28	1.39	1.23	1.25

Table 6.13. Hardwood mill, lumber recovery by small-end diameter (SED) of product output scaled logs (bf lumber/bf log).

SED					
inches	cm	Scribner SL	Scribner LL	Doyle	International ¼-Inch
8	**20**	1.35	1.64	2.28	1.01
9	**23**	1.33	1.59	1.82	1.01
10	**25**	1.20	1.65	1.58	1.00
11	**28**	1.20	1.77	1.45	0.96
12	**30**	1.20	1.48	1.37	0.99
13	**33**	1.20	1.44	1.34	1.01
14	**36**	1.18	1.46	1.27	0.98
15	**38**	1.08	1.35	1.15	0.97
16	**41**	1.12	1.39	1.17	0.99
17	**43**	1.12	1.44	1.19	0.99
18	**46**	1.07	1.31	1.13	0.98
19	**48**	1.07	1.34	1.12	0.97
20	**51**	1.05	1.32	1.13	0.99
21	**53**	1.01	1.14	1.06	0.95
22	**56**	1.05	1.20	1.08	0.98
23	**58**	0.98	1.14	1.01	0.96
24	**61**	1.04	1.22	1.04	0.99
25	**64**	0.93	1.11	0.97	0.93
26	**66**	0.98	1.02	0.99	0.98
27	**69**	0.93	1.10	0.97	0.95
28	**71**	0.90	1.03	0.92	0.92
29	**74**	0.97	1.01	0.95	0.92
30	**76**	0.96	1.05	0.92	0.94

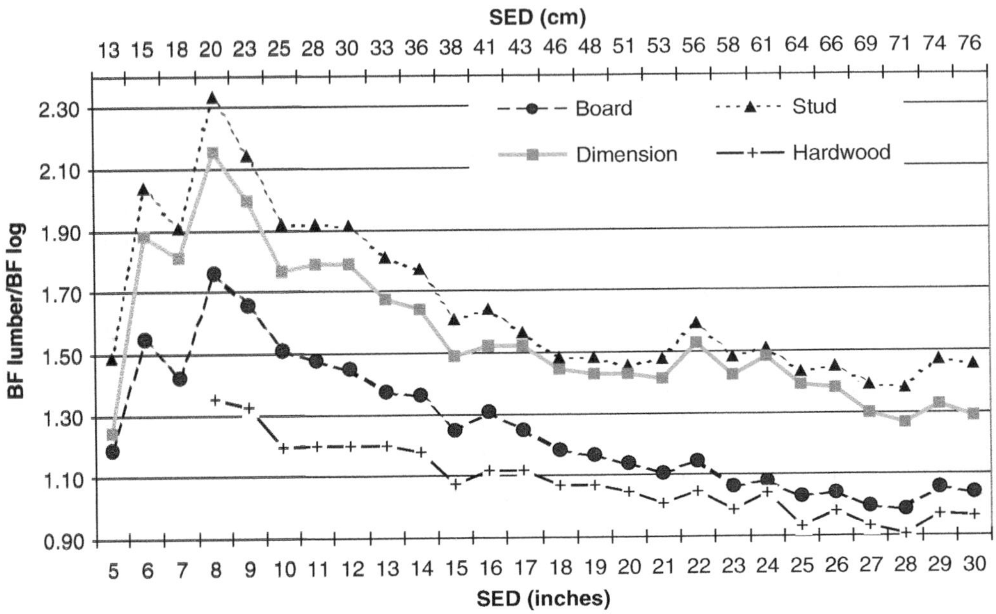

Fig. 6.5. Scribner Short Log Rule recovery by lumber product and log diameter. SED, small-end diameter.

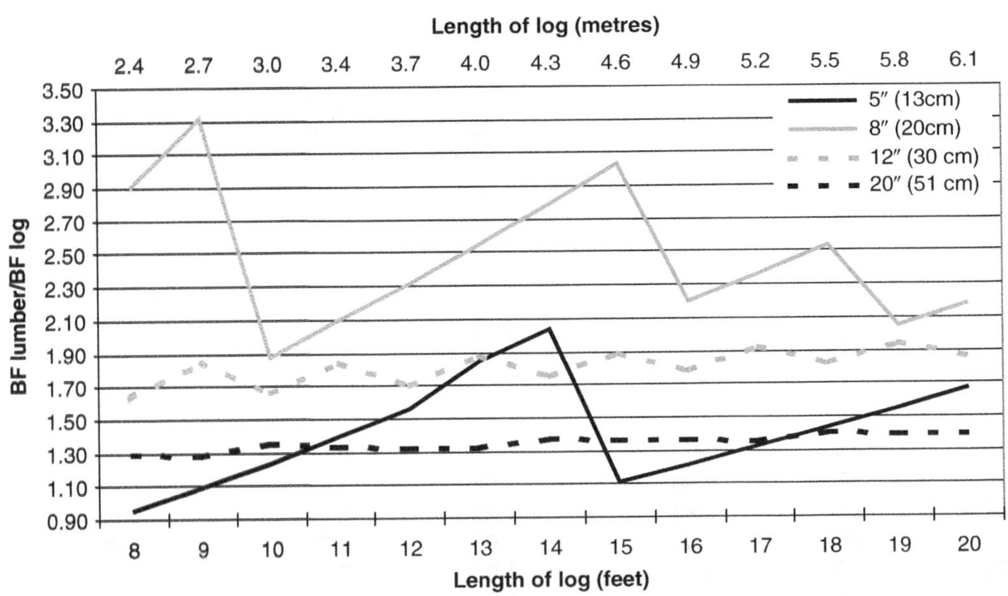

Fig. 6.6. Scribner Short Log Rule recovery trends by log length for selected diameters.

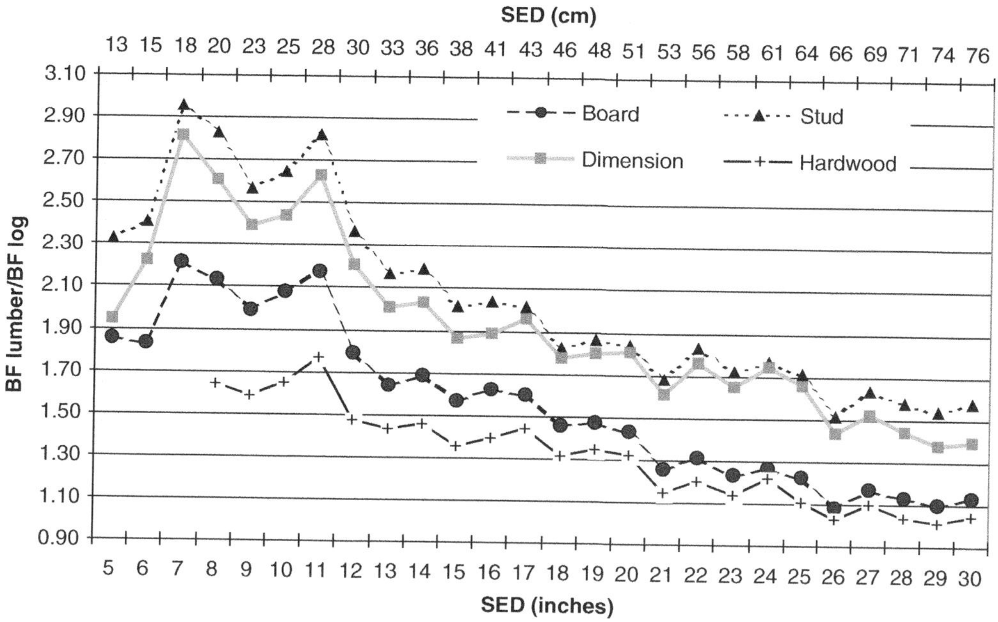

Fig. 6.7. Scribner Long Log Rule recovery by lumber product and log diameter. SED, small-end diameter.

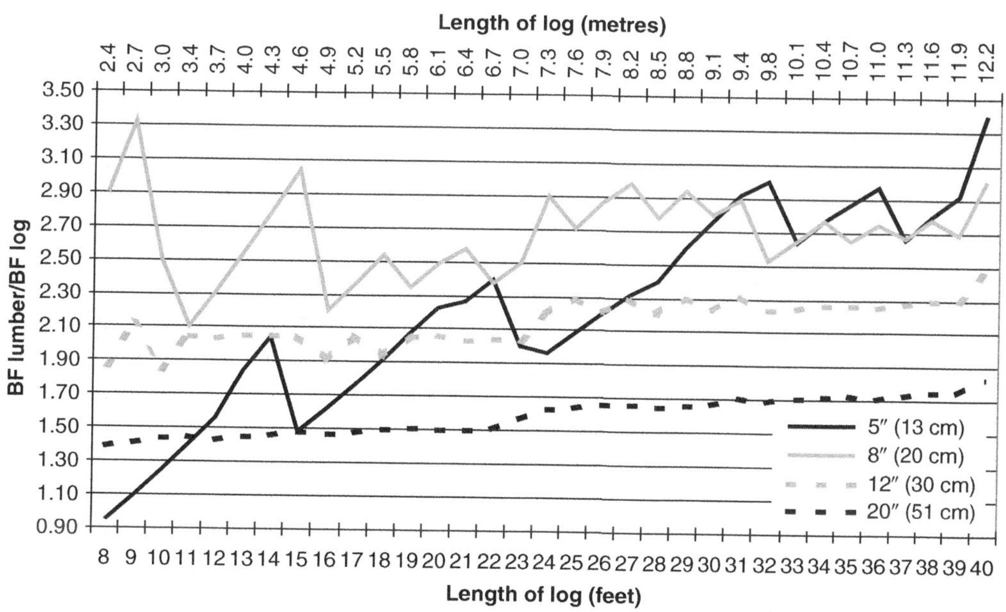

Fig. 6.8. Scribner Long Log Rule recovery trends by log length for selected diameters.

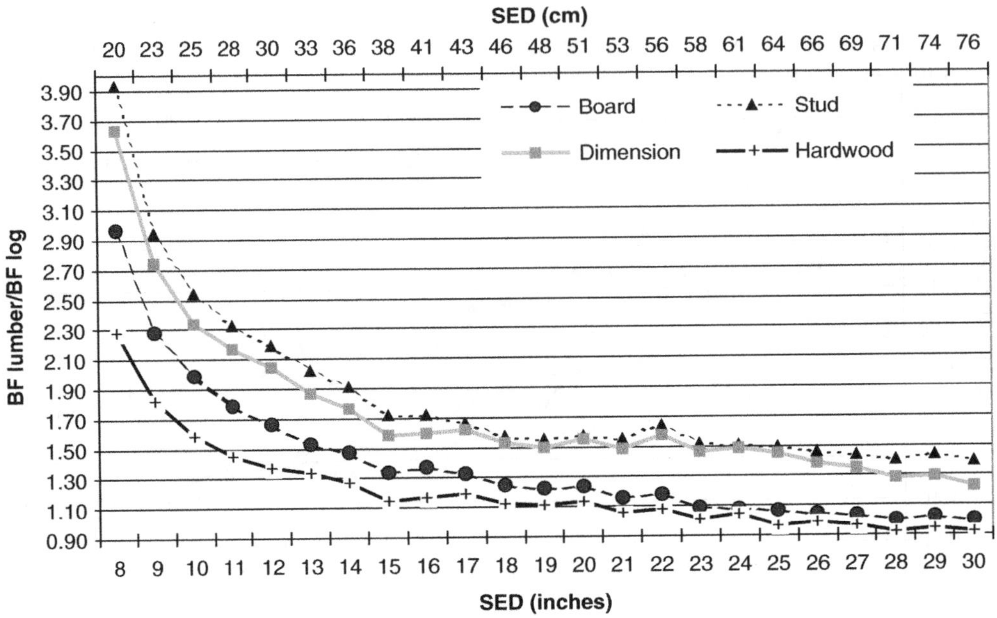

Fig. 6.9. Doyle Log Rule recovery by lumber product and log diameter. SED, small-end diameter.

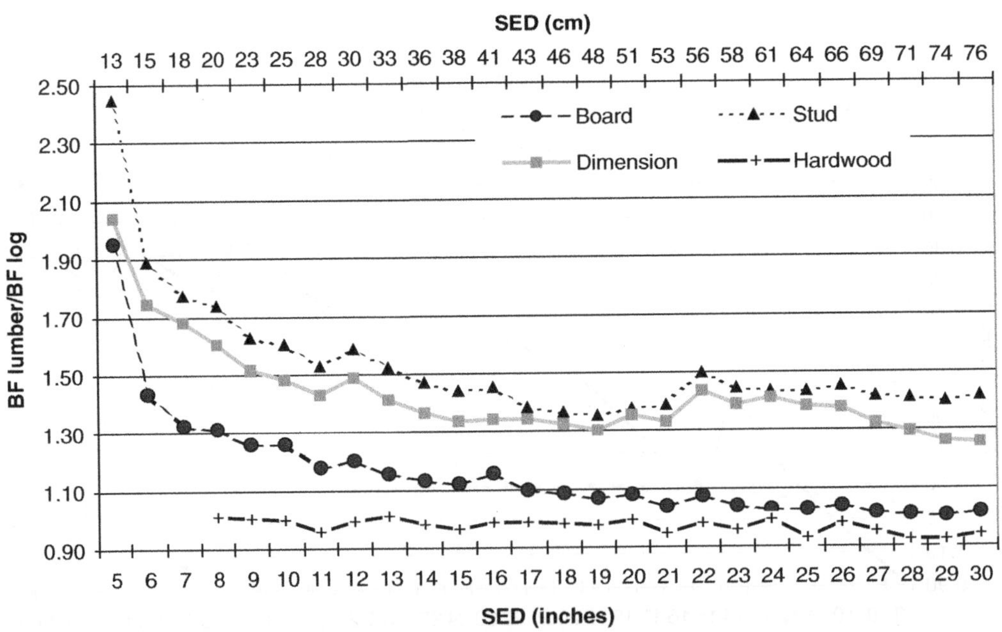

Fig. 6.10. International ¼-Inch Log Rule recovery by lumber product and log diameter. SED, small-end diameter.

7 Metrics of Plywood/Veneer Recovery

Measuring the recovery of solid wood panel products (plywood and veneer) is generally more straightforward than measuring lumber recovery. The concept is the same as lumber – measure product output as a ratio to log input, but the product (veneer or plywood panels) is generally measured with actual measurements, removing at least one significant variable present in many lumber recovery calculations.

When cubic log measure is used, recovery is often measured as a percentage, e.g. 52% recovery means that 52% of the log volume was converted into veneer or plywood. Recovery is also reflected as the units of plywood measure (ft^2 ⅜″ basis, m^2 1 mm basis) vs. units of log measure, which is commonly referred to as the 'veneer recovery factor' (VRF). There are 32 ft^2 ⅜″ basis in a cubic foot, so the equivalent of 52% recovery is: 0.52 $\times$ 32 = a VRF of **16.64**. Since there are 1000 m^2 1 mm basis in a cubic metre, 52% recovery is a VRF of 520 (0.52 $\times$ 1000 = **520**).

When product output measure is used to measure log volume, recovery is always measured via product output in ft^2 ⅜″ basis to bf log scale. If a mill produces 2750 ft^2 ⅜″ basis of plywood from 1000 bf of logs, the recovery is 2.75 (2750 ÷ 1000 = **2.75**).

The focus of this section will be on the measure of recovery for veneer that has been rotary peeled with a lathe and laid up into plywood panels. Veneer is also produced by 'slicing', which is used extensively for expensive face-quality veneer. Slicing can produce very thin veneer (0.2 mm [0.008″]); it is easier to match grain patterns when composing faces, and clear veneer can be made from clear areas of a log even when the other side of the log has knots or blemishes. Often slicing operations utilize flitches, cants and even lumber as raw materials. While we will not be covering the dynamics of veneer recovery from slicing specifically, keep in mind that many of the same basic concepts and trends that apply to rotary peeled veneer also apply to sliced veneer.

7.1 Measuring Plywood and Veneer Volume

Plywood and veneer are measured via surface measure (square area), cubic volume and quantified surface measure (ft^2 ⅜″ basis, or m^2 1 mm basis). Excepting surface measure, all units can be accurately quantified and converted.

Traditionally in North America, plywood and veneer have been measured on what is called surface measure, e.g. a 4′ × 8′ piece of veneer or plywood contains 32 ft^2 surface measure. Because veneer is often peeled in different thicknesses and plywood is made into panels of different thickness, surface measure does not indicate volume, only surface area. In order to put 'surface measure' into a relative unit of measure that indicates volume, measurements are harmonized by putting everything on a relative equivalent thickness of ⅜″. A ft^2 ⅜″ is defined as being the equivalent of a piece of wood 1′ × 1′× ⅜″, or volumetrically speaking it is 0.03125 ft^3 (0.000885 m^3). This concept is similar to the 'bf' used for measuring lumber volume except that rather than being 1″ thick, the ft^2 is 0.375″ thick, and rather than nominal sizing (which can vary greatly in lumber as to the degree of nominalization), measurements are based on actual size when the finished product is measured, with a small plus or minus tolerance. The formula for determining the ft^2 (⅜″ basis) is:

$$ft^2\ ^3/_8'' = \text{width in feet} \times \text{length in feet} \times \text{thickness in inches} \div 0.375$$

For example, ft^2 ⅜″ for a sheet of plywood measuring 4′ × 8′ × ⅝″ is calculated as follows:

$$4 \times 8 \times 0.625 \div 0.375 = \mathbf{53.33\ ft^2}\ ^3/_8''\ \text{basis}$$

In most of the world, outside of North America, plywood and veneer are measured in cubic volume (m^3) or surface measure (m^2 1 mm basis) (Briggs, 1994). A m^2 1 mm basis is equal to a sheet of veneer that is 1 m × 1 m × 1 mm thick, or volumetrically it is 0.001 m^3 (0.035315 ft^3). Note that there are 1000 m^2 on a 1 mm basis in a m^3 and that the volume of a ft^2 ⅜″ basis is 11.5% smaller than that of a m^2 1 mm basis. The formula for determining the m^2 1 mm basis is:

$$m^2\ 1\,mm = \text{width in metres} \times \text{length in metres} \times \text{thickness in millimetres}$$

For example, m^2 1 mm for a sheet of plywood measuring 1.22 m × 2.44 m × 8 mm is calculated as follows:

$$1.22 \times 2.44 \times 8 = \mathbf{23.814\ m^2\ 1\ mm\ basis}$$

7.2 Factors Affecting Plywood and Veneer Recovery

7.2.1 Milling efficiency

Veneer and plywood plants tend to be quite automated. Many of the machines perform their functions with scanners and photocells rather than human command. Listed below is a brief summary of some of the manufacturing functions that commonly affect recovery:

Green target sizing: Like lumber, veneer is dried, resulting in shrinkage. In addition, veneer is glued into plies to form plywood, and thus a certain amount of extra material is required to allow for some misalignments during the lay-up process. Lastly, plywood is often finished via a sander, requiring extra thickness for finishing.

Size control is quite important to ensure proper veneer thickness and thus maximum recovery. There is always some variation in peel thickness, which causes a certain degree of 'oversizing' to prevent undersize.

Core size plays an important role in recovery. Obviously, the smaller the core, the greater the potential amount of wood for making veneer. There is, however, some discussion as to whether the quality of the veneer obtained when the core goes below 3.4″ (8.6 cm) is worth recovering, in lieu of selling the core as a product (post, landscape timber) or chipping the core to obtain chip value.

Centring the peeler block on the lathe chucks is typically accomplished with a piece of equipment called the XY charger. It normally rotates the log, scans its shape and then positions it on the chucks to obtain optimum recovery; obviously if it is not calibrated and performing properly, recovery will suffer.

Clipping the veneer is normally accomplished with a knife that cuts the moving ribbon of veneer into: (i) uniform sized sheets at preset intervals (commonly around 51″ or 1.3 m), or (ii) sections of usable 'strip' veneer which fall between defects that are removed by the clipper. Correct optical readings and calibration of cuts and feed speeds are tantamount to obtaining good recovery, as is the separation and gathering of good strip and fishtail veneer from the defects that were clipped out.

Waste gate operation and outfeed belt speed: The waste gate is normally held open while the peeler block is going through the 'round-up' process or when there is defective veneer not worth saving. Obviously any good veneer that is sent under the waste gate is a loss of recovery, and the waste gate in tandem with the outfeed belt can create wrinkled veneer ribbon, which will normally be clipped out (again wasting good veneer).

7.2.2 Log characteristics

7.2.2.1 Diameter

CUBIC SCALED LOGS. Recovery tends upward steeply in the smaller diameters. In the case of rotary produced veneer, it is not a result of slab loss as it is with lumber (although in sliced veneer it is a factor), but rather due to the fact that core size is generally constant regardless of log size, e.g. a 3.4″ (8.6 cm) diameter core is a bigger percentage of a 6″ (15 cm) log than it is of a 26″ d (66 cm) log. This same principle continues with round-up loss. Given that taper is more a function of log position in the tree, species, growing site conditions and stocking levels (as opposed to diameter), taper rates (inches/ft, cm/m) tend to be similar between small logs and large logs. This means that small logs will have higher percentage of the total volume sent to the waste gate before round-up (causing lower recovery). Further reductions to recovery result from logs that are less round and more elliptical in shape (again creating more round-up loss).

PRODUCT OUTPUT SCALED LOGS. The same general trends that are mentioned in Section 6.2.2.1 concerning the effects of diameter on lumber recovery hold true with plywood and veneer as well.

7.2.2.2 Length

The same trends mentioned in Section 6.2.2.2 regarding the effect of length on lumber recovery are true for veneer and plywood recovery for both cubic and product output scaled logs. However, there are additional factors to consider, such as length which is bucked off logs at the chop saw (and chipped), or which is trimmed off the veneer (as it is extra). Often logs are presented in lengths that are not entirely utilizable for peeling, e.g. a mill may only have a lathe that is capable of peeling log blocks which are 8.33–8.75′ long (2.54–2.67 m), and thus a log which is 12.5′ (3.81 m) will result in a substantial loss of potential recovery. Even when peeler logs are presented in the preferred lengths, there is commonly an additional 9.4% of length and thus volume, which does not make it into the finished product, compared to a typical 3.0% for saw logs. Of course, additional length (trim allowance) is needed, but anything that can be done to reduce it will improve recovery.

7.2.2.3 Taper

Again, the trends outlined in Section 6.2.2.3 regarding lumber recovery and taper are relevant to veneer and plywood productions. Figure 7.1 shows the effects of taper on plywood/veneer recovery.

7.2.2.4 Log defects

While the trends mentioned in Section 6.2.2.4 are applicable to veneer and plywood production as well, there are additional variables introduced by defect. Namely, since most rotary peeled logs need to be grasped firmly on each end in order to spin the log, any unsound or damaged wood in these areas can render the log block unchuckable and therefore unusable. Also, as a result of the nature of rotary peeling, defects that may affect only a small area of the log (scars, heart checks, large knots, knot clusters, etc.) may end up causing severe losses as a result of the fact that the veneer ribbon repeats the defect over and over again, and despite the

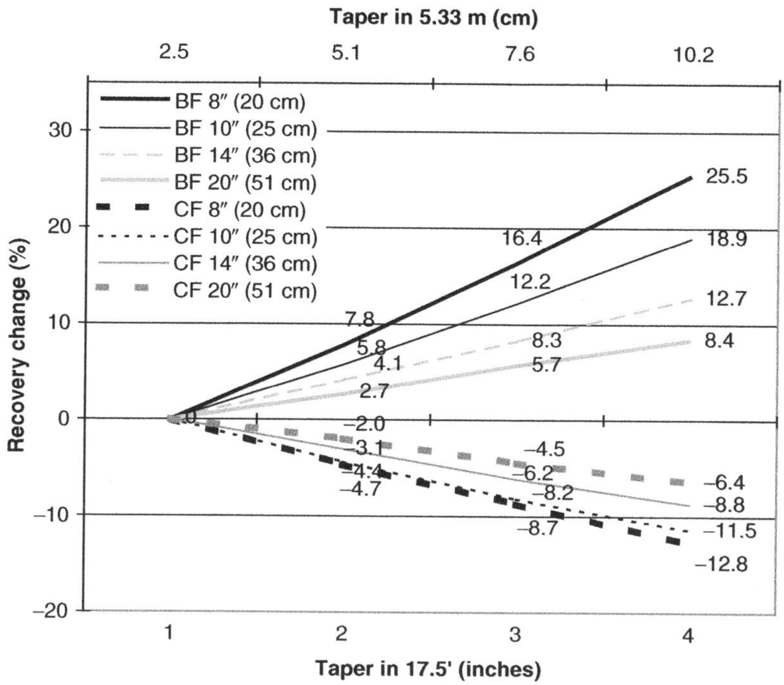

Note: 17.5′ (5.33 m) log length assumed and s.e.d of 8″, 10″ 14″, 20″ (20, 25, 36, 51 cm). CF is cubic foot; BF is Scribner Short Log.
Source: Calculated by the author, using own data.

Fig. 7.1. Effects on plywood recovery from taper in cubic and board foot scaled logs (% change in recovery for taper exceeding 1″ in 17.5″ (2.5 cm in 5.33 m)).

fact that good veneer may reside adjacent to the defect, it may not be possible to salvage it.

Log checks (surface and end checks) result from storage of logs and the resulting drying effect on the wood. This situation can be especially bad given that these defects are not normally accounted for via scaling defect (they normally occur after the log has been scaled). This defect is more common now than in the past, owing to the mechanized log harvesting equipment, which can remove large areas of bark increasing the drying effects on the wood. Keeping logs stored for only short periods or keeping them wet by submersion and/or sprinkling of water are solutions. Neglecting this issue will seriously reduce recovery opportunities.

7.2.2.5 Wood property issues

The wood property dynamics discussed in Section 6.2.2.5 similarly affect veneer and plywood. Shrinkage becomes even more important for veneer and plywood given that veneer is commonly dried to about 6% mc, increasing the amount of shrinkage relative to lumber.

As a result of wood characteristics (moisture content, density, cell structure, etc.), some species are prone to producing veneer that is brittle and breaks in the handling and layout process (especially when rotary peeling), and thus give less recoverable veneer. Some of these wood property issues are minimized through the conditioning of peeler logs with warm water or steam (commonly to temperatures of 30–55°C).

7.3 Plywood and Veneer Recovery Trends by Log Size and Scaling Method

These data were based on numerous 'return to log' studies, and in the author's opinion the mills had above average milling practices. Like the lumber recovery data, it is more important to focus on the relative recovery rates by log size than on the absolute recovery numbers, which vary substantially from one mill to another owing to the variables listed in Section 7.2. The recovery numbers are for finished plywood. If one wants to approximate the percent recovery of just veneer, add the dry fibre chip loss percentage from Table 7.1 to the plywood recovery percentage.

7.3.1 Cubic scaled logs

The data shown in Table 7.1 and illustrated in Fig. 7.2 show a very smooth and predictable progression through the diameter classes. Note that recovery percentages will add up to 100% when using the BC Firmwood percentage of product recovery, but do not when using the USFS Cubic Log Scale.

7.3.2 Product output scaled logs

The Scribner Long Log recovery trends were calculated using the procedures listed in Section 6.3.2. Taper is assumed to be 0.114″/ft. (0.952 cm/m). The recoveries shown in Table 7.2 and Fig. 7.3 are calculated from the cubic recoveries in Table 7.1 using the cubic log scale to product output rule ratios on a segment basis from the logs in Table A.1.M.

When reviewing Table 7.2 and Fig. 7.3, note the predictable recovery trends shown by the International ¼-Inch Rule. Given that at least in principle there are 2.67 ft^2 ⅜″ equivalent in a bf of lumber, it is interesting to note how close the International ¼-Inch comes to this number for most of the diameters. Doyle recovery is only given for 8″ and larger diameters. As with lumber, the two Scribner rules have the most erratic recovery trends, although it should be noted that it is not too common for logs under 8″ to be peeled, and thus when looking at the recovery trend for logs of 8–30″, the trend is fairly linear and predictable for Scribner Short Log, and the Scribner Long Log trend is much improved.

Table 7.1. Plywood recovery by small-end diameter (SED) of cubically scaled logs.

SED		USFS CF Plywood recovery		BC Firmwood Plywood recovery		Log core	Green chip loss	Shrinkage	Dry fibre chip loss
cm	inches	VRF* ft^2/ft^3	%	VRF* ft^2/m^3	%	%	%	%	%
13	**5**	11.3	35.5	368	32.6	30.5	28.9	4.2	3.8
15	**6**	13.3	41.5	431	38.2	22.5	27.8	5.1	6.4
18	**7**	14.6	45.5	472	41.8	17.2	27.0	5.7	8.3
20	**8**	15.5	48.5	509	45.1	13.6	26.4	6.1	8.8
23	**9**	16.2	50.6	531	47.0	11.1	25.8	6.4	9.7
25	**10**	16.8	52.4	551	48.7	9.2	25.3	6.7	10.1
28	**11**	17.2	53.7	563	49.8	7.7	25.0	6.8	10.7
30	**12**	17.6	54.9	586	51.8	6.6	24.8	7.0	9.8
33	**13**	17.8	55.8	596	52.7	5.7	24.4	7.1	10.1
36	**14**	18.0	56.4	602	53.3	4.9	24.1	7.2	10.5
38	**15**	18.3	57.3	612	54.1	4.3	24.1	7.3	10.2
41	**16**	18.5	57.9	631	55.9	3.8	24.0	7.4	8.9
43	**17**	18.7	58.5	638	56.5	3.4	23.8	7.4	8.9
46	**18**	18.8	58.8	641	56.7	3.1	23.7	7.5	9.0
48	**19**	19.0	59.4	648	57.3	2.8	23.5	7.5	8.9
51	**20**	19.1	59.7	651	57.6	2.5	23.4	7.6	8.9
53	**21**	19.2	60.0	655	57.9	2.3	23.3	7.6	8.9
56	**22**	19.4	60.6	661	58.5	2.1	23.2	7.6	8.6
58	**23**	19.5	60.9	664	58.8	1.9	23.1	7.7	8.5
61	**24**	19.6	61.2	668	59.1	1.8	23.0	7.7	8.4
64	**25**	19.7	61.5	671	59.4	1.6	22.9	7.7	8.4
66	**26**	19.8	61.8	674	59.7	1.5	22.9	7.7	8.2
69	**27**	19.9	62.1	678	60.0	1.4	22.8	7.8	8.0
71	**28**	19.9	62.1	678	60.0	1.3	22.7	7.8	8.2
74	**29**	20.0	62.4	681	60.3	1.2	22.7	7.8	8.0
76	**30**	20.1	62.7	684	60.6	1.2	22.7	7.8	7.7

Note: Core assumed to be 3.4″ (8.6 cm) in diameter, log length assumed to be 5.33 m (17′ 6″). *Square feet 3/8″ basis.

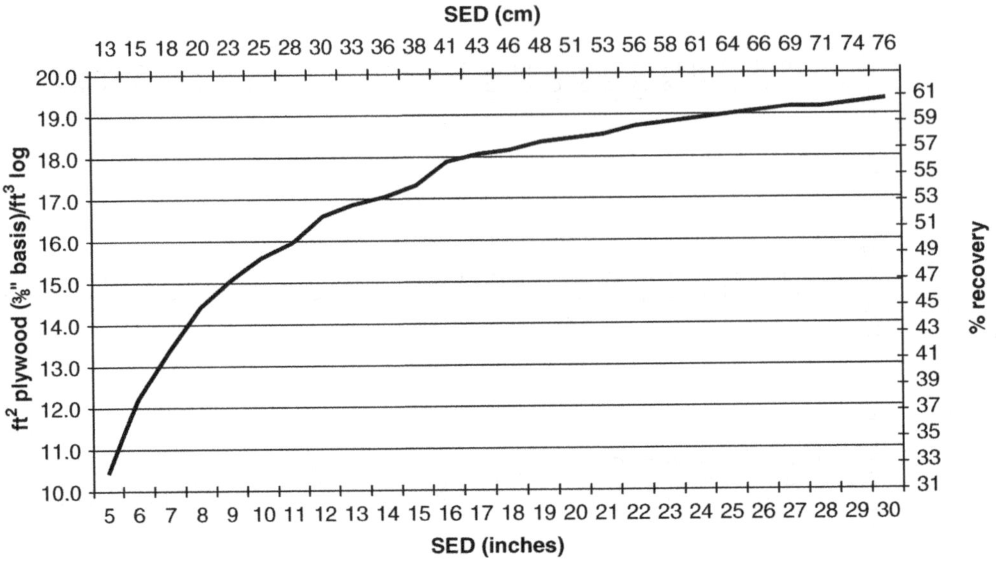

Fig. 7.2. Plywood recovery by small-end diameter (SED) (BC Firmwood).

Table 7.2. Plywood recovery by small-end diameter (SED) of product output scaled logs (ft^2 plywood/bf log).

SED					
inches	cm	Scribner SL	Scribner LL	Doyle	International ¼-Inch
5	**13**	2.12	3.31		3.48
6	**15**	3.10	3.66		2.87
7	**18**	2.95	4.58		2.74
8	**20**	3.74	4.53	6.30	2.79
9	**23**	3.58	4.29	4.92	2.72
10	**25**	3.29	4.53	4.35	2.75
11	**28**	3.30	4.85	4.00	2.63
12	**30**	3.31	4.09	3.79	2.74
13	**33**	3.15	3.77	3.51	2.65
14	**36**	3.12	3.86	3.36	2.59
15	**38**	2.87	3.60	3.06	2.58
16	**41**	2.97	3.69	3.10	2.62
17	**43**	2.86	3.68	3.04	2.52
18	**46**	2.72	3.35	2.88	2.50
19	**48**	2.72	3.42	2.85	2.48
20	**51**	2.66	3.36	2.88	2.52
21	**53**	2.60	2.95	2.74	2.45
22	**56**	2.71	3.12	2.80	2.55
23	**58**	2.54	2.95	2.60	2.48
24	**61**	2.60	3.04	2.60	2.47
25	**64**	2.48	2.96	2.59	2.48
26	**66**	2.52	2.63	2.53	2.52
27	**69**	2.42	2.85	2.51	2.47
28	**71**	2.40	2.75	2.45	2.46
29	**74**	2.58	2.69	2.52	2.46
30	**76**	2.56	2.78	2.45	2.50

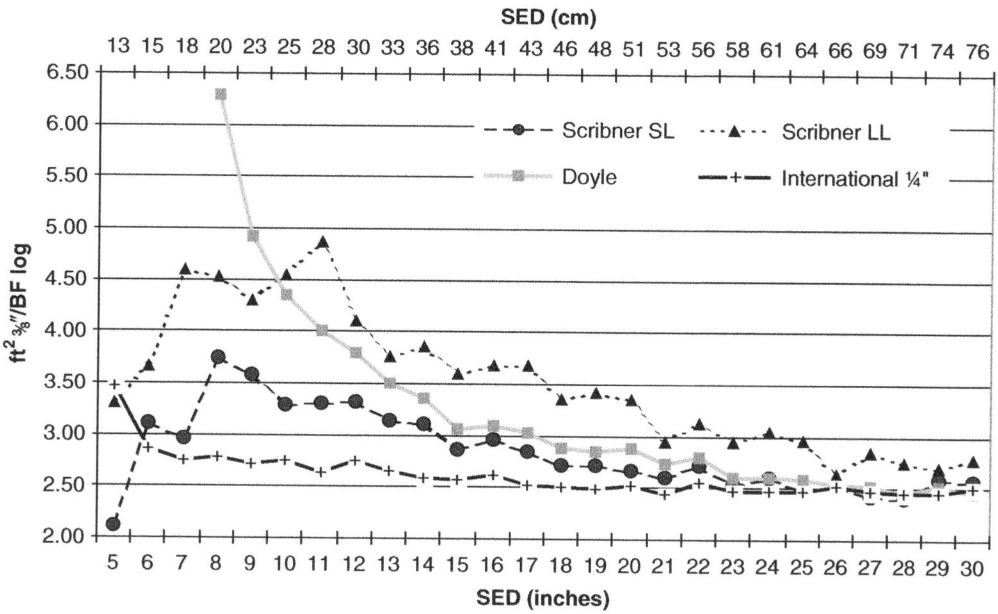

Fig. 7.3. Plywood recovery by small-end diameter (SED), product output scaled logs.

Some of the same trends shown in Figs 6.6 and 6.8 regarding lumber recovery by log length apply to veneer/plywood recovery as well; this trend of recovery volatility is reduced for veneer/plywood recovery (especially for Scribner Short Log) as log lengths tend to be more consistent as a result of common product lengths (peeler logs are commonly manufactured into length multiples of 8.75′ (2.67 m).

8 Metrics of Wood Chips and other Residue Recovery from Logs

Wood chips, sawdust, shavings, bark and 'hog fuel' are all products that are obtained either directly or indirectly from logs. In general, sawdust, shavings and bark are produced as residual products from wood-manufacturing processes. Chips are also commonly produced as a residual product, but wood chips can be made directly from logs as well. Hog fuel is generally a loose classification for combustible wood residues that are used to fire a boiler. At one time, most of these products were burned in incinerators as a means of disposing of them. In recent times, however, the demand for wood chips has reached a point where wood-chipping operations can occasionally compete with sawmills for the smaller lower-valued logs. Some products, namely oriented strand board (OSB), require a particular sized chip, which primarily comes from roundwood as it is difficult to obtain chips that are good for the manufacture of OSB from residual produced chips.

8.1 Units of Measure

As these products start out as solid wood, become loose particles, are dried and are often compressed into products more or less dense than solid wood, it is a challenge to find a relevant unit of measure that is meaningful for all of the changes to volumetric area and weight.

Wood chips, residues and bark can be measured in volumetric units or by weight. Although weight can be straightforward, it is highly affected by contained moisture. Obviously wood residues can have tremendous variability in moisture content due to species variability, heartwood vs. sapwood, seasonality, drying in storage, wood residue from green vs. dry wood, etc. Moisture can even be introduced on stockpiled residuals, and is often added to sawdust with a water spray used to keep the saws cool and the sawdust from becoming airborne. Actual usable wood chip and residue volume vs. displaced volume is also quite variable, depending on compaction (and degree of compaction is often linked to weight). An old unit of volume for measuring wood is the 'unit', which was simply 200 ft^3 of wood chips, residues or bark.

A unit of measure that seems to suit the aforementioned needs is BD weight. In simple terms, it is the weight of the product with all the moisture removed. There are a number of methods for determining BD weight, but normally it is determined by taking a representative sample of the wood fibre, establishing its weight in the green state, heating it in a pan at approximately 103°C (217°F) until the weight stabilizes because all the moisture is gone. The BD weight of the sample is then divided by the green weight to establish a ratio to apply to the non-sampled population.

For example, a truckload of wood chips with a net weight of 32,200 kg (71,000 lb) is sampled; the chip sample has a green weight of 922 g and a BD weight 497 g:

$(497 \div 922) = 53.9\%$ fibre content; $0.539 \times 71{,}000 = 17{,}356$ kg (38,269 lb) of BD weight in the load

Rather than using pounds or kilograms, weight is usually reflected in the following units of measure:

- **Bone-dry unit (BDU),** 2400 lb (1088 kg) of BD fibre.
- **Bone-dry ton (BDT),** 1 t or 2000 lb (907 kg) of BD fibre.
- **Bone-dry metric tonne (BDMT),** 1 MT or 1000 kg (2205 lb) of BD fibre.

In the above example, the load of chips would contain 15.945 BDU, 19.135 BDT, or 17.356 BDMT.

As discussed briefly in Section 5.1.2, SG is another way of presenting the BD weight of material reflected as the decimal representation of its weight (generally measured in volume when green, but weighed at 0% moisture) against the weight of the same volume of water (water weighs 62.4 lb/ft^3 or 1000 kg/m^3). Often in metric countries, it is referred to as 'basic density' rather than SG, with a basic density of 540 kg being the same as SG of 0.54. If volume is measured when dried, wood will have a higher SG than if measured when green, as the volume will be less due to shrinkage, but the BD weight will be the same.

For example, one wants to estimate the weight of lumber that has had the volume measured after it has been dried to 19% mc. The wood is from a tree species listed as having a SG of 0.45; the volumetric shrinkage at the moisture content in question (e.g. 19% for finished lumber) is determined to be 4.5% by using the information and methods in Section 6.2.2.5 and Table 6.5 on p. 144 to determine shrinkage. The formula is: specific gravity $\div$ (1 – shrinkage represented as a decimal) = SG adjusted for shrinkage, i.e. $0.45 \div (1 - 0.045) = 0.471$.

With an SG of 0.471, the BD weight would be 29.4 lb/ft^3 (471 kg/m^3). Since the wood has 19% mc ($0.19 \times 29.4 = 5.6$ lb/ft^3 of moisture (89 kg/m^3)), the weight of the wood is estimated to be 35 lb/ft^3 (560 kg/m^3) or 2.0 lbs/bf using an estimated 17.5 bf/ft^3 for dimension lumber (taken from Table 6.4).

If one knows the SG of a material, it is easy to determine the volume-to-BD measure ratios, such as ft^3/BDU or m^3/BDMT. The green weight-to-volume ratio can be used to determine green weight-to-BD weight ratios. The formulas for converting SG to other BD measures are as follows:

$ft^3/BDU = 2400 \div (SG \times 62.4)$	Green ton/BDU = $ft^3/BDU \times lb/green\ ft^3 \div 2000$
$m^3/BDU = 2400 \div (SG \times 2205)$	Green tonne/BDU = $m^3/BDU \times kg/green\ m^3 \div 1000$
$ft^3/BDT = 2000 \div (SG \times 62.4)$	Green ton/BDT = $ft^3/BDT \times lb/green\ ft^3 \div 2000$
$m^3/BDMT = 1000 \div (SG \times 1000)$	Green tonne/BDMT = $m^3/BDMT \times kg/green\ m^3 \div 1000$

Table 8.1 lists the average BD weight of some common North American timber species by both green weight (including bark) and volume (wood only). A more complete listing of BD volume and weight conversions can be found in Table A.2.B. Note that even amongst conifers, there is great variation in log volume-to-BDU ratios, e.g. Long leaf pine = 71.2 ft^3/BDU vs. Western red cedar = 124.1 ft^3/BDU, and significant variation by green log weight as well, e.g. Western larch = 2.30 t/BDU vs. Ponderosa pine = 3.21 t/BDU.

8.2 Product Recovery

8.2.1 Chips

Chips are wood particles larger and thicker than sawdust or shavings that are produced via a chipper head from larger pieces of wood or whole logs. Chips are used in the manufacture of many products ranging from compost, biofuels, decorative ground cover, paper, OSB and other products. The chips utilized in OSB (strands) tend to be long and thin. As chips require particular characteristics depending upon the end use, many things can affect product recovery. Aside from issues of quality control, such as char, presence of bark, dirt, etc., is the fact that wood chips by definition are large particles of wood, and it is not possible to chip wood without making some smaller particles (fines). Fines are simply undersized particles that are generally removed from the process by the use of screening (the pieces fall through the mesh of a screen). Fines can be generated due to many different causes, some of which are dull chipper knifes, the presence of rot in the wood, overly slow feed speeds of wood fibre into chipping heads, dry wood, frozen wood, and high speed impacts of chips against other objects. Overs (excessively large chips) are also problematic and can lead to reduced recovery if there is not a system for rechipping.

In general, if chipping equipment is set up properly, sound-green logs should yield less than 5% fines, leaving 95%+ available chips. Dry wood or wood containing rot may yield less.

Chips are often segregated into categories based on the characteristics of the source species. Typical categories of chips include softwood, hardwood, brown woods (dark coloration) and white woods (light

Table 8.1. Bone-dry weight and volume conversions for selected tree species of North America.

Common name	Latin name	Green with bark kg/m³	Green with bark lbs/ft³	Specific gravity (green volume)	BD weight kg/m³	BD weight lbs/ft³	ft³/BDU*	m³/BDU*	ft³/BDT*	m³/BDMT*	Ton/BDU**	Tonne/BDU**	Ton/BDT** Tonne/BDMT
North American conifers													
Balsam fir	*Abies balsamea*	864.1	54.0	0.33	330.0	20.6	116.6	3.30	97.1	3.03	3.14	2.85	2.62
White fir	*Abies concolor*	1034.1	64.6	0.37	370.0	23.1	104.0	2.94	86.6	2.70	3.36	3.05	2.80
Grand fir	*Abies grandis*	884.6	55.2	0.35	350.0	21.8	109.9	3.11	91.6	2.86	3.03	2.75	2.53
Subalpine fir	*Abies lasiocarpa*	756.5	47.2	0.31	310.0	19.3	124.1	3.51	103.4	3.23	2.93	2.66	2.44
Tamarack	*Larix laricina*	951.7	59.4	0.49	490.0	30.6	78.5	2.22	65.4	2.04	2.33	2.12	1.94
Western larch	*Larix occidentalis*	921.4	57.5	0.48	480.0	30.0	80.1	2.27	66.8	2.08	2.30	2.09	1.92
Engelmann spruce	*Picea engelmannii*	927.3	57.9	0.33	330.0	20.6	116.6	3.30	97.1	3.03	3.37	3.06	2.81
White spruce	*Picea glauca*	797.7	49.8	0.33	330.0	20.6	116.6	3.30	97.1	3.03	2.90	2.63	2.42
Black spruce	*Picea mariana*	848.8	53.0	0.38	380.0	23.7	101.2	2.86	84.3	2.63	2.68	2.43	2.24
Sitka spruce	*Picea sitchensis*	840.0	52.4	0.37	370.0	23.1	104.0	2.94	86.6	2.70	2.73	2.47	2.27
Jack pine	*Pinus banksiana*	861.7	53.8	0.40	400.0	25.0	96.2	2.72	80.1	2.50	2.59	2.35	2.16
Lodgepole pine	*Pinus contorta*	905.2	56.5	0.38	380.0	23.7	101.2	2.86	84.3	2.63	2.86	2.59	2.38
Shortleaf pine	*Pinus echinata*	1025.0	64.0	0.47	470.0	29.3	81.8	2.32	68.2	2.13	2.62	2.38	2.18
Slash pine	*Pinus elliottii*	1039.5	64.9	0.54	540.0	33.7	71.2	2.02	59.4	1.85	2.31	2.10	1.93
Limber pine	*Pinus flexilis*	930.2	58.1	0.37	370.0	23.1	104.0	2.94	86.6	2.70	3.02	2.74	2.52
Western white pine	*Pinus monticola*	842.5	52.6	0.35	350.0	21.8	109.9	3.11	91.6	2.86	2.89	2.62	2.41
Longleaf pine	*Pinus palustris*	1140.3	71.2	0.54	540.0	33.7	71.2	2.02	59.4	1.85	2.54	2.30	2.11
Ponderosa pine	*Pinus ponderosa*	1015.6	63.4	0.38	380.0	23.7	101.2	2.86	84.3	2.63	3.21	2.91	2.67
Monterey pine	*Pinus radiata*	1042.8	65.1	0.40	400.0	25.0	96.2	2.72	80.1	2.50	3.13	2.84	2.61
Red pine	*Pinus resinosa*	859.1	53.6	0.41	410.0	25.6	93.8	2.65	78.2	2.44	2.52	2.28	2.10
Eastern white pine	*Pinus strobus*	839.9	52.4	0.34	340.0	21.2	113.1	3.20	94.3	2.94	2.97	2.69	2.47
Loblolly pine	*Pinus taeda*	1026.6	64.1	0.47	470.0	29.3	81.8	2.32	68.2	2.13	2.62	2.38	2.19
Douglas fir	*Pseudotsuga menziesii*	950.8	59.4	0.45	450.0	28.1	85.5	2.42	71.2	2.22	2.54	2.30	2.11
Western red cedar	*Thuja plicata*	621.1	38.8	0.31	310.0	19.3	124.1	3.51	103.4	3.23	2.41	2.18	2.00
Eastern hemlock	*Tsuga canadensis*	979.1	61.1	0.38	380.0	23.7	101.2	2.86	84.3	2.63	3.09	2.81	2.58
Western hemlock	*Tsuga heterophylla*	942.8	58.9	0.42	420.0	26.2	91.6	2.59	76.3	2.38	2.70	2.44	2.25
Mountain hemlock	*Tsuga mertensiana*	1073.0	67.0	0.42	420.0	26.2	91.6	2.59	76.3	2.38	3.07	2.78	2.56

North American hardwoods													
Red maple	*Acer rubrum*	993.0	62.0	0.54	540.0	33.7	71.2	2.02	59.4	1.85	2.21	2.00	1.84
Soft maple	*Acer saccharinum*	993.0	62.0	0.49	490.0	30.6	78.5	2.22	65.4	2.04	2.43	2.21	2.03
Hard maple	*Acer saccharum*	1121.1	70.0	0.56	560.0	34.9	68.7	1.94	57.2	1.79	2.40	2.18	2.00
Yellow birch	*Betula alleghaniensis*	1089.1	68.0	0.55	550.0	34.3	69.9	1.98	58.3	1.82	2.38	2.16	1.98
Paper birch	*Betula papyrifera*	1009.0	63.0	0.48	480.0	30.0	80.1	2.27	66.8	2.08	2.52	2.29	2.10
Pecan hickory	*Carya illinoensis*	1057.0	66.0	0.60	600.0	37.4	64.1	1.81	53.4	1.67	2.12	1.92	1.76
Hickory	*Carya* spp.	1201.2	75.0	0.64	640.0	39.9	60.1	1.70	50.1	1.56	2.25	2.04	1.88
Hackberry	*Celtis* spp.	961.0	60.0	0.49	490.0	30.6	78.5	2.22	65.4	2.04	2.35	2.14	1.96
American beech	*Fagus grandifolia*	1025.0	64.0	0.56	560.0	34.9	68.7	1.94	57.2	1.79	2.20	1.99	1.83
White ash	*Fraxinus americana*	961.0	60.0	0.54	540.0	33.7	71.2	2.02	59.4	1.85	2.14	1.94	1.78
Black ash	*Fraxinus nigra*	912.9	57.0	0.45	450.0	28.1	85.5	2.42	71.2	2.22	2.44	2.21	2.03
Green ash	*Fraxinus pennsylvania*	880.9	55.0	0.53	530.0	33.1	72.6	2.05	60.5	1.89	2.00	1.81	1.66
Butternut	*Juglans cinerea*	832.8	52.0	0.36	360.0	22.5	106.8	3.02	89.0	2.78	2.78	2.52	2.31
Black walnut	*Juglans nigra*	993.0	62.0	0.51	510.0	31.8	75.4	2.13	62.8	1.96	2.34	2.12	1.95
Sweetgum	*Liquidambar styraciflua*	1121.1	70.0	0.48	480.0	30.0	80.1	2.27	66.8	2.08	2.80	2.54	2.34
Yellow poplar	*Liriodendron tulipifera*	1025.0	64.0	0.44	440.0	27.5	87.4	2.47	72.8	2.27	2.80	2.54	2.33
Water tupelo	*Nyssa aquatica*	944.9	59.0	0.45	450.0	28.1	85.5	2.42	71.2	2.22	2.52	2.29	2.10
Black gum	*Nyssa sylvatica*	1041.0	65.0	0.48	480.0	30.0	80.1	2.27	66.8	2.08	2.60	2.36	2.17
Sycamore	*Platanus occidentalis*	1089.1	68.0	0.45	450.0	28.1	85.5	2.42	71.2	2.22	2.91	2.64	2.42
Balsam poplar	*Populas balsamifera*	993.0	62.0	0.37	370.0	23.1	104.0	2.94	86.6	2.70	3.22	2.92	2.69
Cottonwood	*Populus* spp.	944.9	59.0	0.37	370.0	23.1	104.0	2.94	86.6	2.70	3.07	2.78	2.56
Aspen	*Populus tremuloides*	944.9	59.0	0.39	390.0	24.3	98.6	2.79	82.2	2.56	2.91	2.64	2.42
Black cherry	*Prunus serotina*	864.9	54.0	0.47	470.0	29.3	81.8	2.32	68.2	2.13	2.21	2.00	1.84
White oak	*Quercus alba*	1185.2	74.0	0.64	640.0	39.9	60.1	1.70	50.1	1.56	2.22	2.02	1.85
Scarlet oak	*Quercus coccinea*	1217.2	76.0	0.57	570.0	35.6	67.5	1.91	56.2	1.75	2.56	2.33	2.14
Laurel oak	*Quercus laurifolia*	1249.2	78.0	0.59	590.0	36.8	65.2	1.84	54.3	1.69	2.54	2.31	2.12
Water oak	*Quercus nigra*	1249.2	78.0	0.57	570.0	35.6	67.5	1.91	56.2	1.75	2.63	2.39	2.19
Chestnut oak	*Quercus prinus*	1153.1	72.0	0.62	620.0	38.7	62.0	1.76	51.7	1.61	2.23	2.03	1.86
Post oak	*Quercus stellata*	1185.2	74.0	0.64	640.0	39.9	60.1	1.70	50.1	1.56	2.22	2.02	1.85
Southern red oak	*Quercus falcata*	1249.2	78.0	0.57	570.0	35.6	67.5	1.91	56.2	1.75	2.63	2.39	2.19
Black willow	*Salix nigra*	896.9	56.0	0.34	340.0	21.2	113.1	3.20	94.3	2.94	3.17	2.87	2.64
American basswood	*Tilia americana*	784.8	49.0	0.32	320.0	20.0	120.2	3.40	100.2	3.13	2.94	2.67	2.45
Elm	*Ulmus* spp.	1089.1	68.0	0.46	460.0	28.7	83.6	2.37	69.7	2.17	2.84	2.58	2.37

Note: All volumetric units (ft^3, m^3) measured in the green state prior to any shrinkage. Green weight to volume ratios for logs includes weight of bark, and volume of wood only. Bone dry weight to volume ratios includes weight and volume of wood only.

*Volume (ft, m^3) is green log volume.

** Ton and Tonne are for green logs weighed with bark.

Source: Calculated by the author from data taken from the sources listed at the end of Table A.2.A on p. 227.

coloration). There are also differentiations made based on density, fibre characteristics and even individual species when there are particular physical, aesthetic or chemical features to account for.

8.2.1.1 Whole log chips

Estimation of chip recovery from whole logs is best accomplished if one knows the BD content of wood fibre contained in the logs that will be chipped. This could be obtained from historical data on timber with similar parameters. Often, however, this type of data is not available, and thus generic information such as that contained in Table 8.1 must be utilized.

For example, assume that one would like to estimate the green tons and cunits of logs needed per BDU (using the data in Table 8.1 or A.2.B), with a mix of thinning logs that will be about 10% Western larch, 30% Douglas fir, 50% Grand fir, and 10% Lodgepole pine.

$$\text{Estimated log tons/BDU:} (0.1 \times 2.3) + (0.3 \times 2.54) + (0.5 \times 3.03) + (0.1 \times 2.86) = 2.79 \text{ log tons/BDU}$$

$$\text{Estimated log ft}^3\text{/BDU:} (0.1 \times 80.1) + (0.3 \times 85.5) + (0.5 \times 109.9) + (0.1 \times 101.2) = 98.73 \text{ ft}^3\text{/BDU}$$

The metric calculation for the above example would be as follows:

$$\text{Estimated log tonnes/BDMT:} (0.1 \times 1.92) + (0.3 \times 2.11) + (0.5 \times 2.53) + (0.1 \times 2.38) = 2.33 \text{ log tonnes/BDMT}$$

$$\text{Estimated log m}^3\text{/BDMT:} (0.1 \times 2.08) + (0.3 \times 2.22) + (0.5 \times 2.86) + (0.1 \times 2.63) = 2.57 \text{ m}^3\text{/BDMT}$$

Adjustments to the above factors may be needed to account for loss of recovery due to log breakage, or fines and overs that are taken out of the system. It should also be kept in mind that all the factors listed in Chapter 5, regarding roundwood weight and physical properties, will affect the accuracy of conversions, including the conversions listed in Tables 8.1 and A.2.B.

8.2.1.2 Residual wood chips

During the manufacture of many wood products (lumber, veneer, poles, etc.) scrap wood is generated, e.g. slabs, edgings, trim ends, pieces of veneer, peeler cores, cull veneer or lumber. During the historical development of the forest products industry, it was discovered that these wood scraps could be chipped and converted into a saleable product. In fact, the revenue from chips alone is often quite substantial; commonly chip revenues can constitute as much as 15% of gross revenues for a sawmill. Obviously, the goal of most mills is to produce as little chips as possible, thereby generating more of the primary products, which are more profitable.

Because of the reasons listed above, many wood manufacturing facilities closely monitor the production of chips, not only for the sake of gaining insight into chip revenues but also as a cross-reference to primary product recovery. When the lumber recovery is lesser or greater than expected, conventional wisdom is to check chip production during the same period to see if the inverse occurred with chip recovery.

The production ratio of residual chips can be estimated by combining data from Tables 8.1 and 8.2; or for more size-specific data, from Tables 6.10–6.13 (if a sawmill) and from Table 7.1 (if a plywood or veneer mill). If Table 8.1 does not list the cubic foot per bone-dry measure for the desired species, it may be listed in Table A.2.B.

For example, assume that one wants to estimate the approximate BDU or BDMT of chips generated per mbf or cubic metre lumber scale in a stud mill with Douglas fir as the raw material:

BDU/mbf = ft^3 of chips from Table 8.2 ÷ ft^3/BDU from Table 8.1, i.e. 38.8 ÷ 85.5 = **0.454 BDU/mbf lumber**

metric calculation is: m^3 chips per m^3 lumber from Table 8.2 ÷ m^3 per BDMT from Table 8.1, i.e. 0.70 ÷ 2.22 m^3 = **0.315 BDMT/m^3 lumber**

Assume that one wants to estimate the approximate BDU of chips generated per mbf lumber scale in a stud mill with subalpine fir as the raw material:

38.8 ÷ 124.1 = **0.313 BDU/mbf lumber**

metric calculation is: 0.70 m^3 chips per m^3 lumber ÷ 3.23 m^3 per BDMT = **0.217 BDMT/m^3 lumber**

As can be seen in the above example, it would appear that while cutting subalpine fir the production of wood chips falls by more than 31%, which could lead to the incorrect assumption that the mill should have better

Table 8.2. Examples of residual product recovery by product type (solid wood equivalent volume by log and products).

	ft^3/mbf lumber					m^3/m^3 lumber				
	Log	Lumber	Chips	Sawdust	Shavings	Log	Lumber	Chips	Sawdust	Shavings
Boards	139.9	56.6	44.6	16.6	18.3	2.47	1	0.79	0.29	0.32
Stud	115.7	55.6	38.8	8.6	9.5	2.08	1	0.70	0.15	0.17
Dimension	119.3	56.6	40.6	8.1	10.6	2.11	1	0.72	0.14	0.19
Hardwood	155.6	69.2	39.0	21.6	15.5	2.25	1	0.56	0.31	0.22
	ft^3/msf plywood					m^3/m^3 plywood				
	Log	Plywood	Chips	Core	Dry fibre	Log	Plywood	Chips	Core	Dry fibre
Plywood	62.5	31.3	15.6	4.7	7.2	2.00	1	0.50	0.15	0.23

Note: These figures are calculated from data in Chapters 6 and 7 and may not apply to any one individual operation. Lumber and residual volume will not add up to log volume on account of lumber shrinkage and rounding. Plywood dry fibre recovery will contain some sanding dust, usable dry chips, and dry chips contaminated by glue; the ratio will be dependent on the operation.
Source: Calculated by the author.

recovery of the primary product (lumber). In fact, the mill is operating at the exact efficiency level with both Douglas fir and subalpine fir. Subalpine fir is a species with very low density, and thus the BD fibre weight of a given volume is very low.

Another good use for the BD measure ratios listed in Table 8.1 or Table A.2.B is to calculate if boards of a given grade are worth further processing or should be chipped.

For example, assume that the price for #5 common grade Western larch lumber is $100 mbf, and the downstream (from the sawmill) manufacturing costs to cover drying, surfacing and handling average to $62/mbf, which leaves a net return from this point forward of about $38/mbf (this does not include any cost for the log or any upstream manufacturing and handling). Appearance boards having a ratio of 12.90 : 13.44 bf/ft^3 for green lumber cut to the GTS (with an approximate 13.1 average) would yield 76.3 ft^3/mbf lumber; dividing 76.3 by the cubic feet per BDU for Western larch (from Table 8.1) gives the number of BDUs in an mbf of lumber. Assuming a chip price of $56/BDU, this example would yield:

$$76.3 \div 80.1 = 0.95 \text{ BDU/mbf}; \; 0.95 \times \$56 = \mathbf{\$53.20}$$

This represents an improvement in value of $15.20/mbf, or 40% over the value if made into lumber. Of course, this is a simplified example. There are some incremental costs in chipping, and there may be a percentage of loss due to fines; nevertheless, given the variables listed in this example, it would seem safe to assume that #5 common larch lumber should be sent to the chipper.

8.2.2 Sawdust

Sawdust is the wood fibre that is removed via the saws in the mill. The particles tend to be quite small with a consistency similar to coffee grounds. The amount of sawdust removed is controlled primarily by the thickness of the saw blade (see Section 6.2.1.1), the number of cuts made and whether or not the mill utilizes slabbers (chipping heads that grind slabs and edgings into chips rather than sawing them).

Sawdust is used for making some paper products (generally as a component in lower grades of paper), as a component in fibreboard and some particleboards, as a source of fuel and raw materials for making heating pellets and as other miscellaneous items such as livestock bedding and compost. Sawdust is generally of significantly lower value than chips. As can be seen in Table 8.2 or Tables 6.6–6.9, board and hardwood mills produce significantly more sawdust than do stud or dimension mills. This is because board mills primarily produce 1″ thick lumber, and hardwood mills, which also commonly favour 1″ lumber, generally use thicker saws.

The procedures for calculating sawdust production per lumber or log volume is the same as for chips: divide volume of sawdust per mbf or m^3

(from Table 8.2) by the desired volume (ft^3 or m^3) per BD measure (from Table 8.1) to calculate approximate ratios per lumber or log volume.

To roughly approximate the solid wood equivalent sawdust production volume, one can estimate the surface area of an mbf or cubic metre of a particular type of lumber, and multiply it by one half the average sawkerf. Unfortunately, this method will underestimate sawdust production in a mill without slabbers, as slabs and edgings have surface area cut by a saw (which will not be accounted for), and may overestimate sawdust volume in mills with slabbers as some of the surface area of lumber was created via the slabber heads (which create chips rather than sawdust). In addition, quite a bit of lumber is trimmed back and even sent to the chipper (in the case of cull). Further complicating the use of a theoretical approach is the fact that small logs have a much higher percentage of volume in slabs relative to large logs, making it difficult to use an assumed ratio for unaccounted for, sawn or chipped slabs.

8.2.3 Shavings

Shavings comprise wood fibre that is removed from lumber via planer knives. The planing process serves two primary functions: (i) it provides a smooth finish to the wood; and (ii) it sizes the wood so that each piece is of the desired thickness and/or width. Planing generally occurs after the wood is dried (to remove the effects of shrinkage).

Planer shavings are used for the same applications as sawdust, but have the added advantage of longer fibre lengths (which can give more strength to composed products), and typically have a lower moisture content, which can be a disadvantage for some products.

If one is using the average dried dimensions out of the kiln vs. the finished size out of the planer to establish the solid wood equivalent cubic volume of shavings to estimate BD quantity, the SG or volume to BD weight measure (BDU, BDT, BDMT) will need to be adjusted to account for the increased density resulting from shrinkage (see Section 8.1).

8.2.4 Bark

Tree bark sees most of its use as fuel for boilers, and as a ground treatment for landscaping and mulch. Most bark is utilized from the debarking systems at mills. Generally, the bark is removed, conveyed into a hog (a machine that pulverizes the bark fragments into smaller, easier-to-convey sizes) and concentrated into a bin for eventual transport or use in a boiler. Roughly, 10% of the weight and volume of softwoods and 13% of the weight and volume of a hardwood log consist of bark. These percentages can vary substantially from one species to another (Table 8.3). Note that the data on bark quantity which is shown in Table 8.3 gives the percentage of total log weight for bark, but lists the percentage of volume relative

Table 8.3. Physical characteristics and heat content of common North American tree species.

Common name	Latin name	Wood lbs/ft³*	Wood specific gravity (green m³)	Wood heating value BTU/lb	Wood gigajoules per tonne	Bark % log weight (bark & wood)	Bark % log wood volume	Bark lbs/ft³**	Bark specific gravity (green m³)	Bark heating value BTU/lb
Conifers										
Balsam fir	*Abies balsamea*	48.2	0.33	—	—	10.7	15.0	53.0	0.38	9339
Grand fir	*Abies grandis*	48.8	0.35	8150	18.99	11.7	14.3	45.1	0.54	—
Subalpine fir	*Abies lasiocarpa*	39.9	0.31	—	—	15.5	15.7	46.7	—	—
Eastern red cedar	*Juniperus virginiana*	43.0	0.44	—	—	10.0	12.0	40.0	0.54	—
Tamarack	*Larix laricina*	55.0	0.49	—	—	7.4	13.0	34.0	0.30	9010
Western larch	*Larix occidentalis*	51.7	0.48	—	—	10.1	19.5	29.9	0.44	8750
Engelmann spruce	*Picea engelmannii*	51.5	0.33	—	—	11.0	13.0	48.8	0.48	8820
White spruce	*Picea glauca*	46.2	0.37	—	—	7.3	10.0	36.0	0.29	8530
Black spruce	*Picea mariana*	49.2	0.38	—	—	7.2	14.0	27.3	0.38	8782
Jack pine	*Pinus banksiana*	47.0	0.40	—	—	12.6	17.0	40.0	0.34	9339
Lodgepole pine	*Pinus contorta*	53.5	0.38	8600	20.04	5.4	6.0	50.7	0.46	9382
Shortleaf pine	*Pinus echinata*	58.0	0.47	—	—	9.4	15.0	40.0	0.32	9319
Slash pine	*Pinus elliottii*	58.9	0.54	—	—	9.2	15.0	40.0	0.32	9327
Western white pine	*Pinus monticola*	46.1	0.35	9610	22.39	12.4	0.0	43.5	0.49	—
Longleaf pine	*Pinus palustris*	65.2	0.54	—	—	8.4	15.0	40.0	0.32	9130
Ponderosa pine	*Pinus ponderosa*	56.0	0.38	9120	21.25	11.7	20.3	33.6	0.34	9616
Red pine	*Pinus resinosa*	49.0	0.41	—	—	8.7	16.0	29.0	0.24	9070
Eastern white pine	*Pinus strobus*	43.0	0.34	—	—	18.0	16.0	59.0	0.37	9647
Loblolly pine	*Pinus taeda*	58.1	0.47	—	—	9.4	15.0	40.0	0.32	9320
Douglas fir	*Pseudotsuga menziesii*	51.2	0.45	8910	20.76	13.7	20.1	40.5	0.44	9962
Bald cypress	*Taxodium distichum*	57.0	0.42	—	—	11.1	13.0	55.0	0.40	—
Northern white cedar	*Thuja occidentalis*	36.0	0.29	9070	21.13	11.2	12.0	38.0	0.30	—
Western red cedar	*Thuja plicata*	34.2	0.31	9700	22.60	11.8	12.6	36.2	0.37	8700
Eastern hemlock	*Tsuga canadensis*	50.0	0.38	8500	19.81	18.2	21.0	53.0	0.40	9348
Western hemlock	*Tsuga heterophylla*	53.5	0.42	—	—	9.2	10.3	52.6	0.50	9297
Mountain hemlock	*Tsuga mertensiana*	56.9	0.42	—	—	15.0	25.1	40.0	—	—

Hardwoods										
Soft maple	*Acer saccharinum*	55.0	0.49	—	—	11.2	12.0	40.0	0.52	8293
Hard maple	*Acer saccharum*	62.7	0.56	—	—	10.5	12.0	61.0	0.54	8230
Yellow birch	*Betula alleghaniensis*	60.4	0.55	—	—	11.1	12.0	63.0	0.56	9548
Paper birch	*Betula papyrifera*	53.7	0.48	9340	21.76	14.7	16.0	58.0	0.51	10,310
Pecan hickory	*Carya illinoensis*	58.8	0.60	—	—	11.6	13.0	59.0	0.60	—
Hickory	*Carya* spp.	62.3	0.64	—	—	17.0	17.0	61.0	0.60	8423
Hackberry	*Celtis* spp.	52.1	0.49	—	—	13.3	15.0	53.0	0.49	—
American beech	*Fagus grandifolia*	59.8	0.56	—	—	6.6	7.0	60.0	0.56	7993
White ash	*Fraxinus americana*	54.6	0.54	9630	22.44	9.1	16.0	34.0	0.34	8453
Black ash	*Fraxinus nigra*	50.0	0.45	—	—	12.3	14.0	50.0	0.45	—
Green ash	*Fraxinus pennsylvania*	48.5	0.53	—	—	11.8	13.0	50.0	0.40	8367
Butternut	*Juglans cinerea*	44.5	0.36	—	—	14.4	15.0	50.0	0.40	—
Black walnut	*Juglans nigra*	57.4	0.51	—	—	7.5	15.0	31.0	0.28	—
Sweetgum	*Liquidambar styraciflua*	61.6	0.48	—	—	12.0	11.0	61.0	0.46	7650
Yellow poplar	*Liriodendron tulipifera*	56.1	0.44	—	—	12.4	15.0	53.0	0.40	8956
Water tupelo	*Nyssa aquatica*	52.5	0.45	—	—	11.0	11.0	—	0.35	—
Black gum	*Nyssa sylvatica*	57.9	0.48	—	—	11.0	12.0	58.0	0.42	8412
Sycamore	*Platanus occidentalis*	65.3	0.45	—	—	4.0	4.6	56.0	0.58	7877
Cottonwood	*Populus* spp.	50.2	0.37	9630	22.44	15.0	15.0	59.0	0.43	8765
Aspen	*Populus tremuloides*	47.8	0.39	—	—	18.9	18.0	62.0	0.50	8712
Black cherry	*Prunus serotina*	48.9	0.47	—	—	9.4	10.0	51.0	0.48	—
White oak	*Quercus alba*	66.6	0.64	9510	22.16	10.0	11.0	53.0	0.61	7450
Water oak	*Quercus nigra*	68.0	0.57	—	—	12.8	13.0	—	0.69	8340
Southern red oak	*Quercus falcata*	64.7	0.57	9360	21.81	17.0	20.0	74.0	0.68	8371
Black willow	*Salix nigra*	46.6	0.34	—	—	16.9	16.0	59.0	0.43	7648
Elm	*Ulmus* spp.	61.1	0.46	—	—	10.1	14.0	34.0	0.45	7385

*Green without bark.

**Green state.

Source: Heat values: Ince, 1979; Hartman *et al.*, 1976.

to wood volume only. This is because log volume is usually given for just the volume of wood. To determine the percentage of bark volume to the entire log volume (wood and bark), use the following formula:

$$\text{Bark \% of wood plus volume} = \text{Bark \% of wood} \div (100 + \text{bark \% of wood})$$

If one is trying to estimate available bark yield at a wood-manufacturing facility, be very careful not to assume that the total percentages listed in Table 8.3 would be available at the debarker. Often, because of log handling, some bark is inadvertently stripped off in the woods and upon arrival at the wood-processing facility. The ratio of missing bark will change by species, log handling equipment and season of the year.

8.2.5 Residual wood fibre-to-product ratios

It is not the intent of this publication to explore the recovery ratios of reconstituted wood products such as composite panels and pulp and paper. However, it is often helpful to have generalized information in order to make approximations. All of the products listed in Table 8.4 can have recovery ratios that can vary significantly from those listed. Variation occurs because specifications on product density differ from the assumptions utilized, quantity of binders and fillers used, efficiency level of the manufacturing process and because SG and physical characteristics of the wood utilized have a great effect on recovery ratios. All of the products listed below assume an SG of about 0.40 for the inputted wood fibre, and the author's interpretation of the middle range of densities for the products

Table 8.4. Composite panels and pulp recovery ratios.

	Inputted solid wood to product output ratios ft³/ft³			
Composite panels	(m³/m³)	BDU/msf ⅜″	BDMT/m³	Approximate product SG
Insulation board	0.66	0.21	0.26	0.26
Particleboard	1.32	0.43	0.53	0.53
MDF	1.73	0.56	0.69	0.69
OSB	1.7	0.55	0.68	0.68
Pulp yields	**ft³/BDT**	**m³/BDMT**	**BDT/BDT**	
Mechanical paper	84.3	2.63	1.05	
Chemi-mechanical process	89.0	2.78	1.11	
Kraft paper (bleached)	186.3	5.82	2.33	
Kraft paper (unbleached)	157.1	4.90	1.96	

Note: These figures are generalized approximations and should not be utilized for anything that requires exactness. Volumetric units are solid wood equivalent (measured in the green state for inputted fibre).
Source: Calculated by the author from Ahmed *et al.*, 2001; Briggs, 1994; Hartman *et al.*, 1976; Spelter, 1996; Spelter *et al.*, 1996a; United Nations Economic Commission for Europe/Food and Agriculture Organization, 1987.

listed, based on available data. Note that in North America, composite panels are often measured via square feet on a standardized thickness like plywood, excepting that particleboard and MDF are generally measured on a ¾″ basis (1 square foot ¾″ basis = 0.0625 ft^3); and insulation board is generally measured on a ½″ basis (1 square foot ½″ basis = 0.04167 ft^3). OSB is measured on a ⅜″ basis like plywood and veneer.

8.2.6 Wood energy

In many regions of the world, wood fibre sees its primary use as a source of heat energy through combustion. Even in regions where wood is valued and utilized for manufacturing products, there are many by-products of timber harvesting and product manufacturing that are burned to generate heat, steam pressure and electricity.

There are different heating values between wood, bark, moisture content, and between species because of chemical make-up and SG. Table 8.3 shows published average heating values per BD pound of wood and bark of some common North American tree species. These values are 'higher heating values' and are not 100% achievable because wood fibre will always have some moisture content, which causes the consumption of energy in order to evaporate the moisture, and heat is always lost out of the exhaust. Table 8.5 shows the heating values for both wood and bark based on solid wood volume (ft^3, m^3) as well as stacked

Table 8.5. Heating value of common North American tree species by volume.

Common name	Latin name	Note	Wood MBTU/ft^3 (1000 BTU)	Bark MBTU/ft^3 (1000 BTU)	Wood and Bark (million BTU/Cord)	Wood (million BTU/m^3)	Bark (million BTU/m^3)	Wood and Bark GJ/stere
Conifers								
Balsam fir	*Abies balsamea*	1	175.0	221.4	16.15	6.2	7.8	4.70
Grand fir	*Abies grandis*	2	178.0	310.0	17.39	6.3	10.9	5.06
Eastern red cedar	*Juniperus virginiana*	3	233.4	310.0	21.68	8.2	10.9	6.31
Tamarack	*Larix laricina*	1	259.9	168.7	22.31	9.2	6.0	6.49
Western larch	*Larix occidentalis*	1	254.6	240.2	22.19	9.0	8.5	6.46
Engelmann spruce	*Picea engelmannii*	1	175.0	264.2	16.60	6.2	9.3	4.83
White spruce	*Picea glauca*	1	196.2	154.4	17.33	6.9	5.5	5.04
Black spruce	*Picea mariana*	1	201.6	208.2	18.08	7.1	7.4	5.26
Jack pine	*Pinus banksiana*	1	212.2	198.1	18.62	7.5	7.0	5.42
Lodgepole pine	*Pinus contorta*		203.9	269.3	18.81	7.2	9.5	5.47
Shortleaf pine	*Pinus echinata*	1	249.3	183.7	21.43	8.8	6.5	6.24
Slash pine	*Pinus elliottii*	1	286.4	183.7	24.30	10.1	6.5	7.07

continued

Table 8.5. Heating value of common North American tree species by volume. *Continued*

Common name	Latin name	Note	Wood MBTU/ft^3 (1000 BTU)	Bark MBTU/ft^3 (1000 BTU)	Wood and Bark (million BTU/Cord)	Wood (million BTU/m^3)	Bark (million BTU/m^3)	Wood and Bark GJ/stere
Western white pine	*Pinus monticola*	2	209.9	281.3	19.55	7.4	9.9	5.69
Longleaf pine	*Pinus palustris*	1	286.4	183.7	24.30	10.1	6.5	7.07
Ponderosa pine	*Pinus ponderosa*		216.3	195.2	18.66	7.6	6.9	5.43
Red pine	*Pinus resinosa*	1	217.5	137.8	18.33	7.7	4.9	5.33
Eastern white pine	*Pinus strobus*	1	180.3	212.4	16.43	6.4	7.5	4.78
Loblolly pine	*Pinus taeda*	1	249.3	183.7	21.43	8.8	6.5	6.24
Douglas fir	*Pseudotsuga menziesii*		250.2	252.6	22.01	8.8	8.9	6.41
Bald cypress	*Taxodium distichum*	3	222.8	229.6	20.01	7.9	8.1	5.82
Northern white cedar	*Thuja occidentalis*	2	164.1	172.2	14.80	5.8	6.1	4.31
Western red cedar	*Thuja plicata*		187.6	212.4	17.06	6.6	7.5	4.97
Eastern hemlock	*Tsuga canadensis*		201.6	229.6	18.09	7.1	8.1	5.27
Western hemlock	*Tsuga heterophylla*	1	222.8	287.0	20.60	7.9	10.1	5.99
Hardwoods								
Soft maple	*Acer saccharinum*	1	259.9	269.1	23.41	9.2	9.5	6.81
Hard maple	*Acer saccharum*	1	297.0	277.3	26.45	10.5	9.8	7.70
Yellow birch	*Betula alleghaniensis*	1	291.7	333.6	26.58	10.3	11.8	7.74
Paper birch	*Betula papyrifera*		279.8	328.1	25.47	9.9	11.6	7.41
Pecan hickory	*Carya illinoensis*	3	318.2	344.4	28.76	11.2	12.2	8.37
Hickory	*Carya* spp.	1	339.5	315.4	29.77	12.0	11.1	8.66
Hackberry	*Celtis* spp.	3	259.9	281.3	23.41	9.2	9.9	6.81
American beech	*Fagus grandifolia*	1	297.0	279.3	26.76	10.5	9.9	7.79
White ash	*Fraxinus americana*		324.5	179.3	27.02	11.5	6.3	7.86
Black ash	*Fraxinus nigra*	3	238.7	258.3	21.54	8.4	9.1	6.27
Green ash	*Fraxinus pennsylvania*	1	281.1	208.8	24.41	9.9	7.4	7.10
Butternut	*Juglans cinerea*	3	190.9	229.6	17.47	6.7	8.1	5.08
Black walnut	*Juglans nigra*	3	270.5	160.7	22.80	9.6	5.7	6.64
Sweetgum	*Liquidambar styraciflua*	1	254.6	219.6	22.57	9.0	7.8	6.57
Yellow poplar	*Liriodendron tulipifera*	1	233.4	223.5	20.68	8.2	7.9	6.02
Water tupelo	*Nyssa aquatica*	3	238.7	200.9	21.11	8.4	7.1	6.15
Black gum	*Nyssa sylvatica*	1	254.6	220.5	22.51	9.0	7.8	6.55
Sycamore	*Platanus occidentalis*	1	238.7	285.1	21.83	8.4	10.1	6.35
Cottonwood	*Populus* spp.		222.3	235.2	19.96	7.9	8.3	5.81
Aspen	*Populus tremuloides*	1	206.9	271.8	19.18	7.3	9.6	5.58
Black cherry	*Prunus serotina*	3	249.3	275.6	22.67	8.8	9.7	6.60
White oak	*Quercus alba*		379.8	283.6	33.27	13.4	10.0	9.68
Water oak	*Quercus nigra*	1	302.3	359.1	27.66	10.7	12.7	8.05
Southern red oak	*Quercus falcata*		332.9	355.2	29.58	11.8	12.5	8.61
Black willow	*Salix nigra*	1	180.3	205.2	16.34	6.4	7.2	4.76
Elm	*Ulmus* spp.	1	244.0	207.4	21.38	8.6	7.3	6.22

Note: 1 = wood assumed to have 8500 BTU/lb if conifer, 9400 BTU/lb if hardwood; 2 = bark assumed to be 9200 BTU/lb if conifer, 8400 BTU/lb if hardwood; 3 = both wood and bark assumed as per 1 and 2.
Source: Calculated by author from data given by Ince, 1979; Hartman *et al.*, 1976.

volume (cord and stere assumes 78% wood and bark content, 22% void space). Heating value is reflected in British thermal units (BTU) and gigajoules (GJ). One million BTU equals 1.055 gigajoules.

Aside from combustion, wood is also used to make ethanol alcohol, which can be used, among other things, as a fuel for internal combustion engines. Ethanol alcohol can be made from virtually all wood fibre including bark and even pulp sludge. According to McCloy and O'Conner (1999), one BDMT will make 350–450 litres of ethanol alcohol with the typical technology being closer to the low side of the range.

References

Cited References

Ahmed, A., Myers, G.C. and AbuBakr, S. (2001) *Packaging Grade Kraft Pulp from Small-Diameter Softwood.* Forest Products Laboratory Forest Service, US Department of Agriculture, Madison, Wisconsin.

Alberta Land & Forest Service (1997) Directive No. 97-09, Off Shore Volume Calculation. Alberta Land & Forest Service, Edmonton, Alberta, Canada.

Alberta Land & Forest Service (1992) *Alberta Scaling Manual.* Alberta Land & Forest Service, Edmonton, Alberta, Canada.

Association pour la rationalisation et la mécanisation de l'exploitation forestière, Centre technique du bois et de l'ameublement, Mutualité sociale agricole (1994) *Manuel d'exploitation forestière, tome 2.* Association pour la rationalisation et la mécanisation de l'exploitation forestière, Centre technique du bois et de l'ameublement, Fontainebleau, France.

Associazioni Regionali Operatori del Legno del Veneto (No year) *Il Legno: conoscerlo amarlo impiegarlo.* Associazioni Regionali Operatori del Legno del Veneto, Milan, Italy.

Brereton, B. (1929) *Lumber and Log Exporters Guide*, 3rd edn. Gateway Printing Company, Seattle, Washington.

Brereton B. (1940) *Practical Lumberman*, 6th edn. Gateway Printing Company, Seattle, Washington.

Briggs, D. (1994) *Forest Products Measurements and Conversion Factors with Special Emphasis on the US Pacific Northwest.* College of Forest Resources, University of Washington, Seattle, Washington.

Briggs, D. and Flora, D. (1991) *Log Scaling Procedures in Japan, Their Relationship to Her Log Sources and Influence on Trade Statistics.* Center for International Trade in Forest Products, College of Forest Resources, University of Washington, Seattle, Washington.

British Columbia Ministry of Forestry (1999) *Scaling Manual.* Crown Publications, Victoria, British Columbia, Canada.

Clark, A. III, Phillips, D.R. and Frederick, D.J. (1986) *Weight, Volume, and Physical Properties of Major Hardwood Species in the Upland-South.* (Research paper SE-257) USDA Southeastern Forest Experiment Station, Ashville, North Carolina.

Clark, A. III, Phillips, D.R. and Frederick, D.J. (1985) *Weight, Volume, and Physical Properties of Major Hardwood Species in the Gulf and Atlantic Coastal Plains.* (Research paper SE-250) USDA Southeastern Forest Experiment Station, Ashville, North Carolina.

Clark, A. III, Phillips, D.R. and Frederick, D.J. (1986) *Weight, Volume, and Physical Properties of Major Hardwood Species in the Piedmont.* (Research paper SE-255) USDA Southeastern Forest Experiment Station, Ashville, North Carolina.

Ellis, J.C. (1994) *Procedures for the Measurement of Roundwood*, 2nd edn. New Zealand Forest Research Institute in association with New Zealand Ministry of Forestry, Rotorua, New Zealand.

Ellis, J.C. and Elliot, D.A. (2001) *Log Scaling Guide for Exporters*, Bulletin No. 221. New Zealand Forest Research Institute, Rotorua, New Zealand.

Entrican, A.R., Hinds, H.V. and Reid, J.S. (1957) *Forest Trees and Timbers of New Zealand.* Government Printer, Wellington, New Zealand.

Freese, F. (1973) *A Collection of Log Rules.* USDA Forest Service General Technical Report FPL 1. Forest Products Laboratory, Madison, Wisconsin.

Giordano, G. (1976) *Tecnologia del legno 3.* Unione Tipografica Torinese, Turin, Italy.

Hallock, H., Steele, P. and Selin, R. (1979) *Comparing Lumber Yields From Board-Foot and Cubically Scaled Logs.* (Research paper FPL-324) Forest Products Laboratory Forest Service, US Department of Agriculture, Madison, Wisconsin.

Hamilton, G.J. (1975) *Forest Mensuration Handbook.* Forestry Commission, Her Majesty's Stationary Office, London.

Hanks, L., Gammon, G., Brisbin, R. and Rast, E. (1980) *Hardwood Log Grades and Lumber Grade Yields for Factory Lumber Logs.* Northeastern Forest Experiment Station, Forest Service US Department of Agriculture, Broomall, Pennsylvania.

Hartman, D.A., Atkinson, W.A., Bryant, B.S. and Woodfin, R.O. Jr. (1976) *Conversion Factors for the Pacific Northwest Forest Industry.* Institute of Forest Resources, College of Forest Resources, University of Washington, Seattle, Washington.

Honer, T.G. (1998) *Without Fear or Favour.* T.G. Honer & Associates, Victoria, British Columbia.

Ince, P.J. (1979) *How to Estimate Recoverable Heat Energy in Wood or Bark Fuels.* Forest Products Laboratory Forest Service, US Department of Agriculture, Madison, Wisconsin.

International Technical Association of Tropical Timber (1982) *Nomenclature générale des bois tropicaux.* International Technical Association of Tropical Timber, Paris, France.

International Technical Association of Tropical Timber (2003) *ATIBT Rules of Mensuration.* International Technical Association of Tropical Timber, Paris, France.

Kuritsyn, A.K., Dmitrenko, O.J. and Korenevitch, L.M. (1997) *Round Timber Delivered for Exports: Method of Measurement of Sizes and Volume.* (RD 13–2–3–97). Centre Lesexpert, Moscow, Russia.

Maine Department of Agriculture, Food & Rural Resources. (No year). *Wood Measurement Rules.* Maine Department of Agriculture, Food & Rural Resources, Bureau of Public Service, Division of Regulations, Augusta, Maine.

McCloy, B.W. and O'Conner, D.V. (1999) *Wood-Ethanol Opportunities and Barriers.* Forest Sector Table, Canada.

National Hardwood Lumber Association (1994) *Rules for the Measurement and Inspection of Hardwood and Cypress.* National Hardwood Lumber Association, Memphis, Tennessee.

New Brunswick Natural Resources and Energy, Forest Management Branch (2003) *New Brunswick Scqling Manual,* 3rd edn. New Brunswick Natural Resources and Energy, Forest Management Branch, New Brunswick, Canada.

Newfoundland Forest Service (No year) *Sawlog Scaling (F.B.M.).* Newfoundland Forest Service, Corner Brook, Newfoundland, Canada.

Northwest Log Advisory Group (1998) *Official Log Scaling and Grading Rules.* Northwest Log Advisory Group, Eugene, Oregon.

Ontario Ministry of Natural Resources (2000) *Scaling Manual,* 2nd edn. Queen's Printer for Canada, Ontario, Canada.

Papua New Guinea Forest Authority (1996) *Procedures for the Identification, Scaling and Reporting (Including Royalty Self-Assessment) on Logs Harvested from Natural Forest Logging Operations.* Papua New Guinea Forest Authority, Hohola, NCD, Papua New Guinea.

Philippines Republic, Department of Environment and Natural Resources (No year) *Measurement Standards for Felled Trees (Round Logs).* Department of Environment and Natural Resources, Quezon City, Philippines.

Québec Ministère des Resources Naturelles (2001) *Méthode de Mesurage des Bois.* Québec Ministère des Resources Naturelles, Québec, Canada.

Sächsisches Staatsministerium (1997) *Messung und Sortierung von Rokholz.* Sächsisches Staatsministerium, Dresden, Germany.

Schmoldt, D.L. (1996) CT imaging, data reduction, and visualization of hardwood logs. In: Meyer, D. (ed.) *Proceedings of the Twenty-Fourth Annual Hardwood Symposium 'Putting Research to Work for the Hardwood Industry: New Technology Available Today'.* National Hardwood Lumber Association, Cashiers, North Carolina, pp. 69–80.

Smith, B.W. (1985) *Factors and Equations to Estimate Forest Biomass in the North Central Region.* USDA North Central Forest Experiment Station, St Paul, Minnesota.

Spelter, H. (1996) *Capacity, Production, and Manufacturing of Wood-Based Panels in the United States and Canada.* Forest Products Laboratory Forest Service, US Department of Agriculture, Madison, Wisconsin.

Spelter, H., Wang, R. and Ince, P. (1996a) *Economic Feasibility of Products from Inland West Small-Diameter Timber.* Forest Products Laboratory Forest Service, US Department of Agriculture, Madison, Wisconsin.

Steele, P.H., Wagner, F.G., Skog, K.E. (1991) *Regional Softwood Sawmill Processing Variables as Influenced by Productive Capacity.* US Department of Agriculture, Madison, Wisconsin.

Swedish Timber Measurement Council (1999) *Regulations for Measuring of Roundwood* (Circular VMR 1-99). Swedish Timber Measurement Council, Sweden.

Tømmermålingforeningenes Fellesorgan (1998) *Grading and Scaling Regulations for Forestry Products.* Tømmermålingforeningenes Fellesorgan, Norway.

US Department of Agriculture Forest Products Laboratory (1999) *Wood Handbook: Wood as an Engineering Material.* Forest Products Laboratory Forest Service, US Department of Agriculture, Washington, DC.

US Forest Services (1985) *National Forest Log Scaling Rules.* USDA Forest Service FSH 2409.11, USDA Forest Service, Washington, DC.

US Forest Services (1991) *National Forest Cubic Scaling Handbook.* USDA Forest Service FSH 2409.11a, USDA Forest Service, Washington, DC.

United Nations Economic Commission for Europe/Food and Agriculture Organization Timber Branch. (1987) *Conversion Factors (Raw Material/Product) for Forest Products.* United Nations Economic Commission for Europe/Food and Agriculture Organization Timber Branch, Geneva, Switzerland.

VMF Nord (1999) *Estimation of the Solid Volume Percentage* (Circular A 13). Swedish Timber Measurement Council, Sweden.

Wade, M.W., Bullard, S.H., Steele, P.H. and Araman, P.A. (1992) *Forest Products Journal* 42, 11/12, Estimating hardwood sawmill conversion efficiency based on sawing machine and log characteristics. Forest Products Research Society.

Western Wood Products Association (1998) *Western Lumber Grading Rules.* Western Wood Products Association, Portland, Oregon.

Background References

APA (1997) *PS 1-95 Construction and Industrial Plywood (With Typical APA Trademarks).* The Engineered Wood Product Association, Tacoma, Washington.

Bell, J. and Dilworth, J.R. (2002) *Log Scaling and Timber Cruising.* Oregon State University Book Stores, Corvallis, Oregon.

Harkin, J.M. (1969) *Uses for Sawdust, Shavings, and Waste Chips* (Research Note FPL-0208). Forest Products Laboratory Forest Service, US Department of Agriculture, Madison, Wisconsin.

Lane, P., Woodfin, R. Jr, Henley, J. and Plank, M. (1973) *Veneer Recovery from Old-growth Coast Douglas-Fir.* Pacific Northwest Forest and Range Experiment Station, Forest Service, US Department of Agriculture, Portland, Oregon.

Marden, R.M., Lothner, D.C. and Kallio, E. (1975) *Wood and Bark Percentages and Moisture Contents of Minnesota Pulpwood Species.* USDA North Central Forest Experiment Station, St Paul, Minnesota.

Spelter, H. (2002) *Conversion of Board Foot Scaled Logs to Cubic Meters in Washington State, 1970–1998.* Forest Products Laboratory Forest Service, US Department of Agriculture, Madison, Wisconsin.

West, S. (2001) *The Application of Modern Methods to Log Measurement for the Purpose of Determining Quantity.* The National Education Trust of the Australian Forest Products Industries, Clayton, South Australia.

Web References

Alberta Scaling Manual. 2004-11-26. http://www3.gov.ab.ca/srd/forests/managing/sustain/ASM/

British Columbia Ministry of Forestry Harvest Billing System. 2004-10-25. http://www.for.gov.bc.ca/hbs/smp/p850.jsp

British Columbia Ministry of Forestry Scaling Manual. 2004-11-26. http://www.for.gov.bc.ca/hva/manuals/scaling/index.htm

Cameroon Policies on Forests Wildlife and Fish. 2004-11-26. http://www.camnet.cm/investir/textes/foret7.htm#titre

Center for Wood Anatomy and Research. 2004-12-05. http://www2.fpl.fs.fed.us

Forest Products Commission of Western Australia. 2004-11-26. http://www.fpc.wa.gov.au/content/plantations/species.asp

Le cubage des bois. 2004-12-05. http://www.iaveyron.com/monde/cube.html

New Brunswick Scaling Manual. 2004-11-26. http://www.gnb.ca/0078/reports/Scaling_manual-e.pdf

Norsk Virkesmåling, *Norwegian log scaling regulations.* 2004-11-26. http://www.tommermaling.no/tms/

Ontario Scaling Manual. 2004-11-26. http://www.mnr.gov.on.ca/mnr/forests/forestdoc/reg_manuals/manuals/scaling/entire_document.pdf

Ontario Woodlot Association. 2003-02-11. http://www.ont-woodlot-assoc.org

Swedish Timber Measurement Council. 2004-11-26. http://www.virkesmatningsradet.org/eng.htm

Appendix 1
Measuring Log Volume

Table A.1.A	Commonly used conversion ratios	193
Table A.1.B	Brief description of other selected log scales in use	196
Table A.1.C	USFS segment length and trim allowance chart	203
Table A.1.D	USFS segment taper distribution chart	204
Table A.1.E	USFS Cubic Log Scale 'half segment' volume chart	205
Table A.1.F	Half segment cubic metre volume chart	207
Table A.1.G	Alberta Cubic Metre Scale 'one diameter regression formula method'	209
Table A.1.H	Full segment cubic metre volume chart	211
Table A.1.I	Swedish cubic volume chart	213
Table A.1.J	New Zealand 3-D volume chart	214
Table A.1.K	Hoppus cubic feet and cubic metre volume chart	216
Table A.1.L	JAS cubic volume	218
Table A.1.M	Summarized volumes of control group of logs used for modeling of conversions used in Section 2.5	220
Table A.1.N	Long Log Scribner volume per Short Log Scribner volume index	222
Table A.1.O	Washington and Oregon mill survey Scribner to BC cubic metre index by length and small-end diameter class	224

Table A.1.A. Commonly used conversion ratios.

To convert from	To	Multiply by
	Length	
millimetre (mm)	centimetre (cm)	0.1
millimetre (mm)	inch (″)	0.0394
millimetre (mm)	foot (′)	0.00328
millimetre (mm)	metre (m)	0.001
centimetre (cm)	millimetre (mm)	10
centimetre (cm)	inch (″)	0.394
centimetre (cm)	foot (′)	0.0328
centimetre (cm)	metre (m)	0.01
inch (″)	millimetre (mm)	25.4
inch (″)	centimetre (cm)	2.54
inch (″)	foot (′)	0.0833
inch (″)	metre (m)	0.0254
foot (′)	millimetre (mm)	304.8
foot (′)	centimetre (cm)	30.48
foot (′)	inch (″)	12
foot (′)	metre (m)	0.3048
metre (m)	millimetre (mm)	1,000
metre (m)	centimetre (cm)	100
metre (m)	inch (″)	39.37
metre (m)	foot (′)	3.281
	Area	
square centimetre (cm^2)	square inch (in^2)	0.155
square centimetre (cm^2)	square foot (ft^2)	0.001076
square centimetre (cm^2)	square metre (m^2)	0.0001
square inch (in^2)	square centimetre (cm^2)	6.452
square inch (in^2)	square foot (ft^2)	0.0069444
square inch (in^2)	square metre (m^2)	0.0006452
square foot (ft^2)	square centimetre (cm^2)	929
square foot (ft^2)	square inch (in^2)	144
square foot (ft^2)	square metre (m^2)	0.0929
square metre (m^2)	square centimetre (cm^2)	10,000
square metre (m^2)	square inch (in^2)	1,550
square metre (m^2)	square foot (ft^2)	10.764
	Density	
specific gravity (sg)	basic density (bd)	1,000
basic density (bd)	specific gravity (sg)	0.001

continued

Table A.1.A. *continued*

To convert from	To	Multiply by
	Stacked measure	
cubic foot (ft^3)	stere	0.02832
cubic foot (ft^3)	cord	0.0078125
cubic metre (m^3)	stere	1
cubic metre (m^3)	cord	0.2759
cord	cubic foot (ft^3)	128
cord	cubic metre (m^3)	3.6245
cord	stere	3.6245
stere	cubic foot (ft^3)	35.315
stere	cubic metre (m^3)	1
stere	cord	0.2759
	Weight	
gram (g)	pounds (lbs)	0.002205
gram (g)	kilogram	0.001
pounds (lbs)	gram (g)	454
pounds (lbs)	kilogram (kg)	0.454
pounds (lbs)	ton (t)	0.0005
pounds (lbs)	tonne (MT)	0.000454
kilogram (kg)	gram (g)	1,000
kilogram (kg)	pounds (lbs)	2.205
kilogram (kg)	ton (t)	0.0011023
kilogram (kg)	tonne (MT)	0.001
ton (t)	kilogram (kg)	907
ton (t)	pounds (lbs)	2,000
ton (t)	tonne (MT)	0.907
tonne (MT)	pounds (lbs)	2,205
tonne (MT)	kilogram	1,000
tonne (MT)	ton (t)	1.1023

continued

Table A.1.A. *continued*

To convert from	To	Multiply by
	Volume	
square foot $\frac{3}{8}$″	square metre^{1mm}	0.885
square foot $\frac{3}{8}$″	cubic foot (ft^3)	0.03125
square foot $\frac{3}{8}$″	cubic metre (m^3)	0.000885
square foot $\frac{3}{8}$″	cunit (ccf)	0.0003125
square metre^{1mm}	square foot $\frac{3}{8}$″	1.13008
square metre^{1mm}	cubic foot (ft^3)	0.35315
square metre^{1mm}	cubic metre (m^3)	0.001
square metre^{1mm}	cunit (ccf)	0.00035315
cubic foot (ft^3)	square foot $\frac{3}{8}$″	32
cubic foot (ft^3)	square metre^{1mm}	28.32
cubic foot (ft^3)	cubic metre (m^3)	0.02832
cubic foot (ft^3)	cunit (ccf)	0.01
cubic metre (m^3)	square foot $\frac{3}{8}$″	1,130.08
cubic metre (m^3)	square metre^{1mm}	1,000
cubic metre (m^3)	cubic foot (ft^3)	35.315
cubic metre (m^3)	cunit (ccf)	2.832
cunit (ccf)	square foot $\frac{3}{8}$″	3,200
cunit (ccf)	square metre^{1mm}	2,832
cunit (ccf)	cubic foot (ft^3)	100
cunit (ccf)	cubic metre (m^3)	2.832
	Heat	
British thermal unit (BTU)	joule (j)	1,055
joule (j)	British thermal unit (BTU)	0.0009481
	Weight to volume ratios	
lbs per cubic foot (lbs/ft^3)	kg per cubic metre (kg/m^3)	16.019
kg per cubic metre (kg/m^3)	lbs per cubic foot (lbs/ft^3)	0.06243

Table A.1.B. Brief description of other selected log scales in use.

British Standard *Source*: Hamilton, 1975.	**Brief description:** this is the standard guideline for measuring roundwood utilized in the United Kingdom. It is overseen by the Forestry Commission. The mid-diameter method is approved for all sawlogs; the top-diameter method approved for conifers only. **Diameters:** the midpoint diameter is measured either via a diameter tape, or with calipers outside bark. The top diameter is obtained by measuring inside bark, the narrow way. Both with the mid- or top-diameter method, round down to the nearest 1-cm class. **Length measurement:** lengths are measured short-end to short-end and rounded down to closest 0.1 m length for lengths up to 10 m long, and rounded down to the nearest 1 m scaling length for lengths over 10 m. **Taper:** assumption for the top-diameter method is 0.833 cm of taper per metre (conifers only). **Gross volume determination:** the formula for mid-diameter method is: m^3 = midpoint in $cm^2 \times$ length in metres $\times$ 0.00007854. The formula for top-diameter method is: m^3 = (((small-end diameter in centimetres + ((length in metres $\times$ 0.5) $\times$ 0.833))^2) $\times$ length) $\times$ 0.00007854. **Defect deductions:** no specific data could be found.
German Log Scaling Standard *Source*: Sächsisches Staatsministerium, 1997.	**Brief description:** the German log scaling standard is based on EU directive (68/89 EEC), which is also the basis for the scaling standards of many other European countries. Apparently, defects, including void, softrot and char, are not taken into account in volume determination, however, the grading structure takes log defect into account. As the Huber formula is used, diameter measurements are taken at the midpoint via calipers. Under bark volume is used by either taking the measurement(s) under bark, or taking over bark diameters minus an allowance for bark from a bark thickness table. **Diameters:** up to a midpoint diameter of 19 cm under bark, only one midpoint diameter is measured and considered. At a midpoint diameter of 20 cm and above under bark, the smallest and widest midpoint diameter should be measured, vertical to each other, and the average calculated. The measured diameters are rounded down to the next lower whole centimetre; the result of the average calculation is also rounded down if fractional. If a branch or other irregularity lies at the midpoint diameter, two diameters, above and below the midpoint diameter, are measured and the average is calculated. **Length measurement:** Length is defined as the shortest possible distance between both ends, however, if there is a beveled cut, measurement starts in the middle of the bevel. The length is measured in metres, to the nearest tenth of a metre (occasionally lengths can be rounded to half metre length increments). A 1% trim allowance is given. **Taper:** not a factor as the midpoint diameter is utilized. **Gross volume determination:** the formula is: m^3 = midpoint in $cm^2 \times$ length in metres $\times$ 0.00007854.

continued

Table A.1.B. *continued*

Hirogoku *Source*: Briggs and Flora, 1991.	**Brief description:** apparently, this is an old method which is used in Japan. This log scale is similar to JAS (see Section 2.2.11 on page 44) in approach, but the procedures for measuring diameters differ and there is no adjustment in diameter made for logs over 6 m in length. **Diameters:** measure the narrow and wide-way and round down to the closest 2 cm class (29.9 is rounded down to 28). Average the narrow and wide dimension and record to the nearest centimetre (24 the narrow-way and 26 the wide-way equals a diameter of 25 cm). Only the small-end diameter is used to calculate volumes. **Length measurement:** lengths are measured short-end to short-end and rounded down to closest 0.2 m length, e.g. 10.78 m rounds down to 10.6 m. **Taper:** as only the small-end diameter is utilized, taper is ignored. **Gross volume determination:** the formula is: $m^3 = $ (scaling diameter in $cm^2 \times$ length in metres) $\div$ 10,000. **Defect deductions:** no specific data could be found.
Norwegian Cubic Log Scale *Source*: Tømmermålingforen-ingenes Fellesorgan, 1998.	**Brief description:** this is the official scaling method for Norway. It is based on a forestry law that came into effect in 1994 and is overseen by a cooperative Joint Committee for Norwegian Grading and Scaling of Roundwood. Logs can be measured by mid-diameter or top-diameter methods. **Diameters:** top diameter measured at 10 cm in from the top-end of the log if using the top measure method, at the midpoint if using the mid-diameter method, or for pulpwood logs that are butt-cut, measure the mid-diameter at a point 40 cm closer to the butt than to the top. Measure under bark (or over bark subtracting the bark thickness), and take two measurements, wide and narrow directions. Round the diameters down to the lower cm class. **Length measurement:** lengths are recorded in multiples of 0.1 m, 0.3 m, or 0.5 m, if lengths are recorded in multiples of 0.1 m, length is rounded down to next 0.1 m length; with 0.3 or 0.5 multiples, lengths are rounded to nearest length class. Maximum scaling length is 5.8 m. **Taper:** Taper for conifers is considered to equal 1 cm per metre when using top-diameter method. **Gross volume determination:** the truncated diameter is compensated for in the formula by adding 0.5 cm to the recorded diameter. The formula is: top diameter $m^3 = 0.00007854 \times (((\text{small-end diameter} + 0.5) + (\text{length in metres} \div 2)^2) \times \text{length})$. mid-diameter $m^3 = 0.00007854 \times (\text{mid-diameter} + 0.5)^2 \times \text{length}$. **Defect deductions:** similar to Swedish National Scale (see Section 2.2.5 on page 26).
Quebec Cubic Metric *Source*: Québec Ministère des Resources Naturelles, 2001.	**Brief description:** the province of Quebec uses a cubic metric method which is similar to the Ontario Cubic Scale (see Section 2.2.4 on page 22) in procedures, although the Smalian formula is used. Short logs can be scaled from one end, like Ontario Cubic, but only if the logs are less than 2.6 m in length and small and large ends are evenly distributed on the end of the log rows or stacks. Only fibre loss defects are deducted (void, soft rot and char).

continued

Table A.1.B. *continued*

	Diameters: measured in 2 cm scaling classes, e.g. 19.01–21.00 cm is the 20 cm scaling class. For elliptical logs (with a 3 or more scaling class difference between the narrow and perpendicular measurement), take two measurements through the true centre of the log (narrow and perpendicular), if the average of the two measurements is an odd diameter class, round down to closest even class, e.g. d1 is 26 cm, d2 is 32 cm average is 29 cm; round down to 28 cm. For logs that have less than three scaling classes between the narrow and perpendicular axis (e.g. narrow way is 26 cm, perpendicular measurement is 30 cm), take one measurement, seeking neither the narrow nor the wide axis. Butt-cut ends are measured via one measurement across the narrow axis and inside of any butt-flare. **Length measurement:** same as Ontario cubic (see section 2.2.4 on page 22) **Taper:** actual taper between the large-end and small-end diameters. **Gross volume determination:** the formula is: $m^3 = (((\text{large-end diameter}^2 + \text{small-end diameter}^2) \div 2) \times \text{length}) \times 0.00007854$. If using the 'one end' method: $m^3 = \text{diameter}^2 \times \text{length} \times 0.00007854$. **Defect deductions:** Similar to Ontario cubic (see Section 2.2.4 on page 22).
New Brunswick Cubic Metre Scale *Source*: New Brunswick Natural Resources and Energy, Forest Management Branch, 2003.	**Brief description:** the New Brunswick Cubic Metre Scale was approved for use in 1979. It is unique in that it is based on sample populations of logs by small-end diameter, length, and whether coniferous or broad-leaved. One must use a volume table (for either conifer or broad-leaved) to look up the volume of the log once a diameter and length are determined. **Diameters:** measure only the small-end, using only one measurement through the true centre and narrow axis. Diameters are rounded to the nearest 2 cm scaling class, e.g. a log which measures 19.01–21 cm is a 20 cm class. **Length measurement:** lengths are measured in specific length classes and are reflected in the metric equivalent of imperial foot classes. Normally logs are manufactured in the metric equivalent of 2′ multiples thus for brevity, the table below only shows the more common even multiple lengths. **Taper:** is not factored in the scaling process, but was accounted for in the sample used to develop the table. **Gross volume determination:** the volume table was originally derived from a large sample of logs scaled via the Smalian formula. To use, find the intersect of the small-end diameter and length class in the hardwoods or softwoods section of the table, e.g. a log which is 4.9 m long (16′ scaling class) and has a small-end diameter of 34 cm contains 0.595 m^3 if a softwood and 0.583 m^3 if a hardwood. **Defect deductions:** deductions are only made for rot and are calculated by deducting the area of the rot.

continued

Table A.1.B. *continued*

New Brunswick Cubic Metre Log Scale Volume (m^3)

	Softwood length in metres						Hardwood length in metres					
sed cm	>2.4 =2.6 8′	>3.0 =3.2 10′	>3.6 =3.8 12′	>4.2 =4.4 14′	>4.8 =5.0 16′	>5.4 =5.6 18′	>2.4 =2.6 8′	>3.0 =3.2 10′	>3.6 =3.8 12′	>4.2 =4.4 14′	>4.8 =5.0 16′	>5.4 =5.6 18′
14	0.048	0.066	0.085	0.100	0.116	0.130	0.050	0.062	0.074	0.091	0.113	0.138
16	0.062	0.085	0.108	0.127	0.146	0.164	0.062	0.081	0.100	0.123	0.146	0.173
18	0.079	0.106	0.133	0.157	0.181	0.202	0.077	0.103	0.128	0.157	0.182	0.212
20	0.097	0.129	0.161	0.190	0.219	0.246	0.094	0.126	0.158	0.192	0.222	0.254
22	0.118	0.155	0.192	0.226	0.261	0.293	0.114	0.152	0.190	0.230	0.264	0.302
24	0.141	0.184	0.226	0.266	0.307	0.345	0.136	0.180	0.224	0.270	0.310	0.353
26	0.166	0.214	0.263	0.310	0.356	0.402	0.160	0.210	0.260	0.313	0.358	0.408
28	0.192	0.248	0.303	0.356	0.410	0.463	0.186	0.242	0.297	0.357	0.410	0.467
30	0.222	0.284	0.346	0.407	0.468	0.528	0.215	0.276	0.337	0.403	0.464	0.531
32	0.253	0.322	0.391	0.460	0.529	0.598	0.246	0.312	0.378	0.451	0.522	0.598
34	0.286	0.363	0.440	0.518	0.595	0.673	0.279	0.351	0.422	0.502	0.583	0.670
36	0.322	0.406	0.491	0.578	0.664	0.752	0.315	0.391	0.467	0.554	0.647	0.745
38	0.360	0.453	0.546	0.642	0.737	0.835	0.353	0.434	0.514	0.609	0.714	0.825
40	0.400	0.502	0.603	0.708	0.814	0.923	0.393	0.478	0.563	0.666	0.784	0.909
42	0.442	0.552	0.663	0.780	0.896	1.015	0.436	0.525	0.614	0.724	0.857	0.997
44	0.486	0.606	0.726	0.853	0.980	1.112	0.481	0.574	0.667	0.785	0.934	1.089
46	0.532	0.662	0.793	0.931	1.069	1.213	0.528	0.625	0.722	0.848	1.013	1.185
48	0.580	0.721	0.862	1.012	1.162	1.318	0.577	0.678	0.779	0.913	1.095	1.285
50	0.631	0.782	0.933	1.096	1.259	1.428	0.629	0.734	0.838	0.980	1.181	1.389
52	0.683	0.846	1.008	1.184	1.359	1.543	0.683	0.791	0.898	1.049	1.269	1.498
54	0.738	0.912	1.086	1.275	1.464	1.662	0.740	0.850	0.961	1.121	1.361	1.610
56	0.795	0.980	1.166	1.369	1.572	1.786	0.798	0.912	1.026	1.194	1.456	1.726
58	0.854	1.052	1.250	1.467	1.684	1.914	0.859	0.976	1.092	1.269	1.554	1.847
60	0.915	1.126	1.336	1.568	1.800	2.046	0.923	1.042	1.160	1.347	1.655	1.972

Maine Rule (also known as the Holland rule) *Source*: Maine Department of Agriculture, Food and Rural Resources, no year; Freese, 1973.

Brief description: the Maine rule is a product output rule (board foot) and like the Scribner is a diagram rule.
Diameters: same as the International ¼″ Log Rule (see Section 2.3.4 on page 67).
Taper: is ignored excepting that logs over 30′ long in nominal length are scaled in two segments.
Length measurement: same as the International ¼″ Log Rule (see section 2.3.4 on page 67), except that segment lengths can be as long as 30′ in nominal length before the log is divided into two segments.
Gross volume determination: the Maine rule is based on a diagram rule, so volume cannot be calculated via a formula, and must be determined via the table below.
Defect deductions: same as International ¼″ Log Rule (see section 2.3.4 on page 67). Logs with less than 33 ⅓% net volume will be culled.

continued

Table A.1.B. *continued*

Maine log rule (*bf*)

Small-end diameter	Segment length in feet: 8	10	12	14	16	18	20	22	24	26	28	30
6	10	12	16	18	20	23	25	28	30	33	36	38
7	16	19	23	27	31	35	39	43	47	50	55	58
8	22	27	33	39	44	50	55	61	66	72	76	82
9	26	32	39	46	52	59	65	72	78	85	92	98
10	34	42	51	59	68	76	85	94	102	111	119	128
11	41	51	62	72	83	93	103	114	124	134	145	155
12	52	65	78	92	105	118	131	144	157	170	183	196
13	60	75	90	105	120	135	150	165	180	195	210	225
14	71	89	107	124	142	160	178	195	213	231	249	266
15	81	101	121	141	161	181	202	222	242	262	282	302
16	89	111	134	157	179	201	223	246	268	291	313	335
17	102	128	154	179	205	231	256	282	307	333	359	384
18	116	145	174	203	232	261	290	319	348	378	407	436
19	136	169	203	237	271	305	339	373	407	441	475	509
20	151	189	227	265	302	340	378	415	454	491	529	567
21	168	210	252	294	336	378	420	462	504	545	587	629
22	182	227	272	318	363	408	454	499	545	590	635	681
23	200	250	300	351	401	451	501	551	601	651	701	751
24	220	274	327	384	439	494	549	604	659	713	768	823
25	238	298	358	417	477	537	596	656	715	775	835	894
26	254	317	380	444	507	570	634	697	761	824	887	951

Newfoundland Board Foot Rule
Source: Newfoundland Forest Service, about 1976.

Brief description: the Newfoundland Board Foot Rule has been in use since the early 1900s. It is measured in imperial units and reflected in board feet. Cubic Measure is starting to take hold in Newfoundland, but the Newfoundland Rule still has a following there.

Diameters: small-end diameters are measured inside of bark through the true centre (seeking neither the narrow nor wide axis), and rounded to the nearest inch, e.g. 7.51″ up to and including 8.5″ are scaled as an 8″ diameter measurement. If a log is significantly elliptical, take two measurements at right angles to each other (round down if fractional).

Taper: is ignored excepting that logs over 24′ long in nominal length are scaled in two segments.

Length measurement: same as the International ¼-Inch Log Rule (see Section 2.3.4 on page 67) except that segment lengths can be as long as 24′ in nominal length before the log is divided into two segments.

Gross volume determination: volume is determined by the following formula:

bf $=$ small-end diameter$^2 \times$ L $\div$ 24; round to the nearest board foot.

Example: 22″ diameter 16′ long; $22^2 \times 16 \div 24 = 322.67$; round to 323 bf.

Defect deductions: similar to the International ¼-Inch Log Rule (see Section 2.3.4 on page 67). Squared defects are deducted by measuring the defect (rounding the width and height up to the next inch if fractional) via the following formula: W″ $\times$ H″ $\times$ L′ $\div$ 16 (round to nearest bf). Defects showing in one end of the log are considered to go through half the log length. Logs with less than 50% net volume will be culled.

continued

Table A.1.B. *continued*

New Brunswick Board Foot Rule
Source: New Brunswick Natural Resources and Energy, Forest Management Branch, 2003; Freese, 1973.

Brief description: the New Brunswick Board Foot Rule was adopted as the statute rule for New Brunswick in 1845. It is still used in the provinces of New Brunswick and Nova Scotia. The rule is at least in part a diagram rule (although logs with diameters of 11–16″ can be calculated via the Newfoundland Board Foot Rule formula).

Diameters: measure through the true centre, under bark. One measurement can be taken on relatively round logs (without trying to favour the narrow or wide measurement), or two measurements can be taken on elliptical logs. Diameter measurements are truncated, e.g. 7.99 rounds to 7″. When averaging two diameters, round down if fractional, e.g. narrow measurement of 10″ and a perpendicular measurement of 13″, round to 11″.

Taper: is ignored excepting that logs over 23′ long in nominal length are scaled in two segments.

Length measurement: same as the International ¼-Inch Log Rule (see Section 2.3.4 on page 67) except that segment lengths can be as long as 23′ in nominal length before the log is divided into two segments.

Gross volume determination: volume is determined via the table below.

Defect deductions: similar to the International ¼-Inch and Newfoundland Log Rules. Squared defects are deducted by measuring the defect (rounding the width and height down to the next inch if fractional) via the following formula: $W'' \times H'' \times L' \div 16$ (round to nearest bf). Defects showing in one end are estimated as to the length affected. Logs with less than 50% net volume will be culled.

New Brunswick Board Foot Log Rule (bf)

Small-end diameter	Segment length in feet															
	8	**9**	**10**	**11**	**12**	**13**	**14**	**15**	**16**	**17**	**18**	**19**	**20**	**21**	**22**	**23**
5	6	7	8	9	10	11	12	13	14	15	16	–	–	–	–	–
6	10	11	12	14	15	17	18	19	20	22	23	–	–	–	–	–
7	15	17	19	21	23	25	27	29	31	33	35	–	–	–	–	–
8	20	23	25	28	30	33	35	38	40	43	45	48	50	35	55	58
9	24	27	30	33	36	39	42	45	48	51	54	57	60	63	66	69
10	32	36	40	44	48	52	56	60	64	68	72	76	80	84	88	92
11	40	45	50	55	60	65	70	75	80	85	90	95	100	105	110	115
12	48	54	60	66	72	78	84	90	96	102	108	114	120	126	132	138
13	56	63	70	77	84	91	98	105	112	119	126	133	140	147	154	161
14	65	74	82	90	98	106	114	122	130	139	147	155	163	171	179	188
15	75	84	93	102	112	121	131	140	150	159	168	177	187	196	205	214
16	85	96	107	117	128	138	149	159	170	181	192	202	213	223	234	245
17	99	111	124	136	149	161	174	186	198	210	223	235	248	261	275	286
18	115	129	143	158	172	186	200	215	229	244	258	272	286	301	315	330
19	131	147	163	180	196	212	228	245	261	278	294	310	326	343	359	376
20	150	168	187	206	225	243	262	281	300	318	337	356	375	393	412	431
21	164	185	206	227	247	268	288	308	327	349	370	391	411	432	453	474
22	181	204	227	250	272	295	317	340	362	385	408	431	453	476	498	521
23	188	223	248	272	297	317	336	356	376	411	445	470	495	519	544	569
24	216	243	270	297	324	352	380	406	432	459	486	513	540	569	594	621
25	238	268	298	328	358	388	419	448	477	507	537	566	596	620	656	685
26	253	285	317	348	380	411	444	475	507	538	570	602	634	665	697	729

continued

Table A.1.B. *continued*

Ontario Board Foot Rule *Source*: Honer, 1998; Ontario Wood Lot Association, 2003.	**Brief description:** the Ontario Rule is a product output rule (board foot). It was authorized for use in the Province of Ontario, Canada in 1869, and is still used for measuring privately owned timber. It is still measured and reflected in imperial units (feet, inches and board feet). **Diameters:** same as the International ¼-Inch Log Rule (see Section 2.3.4 on p. 67). **Taper:** all log segments are considered a cylinder having the diameter of the small-end of the log segment. **Length measurement:** same as the International ¼-Inch Log Rule (see Section 2.3.4 on p. 67), except that the maximum segment length is 18′. **Gross volume determination:** the board foot volume is calculated via the following formula: board foot volume $= ((0.55 \times \text{diameter}^2 - (1.2 \times \text{diameter})) \times (\text{length} \div 12)$; round to the nearest board foot **Defect deductions:** defect deductions are taken for all defects that reduce lumber recovery (similar to the International ¼-Inch Log Rule (see Section 2.3.4 on p. 67). Squared defects are deducted by measuring the defect (rounding the width and height down to the next inch if fractional) via the following formula: $(W'' \times H'' \times L') \div 12$ (round to nearest bf). Logs with less than 33⅓% net volume will be culled.

Table A.1.C. USFS segment length and trim allowance chart (feet).

Measured length	Recorded length	Segment length: Top	Segment length: Segment 2	Measured length	Recorded length	Segment length: Top	Segment length: Segment 2	Segment length: Segment 3
8–8.5	8	8		35.1–36.0	35	17	18	
8.6–9.5	9	9		36.1–37.0	36	18	18	
9.6–10.5	10	10		37.1–38.0	37	18	19	
10.6–11.5	11	11		38.1–39.0	38	18	20	
11.6–12.5	12	12		39.1–40.0	39	19	20	
12.6–13.5	13	13		40.1–41.0	40	20	20	
13.6–14.5	14	14		41.1–42.5	41	13	14	14
14.6–15.5	15	15		42.6–43.5	42	14	14	14
15.6–16.5	16	16		43.6–44.5	43	14	14	15
16.6–17.5	17	17		44.6–45.5	44	14	14	16
17.6–18.5	18	18		45.6–46.5	45	14	15	16
18.6–19.5	19	19		46.6–47.5	46	14	16	16
19.6–20.5	20	20		47.6–48.5	47	15	16	16
20.6–22.0	21	10	11	48.6–49.5	48	16	16	16
22.1–23.0	22	10	12	49.6–50.5	49	16	16	17
23.1–24.0	23	11	12	50.6–51.5	50	16	16	18
24.1–25.0	24	12	12	51.6–52.5	51	16	17	18
25.1–26.0	25	12	13	52.6–53.5	52	16	18	18
26.1–27.0	26	12	14	53.6–54.5	53	17	18	18
27.1–28.0	27	13	14	54.6–55.5	54	18	18	18
28.1–29.0	28	14	14	55.6–56.5	55	18	18	19
29.1–30.0	29	14	15	56.6–57.5	56	18	18	20
30.1–31.0	30	14	16	57.6–58.5	57	18	19	20
31.1–32.0	31	15	16	58.6–59.5	58	18	20	20
32.1–33.0	32	16	16	59.6–60.5	59	19	20	20
33.1–34.0	33	16	17	60.6–61.5	60	20	20	20
34.1–35.0	34	16	18					

Table A.1.D. USFS segment taper distribution chart.

		Total taper in inches																								
		1	2	3	4	5	6	7	8	9	10	11	12	13	14	15	16	17	18	19	20	21	22	23	24	25
Two segment logs 21′–40″	Inches of taper in segment 1	1	1	2	2	3	3	4	4	5	5	6	6	7	7	8	8	9	9	10	10	11	11	12	12	13
	Inches of taper in segment 2	0	1	1	2	2	3	3	4	4	5	5	6	6	7	7	8	8	9	9	10	10	11	11	12	12
		Total taper in inches																								
		1	2	3	4	5	6	7	8	9	10	11	12	13	14	15	16	17	18	19	20	21	22	23	24	25
Three segment logs 41′–60″	Inches of taper in segment 1	1	1	1	2	2	2	3	3	3	4	4	4	5	5	5	6	6	6	7	7	7	8	8	8	9
	Inches of taper in segment 2	0	1	1	1	2	2	2	3	3	3	4	4	4	5	5	5	6	6	6	7	7	7	8	8	8
	Inches of taper in segment 3	0	0	1	1	1	2	2	2	3	3	3	4	4	4	5	5	5	6	6	6	7	7	7	8	8

Note: Segment numbering starts at the small-end of the log, e.g. for a two segment log, segment 1 occurs on the end of the log with the small diameter; segment 2 occurs on the end of the log with the large diameter. To use this table, locate the total taper of the log and find the appropriate taper for the segment dependent on whether a two or three segment log, e.g. for a log which is 44′ long with a small-end diameter of 10″ and a large-end diameter of 17″ (7″ of total taper), the segment 1 l.e.d. and segment 2 s.e.d. will be 13″ (10 + 3 = 13″), the segment 2 l.e.d. and the segment 3 s.e.d. will be 15″ (13 + 2 = 15″).

Table A.1.E. USFS Cubic Log Scale 'half segment' volume chart (ft^3).

Small-end diameter of log segment in inches	Length of log segment in feet																			
	1	**2**	**3**	**4**	**5**	**6**	**7**	**8**	**9**	**10**	**11**	**12**	**13**	**14**	**15**	**16**	**17**	**18**	**19**	**20**
3	0.0	0.0	0.1	0.1	0.1	0.1	0.2	0.2	0.2	0.2	0.3	0.3	0.3	0.3	0.4	0.4	0.4	0.4	0.5	0.5
4	0.0	0.1	0.1	0.2	0.2	0.3	0.3	0.3	0.4	0.4	0.5	0.5	0.6	0.6	0.7	0.7	0.7	0.8	0.8	0.9
5	0.1	0.1	0.2	0.3	0.3	0.4	0.5	0.5	0.6	0.7	0.7	0.8	0.9	1.0	1.0	1.1	1.2	1.2	1.3	1.4
6	0.1	0.2	0.3	0.4	0.5	0.6	0.7	0.8	0.9	1.0	1.1	1.2	1.3	1.4	1.5	1.6	1.7	1.8	1.9	2.0
7	0.1	0.3	0.4	0.5	0.7	0.8	0.9	1.1	1.2	1.3	1.5	1.6	1.7	1.9	2.0	2.1	2.3	2.4	2.5	2.7
8	0.2	0.3	0.5	0.7	0.9	1.0	1.2	1.4	1.6	1.7	1.9	2.1	2.3	2.4	2.6	2.8	3.0	3.1	3.3	3.5
9	0.2	0.4	0.7	0.9	1.1	1.3	1.5	1.8	2.0	2.2	2.4	2.7	2.9	3.1	3.3	3.5	3.8	4.0	4.2	4.4
10	0.3	0.5	0.8	1.1	1.4	1.6	1.9	2.2	2.5	2.7	3.0	3.3	3.5	3.8	4.1	4.4	4.6	4.9	5.2	5.5
11	0.3	0.7	1.0	1.3	1.6	2.0	2.3	2.6	3.0	3.3	3.6	4.0	4.3	4.6	4.9	5.3	5.6	5.9	6.3	6.6
12	0.4	0.8	1.2	1.6	2.0	2.4	2.7	3.1	3.5	3.9	4.3	4.7	5.1	5.5	5.9	6.3	6.7	7.1	7.5	7.9
13	0.5	0.9	1.4	1.8	2.3	2.8	3.2	3.7	4.1	4.6	5.1	5.5	6.0	6.5	6.9	7.4	7.8	8.3	8.8	9.2
14	0.5	1.1	1.6	2.1	2.7	3.2	3.7	4.3	4.8	5.3	5.9	6.4	6.9	7.5	8.0	8.6	9.1	9.6	10.2	10.7
15	0.6	1.2	1.8	2.5	3.1	3.7	4.3	4.9	5.5	6.1	6.7	7.4	8.0	8.6	9.2	9.8	10.4	11.0	11.7	12.3
16	0.7	1.4	2.1	2.8	3.5	4.2	4.9	5.6	6.3	7.0	7.7	8.4	9.1	9.8	10.5	11.2	11.9	12.6	13.3	14.0
17	0.8	1.6	2.4	3.2	3.9	4.7	5.5	6.3	7.1	7.9	8.7	9.5	10.2	11.0	11.8	12.6	13.4	14.2	15.0	15.8
18	0.9	1.8	2.7	3.5	4.4	5.3	6.2	7.1	8.0	8.8	9.7	10.6	11.5	12.4	13.3	14.1	15.0	15.9	16.8	17.7
19	1.0	2.0	3.0	3.9	4.9	5.9	6.9	7.9	8.9	9.8	10.8	11.8	12.8	13.8	14.8	15.8	16.7	17.7	18.7	19.7
20	1.1	2.2	3.3	4.4	5.5	6.5	7.6	8.7	9.8	10.9	12.0	13.1	14.2	15.3	16.4	17.5	18.5	19.6	20.7	21.8
21	1.2	2.4	3.6	4.8	6.0	7.2	8.4	9.6	10.8	12.0	13.2	14.4	15.6	16.8	18.0	19.2	20.4	21.6	22.8	24.1
22	1.3	2.6	4.0	5.3	6.6	7.9	9.2	10.6	11.9	13.2	14.5	15.8	17.2	18.5	19.8	21.1	22.4	23.8	25.1	26.4
23	1.4	2.9	4.3	5.8	7.2	8.7	10.1	11.5	13.0	14.4	15.9	17.3	18.8	20.2	21.6	23.1	24.5	26.0	27.4	28.9
24	1.6	3.1	4.7	6.3	7.9	9.4	11.0	12.6	14.1	15.7	17.3	18.8	20.4	22.0	23.6	25.1	26.7	28.3	29.8	31.4
25	1.7	3.4	5.1	6.8	8.5	10.2	11.9	13.6	15.3	17.0	18.7	20.5	22.2	23.9	25.6	27.3	29.0	30.7	32.4	34.1
26	1.8	3.7	5.5	7.4	9.2	11.1	12.9	14.7	16.6	18.4	20.3	22.1	24.0	25.8	27.7	29.5	31.3	33.2	35.0	36.9
27	2.0	4.0	6.0	8.0	9.9	11.9	13.9	15.9	17.9	19.9	21.9	23.9	25.8	27.8	29.8	31.8	33.8	35.8	37.8	39.8

continued

Table A.1.E. *continued*

Small-end diameter of log segment in inches	Length of log segment in feet																			
	1	2	3	4	5	6	7	8	9	10	11	12	13	14	15	16	17	18	19	20
28	2.1	4.3	6.4	8.6	10.7	12.8	15.0	17.1	19.2	21.4	23.5	25.7	27.8	29.9	32.1	34.2	36.3	38.5	40.6	42.8
29	2.3	4.6	6.9	9.2	11.5	13.8	16.1	18.3	20.6	22.9	25.2	27.5	29.8	32.1	34.4	36.7	39.0	41.3	43.6	45.9
30	2.5	4.9	7.4	9.8	12.3	14.7	17.2	19.6	22.1	24.5	27.0	29.5	31.9	34.4	36.8	39.3	41.7	44.2	46.6	49.1
31	2.6	5.2	7.9	10.5	13.1	15.7	18.3	21.0	23.6	26.2	28.8	31.4	34.1	36.7	39.3	41.9	44.6	47.2	49.8	52.4
32	2.8	5.6	8.4	11.2	14.0	16.8	19.5	22.3	25.1	27.9	30.7	33.5	36.3	39.1	41.9	44.7	47.5	50.3	53.1	55.8
33	3.0	5.9	8.9	11.9	14.8	17.8	20.8	23.8	26.7	29.7	32.7	35.6	38.6	41.6	44.5	47.5	50.5	53.5	56.4	59.4
34	3.2	6.3	9.5	12.6	15.8	18.9	22.1	25.2	28.4	31.5	34.7	37.8	41.0	44.1	47.3	50.4	53.6	56.7	59.9	63.0
35	3.3	6.7	10.0	13.4	16.7	20.0	23.4	26.7	30.1	33.4	36.7	40.1	43.4	46.8	50.1	53.4	56.8	60.1	63.5	66.8
36	3.5	7.1	10.6	14.1	17.7	21.2	24.7	28.3	31.8	35.3	38.9	42.4	45.9	49.5	53.0	56.5	60.1	63.6	67.1	70.7
37	3.7	7.5	11.2	14.9	18.7	22.4	26.1	29.9	33.6	37.3	41.1	44.8	48.5	52.3	56.0	59.7	63.5	67.2	70.9	74.7
38	3.9	7.9	11.8	15.8	19.7	23.6	27.6	31.5	35.4	39.4	43.3	47.3	51.2	55.1	59.1	63.0	66.9	70.9	74.8	78.8
39	4.1	8.3	12.4	16.6	20.7	24.9	29.0	33.2	37.3	41.5	45.6	49.8	53.9	58.1	62.2	66.4	70.5	74.7	78.8	83.0
40	4.4	8.7	13.1	17.5	21.8	26.2	30.5	34.9	39.3	43.6	48.0	52.4	56.7	61.1	65.4	69.8	74.2	78.5	82.9	87.3
41	4.6	9.2	13.8	18.3	22.9	27.5	32.1	36.7	41.3	45.8	50.4	55.0	59.6	64.2	68.8	73.3	77.9	82.5	87.1	91.7
42	4.8	9.6	14.4	19.2	24.1	28.9	33.7	38.5	43.3	48.1	52.9	57.7	62.5	67.3	72.2	77.0	81.8	86.6	91.4	96.2
43	5.0	10.1	15.1	20.2	25.2	30.3	35.3	40.3	45.4	50.4	55.5	60.5	65.5	70.6	75.6	80.7	85.7	90.8	95.8	100.8
44	5.3	10.6	15.8	21.1	26.4	31.7	37.0	42.2	47.5	52.8	58.1	63.4	68.6	73.9	79.2	84.5	89.8	95.0	100.3	105.6
45	5.5	11.0	16.6	22.1	27.6	33.1	38.7	44.2	49.7	55.2	60.7	66.3	71.8	77.3	82.8	88.4	93.9	99.4	104.9	110.4
46	5.8	11.5	17.3	23.1	28.9	34.6	40.4	46.2	51.9	57.7	63.5	69.2	75.0	80.8	86.6	92.3	98.1	103.9	109.6	115.4
47	6.0	12.0	18.1	24.1	30.1	36.1	42.2	48.2	54.2	60.2	66.3	72.3	78.3	84.3	90.4	96.4	102.4	108.4	114.5	120.5
48	6.3	12.6	18.8	25.1	31.4	37.7	44.0	50.3	56.5	62.8	69.1	75.4	81.7	88.0	94.2	100.5	106.8	113.1	119.4	125.7
49	6.5	13.1	19.6	26.2	32.7	39.3	45.8	52.4	58.9	65.5	72.0	78.6	85.1	91.7	98.2	104.8	111.3	117.9	124.4	131.0
50	6.8	13.6	20.5	27.3	34.1	40.9	47.7	54.5	61.4	68.2	75.0	81.8	88.6	95.4	102.3	109.1	115.9	122.7	129.5	136.4

Note: To use this table add the one-half volume shown for the small-end diameter and length combination, with the one-half volume shown for the large-end diameter and length combination, e.g. for a log that is 16′ long, has a small-end diameter of 13″ and a large-end diameter of 14″, a value of 7.4 ft^3 is shown for 13″ and 8.6 ft^3 for 14″ ($7.4 + 8.6 = 16.0$); log volume is **16.0 ft^3**.

Table A.1.F. Half segment cubic metre volume chart (m^3).

Small-end diameter of log in cm or rads

		Length of log segment in metres																			
Rads	cm	1	2	3	4	5	6	7	8	9	10	11	12	13	14	15	16	17	18	19	20
4	8	0.003	0.005	0.008	0.010	0.013	0.015	0.018	0.020	0.023	0.025	0.028	0.030	0.033	0.035	0.038	0.040	0.043	0.045	0.048	0.050
	9	0.003	0.006	0.010	0.013	0.016	0.019	0.022	0.025	0.029	0.032	0.035	0.038	0.041	0.045	0.048	0.051	0.054	0.057	0.060	0.064
5	10	0.004	0.008	0.012	0.016	0.020	0.024	0.027	0.031	0.035	0.039	0.043	0.047	0.051	0.055	0.059	0.063	0.067	0.071	0.075	0.079
	11	0.005	0.010	0.014	0.019	0.024	0.029	0.033	0.038	0.043	0.048	0.052	0.057	0.062	0.067	0.071	0.076	0.081	0.086	0.090	0.095
6	12	0.006	0.011	0.017	0.023	0.028	0.034	0.040	0.045	0.051	0.057	0.062	0.068	0.074	0.079	0.085	0.090	0.096	0.102	0.107	0.113
	13	0.007	0.013	0.020	0.027	0.033	0.040	0.046	0.053	0.060	0.066	0.073	0.080	0.086	0.093	0.100	0.106	0.113	0.119	0.126	0.133
7	14	0.008	0.015	0.023	0.031	0.038	0.046	0.054	0.062	0.069	0.077	0.085	0.092	0.100	0.108	0.115	0.123	0.131	0.139	0.146	0.154
	15	0.009	0.018	0.027	0.035	0.044	0.053	0.062	0.071	0.080	0.088	0.097	0.106	0.115	0.124	0.133	0.141	0.150	0.159	0.168	0.177
8	16	0.010	0.020	0.030	0.040	0.050	0.060	0.070	0.080	0.090	0.101	0.111	0.121	0.131	0.141	0.151	0.161	0.171	0.181	0.191	0.201
	17	0.011	0.023	0.034	0.045	0.057	0.068	0.079	0.091	0.102	0.113	0.125	0.136	0.148	0.159	0.170	0.182	0.193	0.204	0.216	0.227
9	18	0.013	0.025	0.038	0.051	0.064	0.076	0.089	0.102	0.115	0.127	0.140	0.153	0.165	0.178	0.191	0.204	0.216	0.229	0.242	0.254
	19	0.014	0.028	0.043	0.057	0.071	0.085	0.099	0.113	0.128	0.142	0.156	0.170	0.184	0.198	0.213	0.227	0.241	0.255	0.269	0.284
10	20	0.016	0.031	0.047	0.063	0.079	0.094	0.110	0.126	0.141	0.157	0.173	0.188	0.204	0.220	0.236	0.251	0.267	0.283	0.298	0.314
	21	0.017	0.035	0.052	0.069	0.087	0.104	0.121	0.139	0.156	0.173	0.190	0.208	0.225	0.242	0.260	0.277	0.294	0.312	0.329	0.346
11	22	0.019	0.038	0.057	0.076	0.095	0.114	0.133	0.152	0.171	0.190	0.209	0.228	0.247	0.266	0.285	0.304	0.323	0.342	0.361	0.380
	23	0.021	0.042	0.062	0.083	0.104	0.125	0.145	0.166	0.187	0.208	0.229	0.249	0.270	0.291	0.312	0.332	0.353	0.374	0.395	0.415
12	24	0.023	0.045	0.068	0.090	0.113	0.136	0.158	0.181	0.204	0.226	0.249	0.271	0.294	0.317	0.339	0.362	0.385	0.407	0.430	0.452
	25	0.025	0.049	0.074	0.098	0.123	0.147	0.172	0.196	0.221	0.245	0.270	0.295	0.319	0.344	0.368	0.393	0.417	0.442	0.466	0.491
13	26	0.027	0.053	0.080	0.106	0.133	0.159	0.186	0.212	0.239	0.265	0.292	0.319	0.345	0.372	0.398	0.425	0.451	0.478	0.504	0.531
	27	0.029	0.057	0.086	0.115	0.143	0.172	0.200	0.229	0.258	0.286	0.315	0.344	0.372	0.401	0.429	0.458	0.487	0.515	0.544	0.573
14	28	0.031	0.062	0.092	0.123	0.154	0.185	0.216	0.246	0.277	0.308	0.339	0.369	0.400	0.431	0.462	0.493	0.523	0.554	0.585	0.616
	29	0.033	0.066	0.099	0.132	0.165	0.198	0.231	0.264	0.297	0.330	0.363	0.396	0.429	0.462	0.495	0.528	0.561	0.594	0.627	0.661
15	30	0.035	0.071	0.106	0.141	0.177	0.212	0.247	0.283	0.318	0.353	0.389	0.424	0.459	0.495	0.530	0.565	0.601	0.636	0.672	0.707
	31	0.038	0.075	0.113	0.151	0.189	0.226	0.264	0.302	0.340	0.377	0.415	0.453	0.491	0.528	0.566	0.604	0.642	0.679	0.717	0.755
16	32	0.040	0.080	0.121	0.161	0.201	0.241	0.281	0.322	0.362	0.402	0.442	0.483	0.523	0.563	0.603	0.643	0.684	0.724	0.764	0.804
	33	0.043	0.086	0.128	0.171	0.214	0.257	0.299	0.342	0.385	0.428	0.470	0.513	0.556	0.599	0.641	0.684	0.727	0.770	0.813	0.855
17	34	0.045	0.091	0.136	0.182	0.227	0.272	0.318	0.363	0.409	0.454	0.499	0.545	0.590	0.636	0.681	0.726	0.772	0.817	0.863	0.908
	35	0.048	0.096	0.144	0.192	0.241	0.289	0.337	0.385	0.433	0.481	0.529	0.577	0.625	0.673	0.722	0.770	0.818	0.866	0.914	0.962
18	36	0.051	0.102	0.153	0.204	0.254	0.305	0.356	0.407	0.458	0.509	0.560	0.611	0.662	0.713	0.763	0.814	0.865	0.916	0.967	1.018
	37	0.054	0.108	0.161	0.215	0.269	0.323	0.376	0.430	0.484	0.538	0.591	0.645	0.699	0.753	0.806	0.860	0.914	0.968	1.021	1.075
19	38	0.057	0.113	0.170	0.227	0.284	0.340	0.397	0.454	0.510	0.567	0.624	0.680	0.737	0.794	0.851	0.907	0.964	1.021	1.077	1.134
	39	0.060	0.119	0.179	0.239	0.299	0.358	0.418	0.478	0.538	0.597	0.657	0.717	0.776	0.836	0.896	0.956	1.015	1.075	1.135	1.195
20	40	0.063	0.126	0.188	0.251	0.314	0.377	0.440	0.503	0.565	0.628	0.691	0.754	0.817	0.880	0.942	1.005	1.068	1.131	1.194	1.257
	41	0.066	0.132	0.198	0.264	0.330	0.396	0.462	0.528	0.594	0.660	0.726	0.792	0.858	0.924	0.990	1.056	1.122	1.188	1.254	1.320

continued

Table A.1.F. *continued*

Small-end diameter of log in cm or rads

		Length of log segment in metres																			
Rads	**cm**	**1**	**2**	**3**	**4**	**5**	**6**	**7**	**8**	**9**	**10**	**11**	**12**	**13**	**14**	**15**	**16**	**17**	**18**	**19**	**20**
21	42	0.069	0.139	0.208	0.277	0.346	0.416	0.485	0.554	0.623	0.693	0.762	0.831	0.901	0.970	1.039	1.108	1.178	1.247	1.316	1.385
	43	0.073	0.145	0.218	0.290	0.363	0.436	0.508	0.581	0.653	0.726	0.799	0.871	0.944	1.017	1.089	1.162	1.234	1.307	1.380	1.452
22	44	0.076	0.152	0.228	0.304	0.380	0.456	0.532	0.608	0.684	0.760	0.836	0.912	0.988	1.064	1.140	1.216	1.292	1.368	1.445	1.521
	45	0.080	0.159	0.239	0.318	0.398	0.477	0.557	0.636	0.716	0.795	0.875	0.954	1.034	1.113	1.193	1.272	1.352	1.431	1.511	1.590
23	46	0.083	0.166	0.249	0.332	0.415	0.499	0.582	0.665	0.748	0.831	0.914	0.997	1.080	1.163	1.246	1.330	1.413	1.496	1.579	1.662
	47	0.087	0.173	0.260	0.347	0.434	0.520	0.607	0.694	0.781	0.867	0.954	1.041	1.128	1.214	1.301	1.388	1.475	1.561	1.648	1.735
24	48	0.090	0.181	0.271	0.362	0.452	0.543	0.633	0.724	0.814	0.905	0.995	1.086	1.176	1.267	1.357	1.448	1.538	1.629	1.719	1.810
	49	0.094	0.189	0.283	0.377	0.471	0.566	0.660	0.754	0.849	0.943	1.037	1.131	1.226	1.320	1.414	1.509	1.603	1.697	1.791	1.886
25	50	0.098	0.196	0.295	0.393	0.491	0.589	0.687	0.785	0.884	0.982	1.080	1.178	1.276	1.374	1.473	1.571	1.669	1.767	1.865	1.964
	51	0.102	0.204	0.306	0.409	0.511	0.613	0.715	0.817	0.919	1.021	1.124	1.226	1.328	1.430	1.532	1.634	1.736	1.839	1.941	2.043
26	52	0.106	0.212	0.319	0.425	0.531	0.637	0.743	0.849	0.956	1.062	1.168	1.274	1.380	1.487	1.593	1.699	1.805	1.911	2.018	2.124
	53	0.110	0.221	0.331	0.441	0.552	0.662	0.772	0.882	0.993	1.103	1.213	1.324	1.434	1.544	1.655	1.765	1.875	1.986	2.096	2.206
27	54	0.115	0.229	0.344	0.458	0.573	0.687	0.802	0.916	1.031	1.145	1.260	1.374	1.489	1.603	1.718	1.832	1.947	2.061	2.176	2.290
	55	0.119	0.238	0.356	0.475	0.594	0.713	0.832	0.950	1.069	1.188	1.307	1.426	1.544	1.663	1.782	1.901	2.019	2.138	2.257	2.376
28	56	0.123	0.246	0.369	0.493	0.616	0.739	0.862	0.985	1.108	1.232	1.355	1.478	1.601	1.724	1.847	1.970	2.094	2.217	2.340	2.463
	57	0.128	0.255	0.383	0.510	0.638	0.766	0.893	1.021	1.148	1.276	1.403	1.531	1.659	1.786	1.914	2.041	2.169	2.297	2.424	2.552
29	58	0.132	0.264	0.396	0.528	0.661	0.793	0.925	1.057	1.189	1.321	1.453	1.585	1.717	1.849	1.982	2.114	2.246	2.378	2.510	2.642
	59	0.137	0.273	0.410	0.547	0.683	0.820	0.957	1.094	1.230	1.367	1.504	1.640	1.777	1.914	2.050	2.187	2.324	2.461	2.597	2.734
30	60	0.141	0.283	0.424	0.565	0.707	0.848	0.990	1.131	1.272	1.414	1.555	1.696	1.838	1.979	2.121	2.262	2.403	2.545	2.686	2.827
	61	0.146	0.292	0.438	0.584	0.731	0.877	1.023	1.169	1.315	1.461	1.607	1.753	1.900	2.046	2.192	2.338	2.484	2.630	2.776	2.922
31	62	0.151	0.302	0.453	0.604	0.755	0.906	1.057	1.208	1.359	1.510	1.660	1.811	1.962	2.113	2.264	2.415	2.566	2.717	2.868	3.019
	63	0.156	0.312	0.468	0.623	0.779	0.935	1.091	1.247	1.403	1.559	1.714	1.870	2.026	2.182	2.338	2.494	2.650	2.806	2.961	3.117
32	64	0.161	0.322	0.483	0.643	0.804	0.965	1.126	1.287	1.448	1.608	1.769	1.930	2.091	2.252	2.413	2.574	2.734	2.895	3.056	3.217
	65	0.166	0.332	0.498	0.664	0.830	0.995	1.161	1.327	1.493	1.659	1.825	1.991	2.157	2.323	2.489	2.655	2.821	2.986	3.152	3.318
33	66	0.171	0.342	0.513	0.684	0.855	1.026	1.197	1.368	1.540	1.711	1.882	2.053	2.224	2.395	2.566	2.737	2.908	3.079	3.250	3.421
	67	0.176	0.353	0.529	0.705	0.881	1.058	1.234	1.410	1.587	1.763	1.939	2.115	2.292	2.468	2.644	2.821	2.997	3.173	3.349	3.526
34	68	0.182	0.363	0.545	0.726	0.908	1.090	1.271	1.453	1.634	1.816	1.997	2.179	2.361	2.542	2.724	2.905	3.087	3.269	3.450	3.632
	69	0.187	0.374	0.561	0.748	0.935	1.122	1.309	1.496	1.683	1.870	2.057	2.244	2.431	2.618	2.804	2.991	3.178	3.365	3.552	3.739
35	70	0.192	0.385	0.577	0.770	0.962	1.155	1.347	1.539	1.732	1.924	2.117	2.309	2.501	2.694	2.886	3.079	3.271	3.464	3.656	3.848
	71	0.198	0.396	0.594	0.792	0.990	1.188	1.386	1.584	1.782	1.980	2.178	2.376	2.573	2.771	2.969	3.167	3.365	3.563	3.761	3.959
36	72	0.204	0.407	0.611	0.814	1.018	1.221	1.425	1.629	1.832	2.036	2.239	2.443	2.646	2.850	3.054	3.257	3.461	3.664	3.868	4.072
	73	0.209	0.419	0.628	0.837	1.046	1.256	1.465	1.674	1.883	2.093	2.302	2.511	2.721	2.930	3.139	3.348	3.558	3.767	3.976	4.185
37	74	0.215	0.430	0.645	0.860	1.075	1.290	1.505	1.720	1.935	2.150	2.365	2.581	2.796	3.011	3.226	3.441	3.656	3.871	4.086	4.301

Note: To use this table, add the volume shown for the small-end diameter and length combination, with the volume shown for the large-end diameter and length combination, e.g. for a log that is 5.2 m long, with a small-end diameter of 46 cm (23 rads) and a large-end diameter of 48 cm (24 rads), a value of 0.415 m^3 is shown for 46 cm – 5 m and 0.452 m^3 for 48 cm – 5 m; for the remaining 0.2 m use 10% of 2 m volume, e.g. 0.017 m^3 for 46 cm – 0.2 m and 0.018 m^3 for 48 cm (0.415 + 0.452 + 0.017 + 0.018 = 0.902); log volume is **0.902 m^3**.

Table A.1.G. Alberta Cubic Metre Scale 'one diameter regression formula method' (m^3).

Small-end diameter in cm (conifer)	Length of log segment in metres																								
	2.4	2.6	2.8	3.0	3.2	3.4	3.6	3.8	4.0	4.2	4.4	4.6	4.8	5.0	5.2	5.4	5.6	5.8	6.0	6.2	6.4	6.6	6.8	7.0	7.2
10	0.026	0.028	0.031	0.034	0.038	0.041	0.044	0.048	0.052	0.056	0.060	0.064	0.068	0.073	0.077	0.082	0.087	0.092	0.098	0.103	0.109	0.115	0.121	0.127	0.134
12	0.034	0.038	0.042	0.045	0.049	0.053	0.058	0.062	0.066	0.071	0.076	0.081	0.086	0.091	0.096	0.102	0.108	0.114	0.120	0.126	0.132	0.139	0.145	0.152	0.159
14	0.044	0.049	0.053	0.058	0.063	0.068	0.073	0.078	0.084	0.089	0.095	0.101	0.106	0.113	0.119	0.125	0.132	0.138	0.145	0.152	0.159	0.167	0.174	0.182	0.190
16	0.056	0.062	0.067	0.073	0.079	0.085	0.091	0.097	0.103	0.110	0.116	0.123	0.130	0.137	0.145	0.152	0.159	0.167	0.175	0.183	0.191	0.200	0.208	0.217	0.226
18	0.069	0.076	0.083	0.089	0.096	0.104	0.111	0.118	0.126	0.133	0.141	0.149	0.157	0.166	0.174	0.183	0.191	0.200	0.209	0.218	0.228	0.237	0.247	0.257	0.267
20	0.084	0.092	0.100	0.108	0.117	0.125	0.133	0.142	0.151	0.160	0.169	0.179	0.188	0.198	0.207	0.217	0.227	0.238	0.248	0.259	0.269	0.280	0.291	0.303	0.314
22	0.101	0.110	0.120	0.129	0.139	0.149	0.159	0.169	0.179	0.190	0.201	0.211	0.222	0.234	0.245	0.256	0.268	0.280	0.292	0.304	0.316	0.329	0.342	0.355	0.368
24	0.119	0.130	0.141	0.152	0.164	0.175	0.187	0.199	0.211	0.223	0.235	0.248	0.261	0.274	0.287	0.300	0.313	0.327	0.341	0.355	0.369	0.384	0.398	0.413	0.428
26	0.139	0.152	0.165	0.178	0.191	0.204	0.218	0.232	0.246	0.260	0.274	0.289	0.303	0.318	0.333	0.349	0.364	0.380	0.396	0.412	0.428	0.445	0.461	0.478	0.496
28	0.161	0.176	0.190	0.205	0.221	0.236	0.252	0.268	0.284	0.300	0.317	0.333	0.350	0.367	0.385	0.402	0.420	0.438	0.457	0.475	0.494	0.513	0.532	0.551	0.571
30	0.185	0.201	0.219	0.236	0.253	0.271	0.289	0.307	0.326	0.344	0.363	0.382	0.402	0.422	0.441	0.462	0.482	0.503	0.524	0.545	0.566	0.588	0.610	0.633	0.655
32	0.210	0.230	0.249	0.269	0.289	0.309	0.329	0.350	0.371	0.393	0.414	0.436	0.458	0.481	0.504	0.527	0.550	0.574	0.598	0.622	0.647	0.672	0.697	0.722	0.748
34	0.238	0.260	0.282	0.304	0.327	0.350	0.373	0.397	0.421	0.445	0.470	0.495	0.520	0.546	0.572	0.598	0.625	0.652	0.679	0.707	0.735	0.763	0.792	0.821	0.851
36	0.268	0.292	0.317	0.343	0.368	0.394	0.421	0.448	0.475	0.502	0.530	0.558	0.587	0.616	0.646	0.676	0.706	0.737	0.768	0.800	0.832	0.864	0.897	0.930	0.964
38	0.300	0.327	0.355	0.384	0.413	0.442	0.472	0.502	0.533	0.564	0.596	0.628	0.660	0.693	0.727	0.761	0.795	0.830	0.865	0.901	0.938	0.975	1.012	1.050	1.088
40	0.334	0.365	0.396	0.428	0.461	0.494	0.527	0.561	0.596	0.631	0.666	0.702	0.739	0.776	0.814	0.853	0.891	0.931	0.971	1.012	1.053	1.095	1.138	1.181	1.225
42	0.370	0.405	0.440	0.476	0.512	0.549	0.586	0.624	0.663	0.702	0.742	0.783	0.824	0.866	0.909	0.952	0.996	1.041	1.086	1.132	1.179	1.227	1.275	1.324	1.373
44	0.409	0.447	0.486	0.526	0.567	0.608	0.650	0.692	0.735	0.780	0.824	0.870	0.916	0.963	1.011	1.060	1.110	1.160	1.211	1.263	1.316	1.370	1.424	1.479	1.536
46	0.450	0.493	0.536	0.580	0.625	0.671	0.717	0.765	0.813	0.862	0.912	0.963	1.015	1.068	1.122	1.176	1.232	1.289	1.346	1.405	1.464	1.525	1.586	1.649	1.712
48	0.494	0.541	0.589	0.637	0.687	0.738	0.790	0.842	0.896	0.951	1.007	1.064	1.121	1.180	1.241	1.302	1.364	1.428	1.492	1.558	1.625	1.693	1.762	1.833	1.904
50	0.540	0.592	0.644	0.698	0.753	0.809	0.867	0.925	0.985	1.046	1.108	1.171	1.235	1.301	1.368	1.437	1.506	1.577	1.650	1.723	1.798	1.875	1.952	2.032	2.112
52	0.589	0.646	0.704	0.763	0.823	0.885	0.949	1.013	1.079	1.147	1.215	1.286	1.357	1.431	1.505	1.581	1.659	1.738	1.819	1.901	1.985	2.071	2.158	2.247	2.338
54	0.640	0.702	0.766	0.831	0.898	0.966	1.036	1.107	1.180	1.254	1.330	1.408	1.488	1.569	1.652	1.736	1.823	1.911	2.001	2.093	2.187	2.282	2.380	2.480	2.581
56	0.695	0.763	0.832	0.904	0.977	1.051	1.128	1.207	1.287	1.369	1.453	1.539	1.627	1.717	1.809	1.903	1.999	2.097	2.197	2.299	2.404	2.510	2.619	2.730	2.844
58	0.752	0.826	0.902	0.980	1.060	1.142	1.226	1.312	1.400	1.491	1.583	1.678	1.775	1.875	1.976	2.080	2.187	2.295	2.406	2.520	2.636	2.755	2.876	3.000	3.126
60	0.812	0.892	0.975	1.060	1.148	1.237	1.330	1.424	1.521	1.620	1.722	1.826	1.933	2.043	2.155	2.270	2.387	2.508	2.631	2.757	2.886	3.017	3.152	3.290	3.431
62	0.875	0.962	1.052	1.145	1.241	1.338	1.439	1.542	1.648	1.757	1.869	1.984	2.101	2.222	2.345	2.472	2.602	2.735	2.871	3.010	3.153	3.299	3.448	3.601	3.757
64	0.941	1.036	1.134	1.234	1.338	1.445	1.555	1.667	1.783	1.902	2.025	2.150	2.280	2.412	2.548	2.687	2.830	2.977	3.127	3.281	3.438	3.600	3.765	3.935	4.108
66	1.010	1.113	1.219	1.328	1.441	1.557	1.677	1.799	1.926	2.056	2.190	2.327	2.469	2.614	2.763	2.916	3.073	3.234	3.400	3.569	3.743	3.922	4.104	4.292	4.483
68	1.082	1.194	1.308	1.427	1.549	1.675	1.805	1.939	2.077	2.218	2.364	2.515	2.669	2.828	2.991	3.159	3.332	3.509	3.691	3.877	4.069	4.265	4.467	4.673	4.885
70	1.158	1.278	1.402	1.530	1.663	1.799	1.940	2.086	2.235	2.390	2.549	2.713	2.882	3.055	3.234	3.418	3.606	3.801	4.000	4.205	4.415	4.632	4.853	5.081	5.314
72	1.237	1.367	1.501	1.639	1.782	1.930	2.082	2.240	2.403	2.571	2.744	2.922	3.106	3.295	3.490	3.691	3.898	4.110	4.329	4.554	4.784	5.022	5.265	5.515	5.772
74	1.320	1.459	1.603	1.753	1.907	2.067	2.232	2.403	2.579	2.761	2.949	3.143	3.343	3.550	3.762	3.981	4.207	4.439	4.678	4.924	5.177	5.437	5.704	5.978	6.260
76	1.406	1.556	1.711	1.872	2.038	2.210	2.389	2.573	2.764	2.962	3.166	3.376	3.594	3.818	4.049	4.288	4.534	4.787	5.048	5.317	5.593	5.878	6.170	6.471	6.780
78	1.496	1.657	1.823	1.996	2.175	2.361	2.553	2.753	2.959	3.173	3.394	3.622	3.858	4.101	4.353	4.612	4.880	5.156	5.440	5.733	6.035	6.346	6.665	6.994	7.332

continued

Table A.1.G. *continued*

Small-end diameter in cm (hardwood)	Length of log segment in metres																							
	2.4	2.6	2.8	3.0	3.2	3.4	3.6	3.8	4.0	4.2	4.4	4.6	4.8	5.0	5.2	5.4	5.6	5.8	6.0	6.2	6.4	6.6	6.8	7.0	7.2
10	0.025	0.028	0.031	0.034	0.037	0.040	0.043	0.047	0.050	0.054	0.058	0.062	0.066	0.070	0.075	0.079	0.084	0.089	0.094	0.099	0.104	0.110	0.115	0.121	0.127
12	0.034	0.037	0.041	0.044	0.048	0.052	0.056	0.060	0.064	0.069	0.073	0.078	0.083	0.088	0.093	0.098	0.103	0.108	0.114	0.120	0.126	0.132	0.138	0.144	0.151
14	0.044	0.048	0.052	0.057	0.061	0.066	0.071	0.076	0.081	0.086	0.091	0.097	0.102	0.108	0.114	0.120	0.126	0.132	0.138	0.145	0.152	0.158	0.165	0.172	0.179
16	0.055	0.060	0.066	0.071	0.077	0.082	0.088	0.094	0.100	0.106	0.113	0.119	0.125	0.132	0.139	0.146	0.153	0.160	0.167	0.174	0.182	0.190	0.198	0.205	0.214
18	0.068	0.074	0.081	0.088	0.094	0.101	0.108	0.115	0.122	0.129	0.137	0.144	0.152	0.160	0.168	0.176	0.184	0.192	0.200	0.209	0.217	0.226	0.235	0.244	0.253
20	0.083	0.090	0.098	0.106	0.114	0.122	0.130	0.139	0.147	0.156	0.164	0.173	0.182	0.191	0.200	0.210	0.219	0.229	0.238	0.248	0.258	0.268	0.279	0.289	0.300
22	0.099	0.108	0.118	0.127	0.136	0.146	0.155	0.165	0.175	0.185	0.195	0.206	0.216	0.227	0.238	0.248	0.259	0.271	0.282	0.293	0.305	0.317	0.329	0.341	0.353
24	0.118	0.128	0.139	0.150	0.161	0.172	0.184	0.195	0.207	0.218	0.230	0.242	0.255	0.267	0.280	0.292	0.305	0.318	0.331	0.344	0.358	0.372	0.385	0.399	0.413
26	0.138	0.150	0.163	0.176	0.188	0.202	0.215	0.228	0.242	0.255	0.269	0.283	0.298	0.312	0.327	0.341	0.356	0.371	0.387	0.402	0.418	0.434	0.449	0.466	0.482
28	0.160	0.174	0.189	0.204	0.219	0.234	0.249	0.265	0.281	0.296	0.313	0.329	0.345	0.362	0.379	0.396	0.414	0.431	0.449	0.467	0.485	0.503	0.522	0.541	0.560
30	0.184	0.201	0.217	0.234	0.252	0.269	0.287	0.305	0.323	0.342	0.360	0.379	0.398	0.418	0.437	0.457	0.477	0.498	0.518	0.539	0.560	0.581	0.603	0.625	0.647
32	0.210	0.229	0.248	0.268	0.288	0.308	0.329	0.349	0.370	0.391	0.413	0.435	0.457	0.479	0.502	0.525	0.548	0.571	0.595	0.619	0.644	0.669	0.694	0.719	0.744
34	0.238	0.260	0.282	0.305	0.327	0.350	0.374	0.398	0.422	0.446	0.471	0.496	0.521	0.547	0.573	0.599	0.626	0.653	0.681	0.709	0.737	0.765	0.794	0.824	0.854
36	0.269	0.294	0.319	0.344	0.370	0.396	0.423	0.450	0.478	0.506	0.534	0.563	0.592	0.621	0.651	0.681	0.712	0.744	0.775	0.807	0.840	0.873	0.907	0.940	0.975
38	0.302	0.330	0.358	0.387	0.416	0.446	0.477	0.507	0.539	0.570	0.603	0.635	0.669	0.703	0.737	0.772	0.807	0.843	0.879	0.916	0.954	0.992	1.031	1.070	1.110
40	0.337	0.369	0.401	0.433	0.466	0.500	0.534	0.569	0.605	0.641	0.678	0.715	0.753	0.791	0.831	0.870	0.911	0.952	0.994	1.036	1.079	1.123	1.168	1.213	1.259
42	0.375	0.410	0.446	0.483	0.520	0.558	0.597	0.636	0.677	0.717	0.759	0.801	0.845	0.888	0.933	0.978	1.025	1.071	1.119	1.168	1.217	1.267	1.318	1.370	1.423
44	0.415	0.455	0.495	0.536	0.578	0.621	0.664	0.709	0.754	0.800	0.847	0.895	0.944	0.994	1.044	1.096	1.149	1.202	1.257	1.312	1.368	1.426	1.484	1.544	1.604
46	0.458	0.502	0.547	0.593	0.640	0.688	0.737	0.787	0.838	0.890	0.943	0.997	1.052	1.108	1.165	1.224	1.284	1.344	1.406	1.469	1.534	1.599	1.666	1.734	1.803
48	0.504	0.553	0.603	0.654	0.707	0.760	0.815	0.870	0.928	0.986	1.045	1.106	1.168	1.232	1.297	1.363	1.430	1.499	1.570	1.641	1.714	1.789	1.865	1.943	2.022
50	0.553	0.607	0.663	0.719	0.778	0.837	0.898	0.960	1.024	1.089	1.156	1.225	1.294	1.366	1.439	1.513	1.590	1.667	1.747	1.828	1.911	1.996	2.082	2.170	2.260
52	0.605	0.665	0.726	0.789	0.853	0.919	0.987	1.057	1.128	1.201	1.275	1.352	1.430	1.510	1.592	1.676	1.762	1.850	1.940	2.031	2.125	2.221	2.319	2.419	2.521
54	0.660	0.725	0.793	0.863	0.934	1.007	1.082	1.160	1.239	1.320	1.403	1.489	1.576	1.666	1.758	1.852	1.948	2.047	2.148	2.252	2.357	2.466	2.576	2.690	2.805
56	0.718	0.790	0.864	0.941	1.020	1.101	1.184	1.270	1.357	1.448	1.540	1.636	1.733	1.833	1.936	2.042	2.150	2.260	2.374	2.490	2.609	2.731	2.856	2.984	3.114
58	0.779	0.858	0.940	1.024	1.111	1.200	1.292	1.387	1.484	1.584	1.687	1.793	1.902	2.013	2.128	2.246	2.366	2.490	2.618	2.748	2.882	3.018	3.159	3.303	3.450
60	0.844	0.930	1.020	1.112	1.208	1.306	1.407	1.512	1.619	1.730	1.844	1.961	2.082	2.206	2.334	2.465	2.600	2.738	2.880	3.026	3.176	3.329	3.487	3.648	3.813
62	0.912	1.006	1.104	1.206	1.310	1.418	1.530	1.645	1.763	1.886	2.012	2.141	2.275	2.413	2.555	2.700	2.850	3.004	3.163	3.326	3.493	3.664	3.841	4.021	4.207
64	0.983	1.087	1.194	1.304	1.419	1.537	1.659	1.786	1.916	2.051	2.190	2.334	2.481	2.634	2.791	2.953	3.119	3.290	3.467	3.648	3.834	4.026	4.222	4.424	4.632
66	1.059	1.171	1.288	1.408	1.534	1.663	1.797	1.936	2.079	2.227	2.380	2.538	2.702	2.870	3.043	3.222	3.407	3.597	3.793	3.994	4.201	4.414	4.633	4.859	5.090
68	1.138	1.260	1.387	1.518	1.655	1.796	1.943	2.095	2.252	2.414	2.583	2.756	2.936	3.122	3.313	3.511	3.715	3.925	4.142	4.365	4.595	4.832	5.076	5.326	5.584
70	1.221	1.354	1.491	1.634	1.783	1.937	2.097	2.263	2.435	2.613	2.797	2.988	3.186	3.390	3.601	3.819	4.044	4.276	4.516	4.763	5.018	5.280	5.550	5.828	6.115
72	1.309	1.452	1.601	1.756	1.918	2.085	2.260	2.441	2.629	2.824	3.026	3.235	3.451	3.676	3.908	4.147	4.395	4.651	4.916	5.188	5.470	5.760	6.059	6.367	6.684
74	1.400	1.555	1.716	1.885	2.060	2.242	2.432	2.629	2.834	3.047	3.267	3.496	3.734	3.979	4.234	4.497	4.770	5.051	5.342	5.643	5.953	6.274	6.604	6.944	7.295
76	1.496	1.663	1.837	2.019	2.209	2.407	2.613	2.828	3.051	3.283	3.523	3.773	4.033	4.302	4.581	4.869	5.168	5.478	5.797	6.128	6.470	6.822	7.186	7.562	7.949
78	1.596	1.776	1.964	2.161	2.367	2.581	2.804	3.037	3.280	3.532	3.794	4.067	4.350	4.644	4.949	5.265	5.592	5.931	6.282	6.645	7.020	7.408	7.809	8.222	8.649

Note: To use this table find the volume at the intersect of the small-end diameter and log length for the appropriate category (conifer or hardwood).

Table A.1.H. Full segment cubic metre volume chart (m^3)

Diameter in cm	Length of log segment in metres																			
	1	2	3	4	5	6	7	8	9	10	11	12	13	14	15	16	17	18	19	20
8	0.005	0.010	0.015	0.020	0.025	0.030	0.035	0.040	0.045	0.050	0.055	0.060	0.065	0.070	0.075	0.080	0.085	0.090	0.096	0.101
9	0.006	0.013	0.019	0.025	0.032	0.038	0.045	0.051	0.057	0.064	0.070	0.076	0.083	0.089	0.095	0.102	0.108	0.115	0.121	0.127
10	0.008	0.016	0.024	0.031	0.039	0.047	0.055	0.063	0.071	0.079	0.086	0.094	0.102	0.110	0.118	0.126	0.134	0.141	0.149	0.157
11	0.010	0.019	0.029	0.038	0.048	0.057	0.067	0.076	0.086	0.095	0.105	0.114	0.124	0.133	0.143	0.152	0.162	0.171	0.181	0.190
12	0.011	0.023	0.034	0.045	0.057	0.068	0.079	0.090	0.102	0.113	0.124	0.136	0.147	0.158	0.170	0.181	0.192	0.204	0.215	0.226
13	0.013	0.027	0.040	0.053	0.066	0.080	0.093	0.106	0.119	0.133	0.146	0.159	0.173	0.186	0.199	0.212	0.226	0.239	0.252	0.265
14	0.015	0.031	0.046	0.062	0.077	0.092	0.108	0.123	0.139	0.154	0.169	0.185	0.200	0.216	0.231	0.246	0.262	0.277	0.292	0.308
15	0.018	0.035	0.053	0.071	0.088	0.106	0.124	0.141	0.159	0.177	0.194	0.212	0.230	0.247	0.265	0.283	0.300	0.318	0.336	0.353
16	0.020	0.040	0.060	0.080	0.101	0.121	0.141	0.161	0.181	0.201	0.221	0.241	0.261	0.281	0.302	0.322	0.342	0.362	0.382	0.402
17	0.023	0.045	0.068	0.091	0.113	0.136	0.159	0.182	0.204	0.227	0.250	0.272	0.295	0.318	0.340	0.363	0.386	0.409	0.431	0.454
18	0.025	0.051	0.076	0.102	0.127	0.153	0.178	0.204	0.229	0.254	0.280	0.305	0.331	0.356	0.382	0.407	0.433	0.458	0.483	0.509
19	0.028	0.057	0.085	0.113	0.142	0.170	0.198	0.227	0.255	0.284	0.312	0.340	0.369	0.397	0.425	0.454	0.482	0.510	0.539	0.567
20	0.031	0.063	0.094	0.126	0.157	0.188	0.220	0.251	0.283	0.314	0.346	0.377	0.408	0.440	0.471	0.503	0.534	0.565	0.597	0.628
21	0.035	0.069	0.104	0.139	0.173	0.208	0.242	0.277	0.312	0.346	0.381	0.416	0.450	0.485	0.520	0.554	0.589	0.623	0.658	0.693
22	0.038	0.076	0.114	0.152	0.190	0.228	0.266	0.304	0.342	0.380	0.418	0.456	0.494	0.532	0.570	0.608	0.646	0.684	0.722	0.760
23	0.042	0.083	0.125	0.166	0.208	0.249	0.291	0.332	0.374	0.415	0.457	0.499	0.540	0.582	0.623	0.665	0.706	0.748	0.789	0.831
24	0.045	0.090	0.136	0.181	0.226	0.271	0.317	0.362	0.407	0.452	0.498	0.543	0.588	0.633	0.679	0.724	0.769	0.814	0.860	0.905
25	0.049	0.098	0.147	0.196	0.245	0.295	0.344	0.393	0.442	0.491	0.540	0.589	0.638	0.687	0.736	0.785	0.834	0.884	0.933	0.982
26	0.053	0.106	0.159	0.212	0.265	0.319	0.372	0.425	0.478	0.531	0.584	0.637	0.690	0.743	0.796	0.849	0.903	0.956	1.009	1.062
27	0.057	0.115	0.172	0.229	0.286	0.344	0.401	0.458	0.515	0.573	0.630	0.687	0.744	0.802	0.859	0.916	0.973	1.031	1.088	1.145
28	0.062	0.123	0.185	0.246	0.308	0.369	0.431	0.493	0.554	0.616	0.677	0.739	0.800	0.862	0.924	0.985	1.047	1.108	1.170	1.232
29	0.066	0.132	0.198	0.264	0.330	0.396	0.462	0.528	0.594	0.661	0.727	0.793	0.859	0.925	0.991	1.057	1.123	1.189	1.255	1.321
30	0.071	0.141	0.212	0.283	0.353	0.424	0.495	0.565	0.636	0.707	0.778	0.848	0.919	0.990	1.060	1.131	1.202	1.272	1.343	1.414
31	0.075	0.151	0.226	0.302	0.377	0.453	0.528	0.604	0.679	0.755	0.830	0.906	0.981	1.057	1.132	1.208	1.283	1.359	1.434	1.510
32	0.080	0.161	0.241	0.322	0.402	0.483	0.563	0.643	0.724	0.804	0.885	0.965	1.046	1.126	1.206	1.287	1.367	1.448	1.528	1.608
33	0.086	0.171	0.257	0.342	0.428	0.513	0.599	0.684	0.770	0.855	0.941	1.026	1.112	1.197	1.283	1.368	1.454	1.540	1.625	1.711
34	0.091	0.182	0.272	0.363	0.454	0.545	0.636	0.726	0.817	0.908	0.999	1.090	1.180	1.271	1.362	1.453	1.543	1.634	1.725	1.816
35	0.096	0.192	0.289	0.385	0.481	0.577	0.673	0.770	0.866	0.962	1.058	1.155	1.251	1.347	1.443	1.539	1.636	1.732	1.828	1.924
36	0.102	0.204	0.305	0.407	0.509	0.611	0.713	0.814	0.916	1.018	1.120	1.221	1.323	1.425	1.527	1.629	1.730	1.832	1.934	2.036
37	0.108	0.215	0.323	0.430	0.538	0.645	0.753	0.860	0.968	1.075	1.183	1.290	1.398	1.505	1.613	1.720	1.828	1.935	2.043	2.150
38	0.113	0.227	0.340	0.454	0.567	0.680	0.794	0.907	1.021	1.134	1.248	1.361	1.474	1.588	1.701	1.815	1.928	2.041	2.155	2.268
39	0.119	0.239	0.358	0.478	0.597	0.717	0.836	0.956	1.075	1.195	1.314	1.434	1.553	1.672	1.792	1.911	2.031	2.150	2.270	2.389
40	0.126	0.251	0.377	0.503	0.628	0.754	0.880	1.005	1.131	1.257	1.382	1.508	1.634	1.759	1.885	2.011	2.136	2.262	2.388	2.513
41	0.132	0.264	0.396	0.528	0.660	0.792	0.924	1.056	1.188	1.320	1.452	1.584	1.716	1.848	1.980	2.112	2.244	2.376	2.508	2.641
42	0.139	0.277	0.416	0.554	0.693	0.831	0.970	1.108	1.247	1.385	1.524	1.663	1.801	1.940	2.078	2.217	2.355	2.494	2.632	2.771

continued

Table A.1.H. *continued*

Diameter in cm	Length of log segment in metres																			
	1	2	3	4	5	6	7	8	9	10	11	12	13	14	15	16	17	18	19	20
43	0.145	0.290	0.436	0.581	0.726	0.871	1.017	1.162	1.307	1.452	1.597	1.743	1.888	2.033	2.178	2.324	2.469	2.614	2.759	2.904
44	0.152	0.304	0.456	0.608	0.760	0.912	1.064	1.216	1.368	1.521	1.673	1.825	1.977	2.129	2.281	2.433	2.585	2.737	2.889	3.041
45	0.159	0.318	0.477	0.636	0.795	0.954	1.113	1.272	1.431	1.590	1.749	1.909	2.068	2.227	2.386	2.545	2.704	2.863	3.022	3.181
46	0.166	0.332	0.499	0.665	0.831	0.997	1.163	1.330	1.496	1.662	1.828	1.994	2.160	2.327	2.493	2.659	2.825	2.991	3.158	3.324
47	0.173	0.347	0.520	0.694	0.867	1.041	1.214	1.388	1.561	1.735	1.908	2.082	2.255	2.429	2.602	2.776	2.949	3.123	3.296	3.470
48	0.181	0.362	0.543	0.724	0.905	1.086	1.267	1.448	1.629	1.810	1.991	2.171	2.352	2.533	2.714	2.895	3.076	3.257	3.438	3.619
49	0.189	0.377	0.566	0.754	0.943	1.131	1.320	1.509	1.697	1.886	2.074	2.263	2.451	2.640	2.829	3.017	3.206	3.394	3.583	3.771
50	0.196	0.393	0.589	0.785	0.982	1.178	1.374	1.571	1.767	1.964	2.160	2.356	2.553	2.749	2.945	3.142	3.338	3.534	3.731	3.927
51	0.204	0.409	0.613	0.817	1.021	1.226	1.430	1.634	1.839	2.043	2.247	2.451	2.656	2.860	3.064	3.269	3.473	3.677	3.881	4.086
52	0.212	0.425	0.637	0.849	1.062	1.274	1.487	1.699	1.911	2.124	2.336	2.548	2.761	2.973	3.186	3.398	3.610	3.823	4.035	4.247
53	0.221	0.441	0.662	0.882	1.103	1.324	1.544	1.765	1.986	2.206	2.427	2.647	2.868	3.089	3.309	3.530	3.751	3.971	4.192	4.412
54	0.229	0.458	0.687	0.916	1.145	1.374	1.603	1.832	2.061	2.290	2.519	2.748	2.977	3.206	3.435	3.664	3.893	4.122	4.351	4.580
55	0.238	0.475	0.713	0.950	1.188	1.426	1.663	1.901	2.138	2.376	2.613	2.851	3.089	3.326	3.564	3.801	4.039	4.277	4.514	4.752
56	0.246	0.493	0.739	0.985	1.232	1.478	1.724	1.970	2.217	2.463	2.709	2.956	3.202	3.448	3.695	3.941	4.187	4.433	4.680	4.926
57	0.255	0.510	0.766	1.021	1.276	1.531	1.786	2.041	2.297	2.552	2.807	3.062	3.317	3.572	3.828	4.083	4.338	4.593	4.848	5.104
58	0.264	0.528	0.793	1.057	1.321	1.585	1.849	2.114	2.378	2.642	2.906	3.171	3.435	3.699	3.963	4.227	4.492	4.756	5.020	5.284
59	0.273	0.547	0.820	1.094	1.367	1.640	1.914	2.187	2.461	2.734	3.007	3.281	3.554	3.828	4.101	4.374	4.648	4.921	5.195	5.468
60	0.283	0.565	0.848	1.131	1.414	1.696	1.979	2.262	2.545	2.827	3.110	3.393	3.676	3.958	4.241	4.524	4.807	5.089	5.372	5.655
61	0.292	0.584	0.877	1.169	1.461	1.753	2.046	2.338	2.630	2.922	3.215	3.507	3.799	4.091	4.384	4.676	4.968	5.260	5.553	5.845
62	0.302	0.604	0.906	1.208	1.510	1.811	2.113	2.415	2.717	3.019	3.321	3.623	3.925	4.227	4.529	4.831	5.132	5.434	5.736	6.038
63	0.312	0.623	0.935	1.247	1.559	1.870	2.182	2.494	2.806	3.117	3.429	3.741	4.052	4.364	4.676	4.988	5.299	5.611	5.923	6.235
64	0.322	0.643	0.965	1.287	1.608	1.930	2.252	2.574	2.895	3.217	3.539	3.860	4.182	4.504	4.825	5.147	5.469	5.791	6.112	6.434
65	0.332	0.664	0.995	1.327	1.659	1.991	2.323	2.655	2.986	3.318	3.650	3.982	4.314	4.646	4.977	5.309	5.641	5.973	6.305	6.637
66	0.342	0.684	1.026	1.368	1.711	2.053	2.395	2.737	3.079	3.421	3.763	4.105	4.448	4.790	5.132	5.474	5.816	6.158	6.500	6.842
67	0.353	0.705	1.058	1.410	1.763	2.115	2.468	2.821	3.173	3.526	3.878	4.231	4.583	4.936	5.288	5.641	5.994	6.346	6.699	7.051
68	0.363	0.726	1.090	1.453	1.816	2.179	2.542	2.905	3.269	3.632	3.995	4.358	4.721	5.084	5.448	5.811	6.174	6.537	6.900	7.263
69	0.374	0.748	1.122	1.496	1.870	2.244	2.618	2.991	3.365	3.739	4.113	4.487	4.861	5.235	5.609	5.983	6.357	6.731	7.105	7.479
70	0.385	0.770	1.155	1.539	1.924	2.309	2.694	3.079	3.464	3.848	4.233	4.618	5.003	5.388	5.773	6.158	6.542	6.927	7.312	7.697
71	0.396	0.792	1.188	1.584	1.980	2.376	2.771	3.167	3.563	3.959	4.355	4.751	5.147	5.543	5.939	6.335	6.731	7.127	7.522	7.918
72	0.407	0.814	1.221	1.629	2.036	2.443	2.850	3.257	3.664	4.072	4.479	4.886	5.293	5.700	6.107	6.514	6.922	7.329	7.736	8.143
73	0.419	0.837	1.256	1.674	2.093	2.511	2.930	3.348	3.767	4.185	4.604	5.022	5.441	5.860	6.278	6.697	7.115	7.534	7.952	8.371
74	0.430	0.860	1.290	1.720	2.150	2.581	3.011	3.441	3.871	4.301	4.731	5.161	5.591	6.021	6.451	6.881	7.311	7.742	8.172	8.602
75	0.442	0.884	1.325	1.767	2.209	2.651	3.093	3.534	3.976	4.418	4.860	5.301	5.743	6.185	6.627	7.069	7.510	7.952	8.394	8.836

Note: To use this table find the volume at the intersect of the mean diameter and log length. For lengths which are not on the full metre, use a percentage of the full metre lengths. Example: for a log which is 12.4 m long with a mean diameter of 40 cm, m^3 is determined as follows: 1.508 m^3 for 12 m + 0.050 (which is 1/10th of a 4 m long log) = **1.558 m^3**.

Table A.1.I. Swedish cubic volume chart (m^3).

Length in m	Small-end diameter of log in cm																															
	14	15	16	17	18	19	20	21	22	23	24	25	26	27	28	29	30	31	32	33	34	35	36	37	38	39	40	41	42	43	44	45
2.0	0.034	0.039	0.044	0.049	0.055	0.061	0.068	0.074	0.082	0.089	0.097	0.105	0.113	0.122	0.131	0.140	0.150	0.160	0.170	0.181	0.192	0.203	0.215	0.226	0.239	0.251	0.264	0.277	0.291	0.305	0.319	0.333
2.1	0.036	0.041	0.046	0.052	0.058	0.064	0.071	0.078	0.085	0.093	0.101	0.110	0.119	0.128	0.137	0.147	0.157	0.168	0.178	0.190	0.201	0.213	0.225	0.237	0.250	0.263	0.277	0.291	0.305	0.320	0.334	0.350
2.2	0.037	0.042	0.048	0.054	0.060	0.067	0.074	0.082	0.089	0.098	0.106	0.115	0.124	0.134	0.144	0.154	0.164	0.175	0.187	0.198	0.210	0.223	0.235	0.249	0.262	0.276	0.290	0.304	0.319	0.334	0.350	0.366
2.3	0.039	0.044	0.050	0.057	0.063	0.070	0.078	0.085	0.093	0.102	0.111	0.120	0.130	0.140	0.150	0.161	0.172	0.183	0.195	0.207	0.220	0.233	0.246	0.260	0.274	0.288	0.303	0.318	0.333	0.349	0.365	0.382
2.4	0.040	0.046	0.052	0.059	0.066	0.073	0.081	0.089	0.097	0.106	0.116	0.125	0.135	0.146	0.156	0.167	0.179	0.191	0.203	0.216	0.229	0.243	0.256	0.271	0.285	0.300	0.316	0.331	0.348	0.364	0.381	0.398
2.5	0.042	0.048	0.055	0.061	0.069	0.076	0.084	0.093	0.101	0.111	0.120	0.130	0.141	0.151	0.163	0.174	0.186	0.199	0.212	0.225	0.238	0.252	0.267	0.282	0.297	0.312	0.329	0.345	0.362	0.379	0.397	0.415
2.6	0.044	0.050	0.057	0.064	0.071	0.079	0.087	0.096	0.105	0.115	0.125	0.135	0.146	0.157	0.169	0.181	0.194	0.207	0.220	0.234	0.248	0.262	0.277	0.293	0.309	0.325	0.341	0.358	0.376	0.394	0.412	0.431
2.7	0.045	0.052	0.059	0.066	0.074	0.082	0.091	0.100	0.109	0.119	0.130	0.140	0.152	0.163	0.175	0.188	0.201	0.214	0.228	0.242	0.257	0.272	0.288	0.304	0.320	0.337	0.354	0.372	0.390	0.409	0.428	0.447
2.8	0.047	0.054	0.061	0.069	0.077	0.085	0.094	0.103	0.113	0.124	0.134	0.146	0.157	0.169	0.182	0.195	0.208	0.222	0.236	0.251	0.266	0.282	0.298	0.315	0.332	0.349	0.367	0.386	0.404	0.424	0.443	0.463
2.9	0.049	0.056	0.063	0.071	0.079	0.088	0.097	0.107	0.117	0.128	0.139	0.151	0.163	0.175	0.188	0.202	0.216	0.230	0.245	0.260	0.276	0.292	0.309	0.326	0.343	0.361	0.380	0.399	0.418	0.438	0.459	0.480
3.0	0.050	0.058	0.065	0.073	0.082	0.091	0.101	0.111	0.121	0.132	0.144	0.156	0.168	0.181	0.195	0.208	0.223	0.238	0.253	0.269	0.285	0.302	0.319	0.337	0.355	0.374	0.393	0.413	0.433	0.453	0.474	0.496
3.1	0.052	0.059	0.067	0.076	0.085	0.094	0.104	0.114	0.125	0.137	0.149	0.161	0.174	0.187	0.201	0.215	0.230	0.245	0.261	0.278	0.294	0.312	0.330	0.348	0.367	0.386	0.406	0.426	0.447	0.468	0.490	0.512
3.2	0.054	0.061	0.069	0.078	0.087	0.097	0.107	0.118	0.129	0.141	0.153	0.166	0.179	0.193	0.207	0.222	0.237	0.253	0.270	0.286	0.304	0.322	0.340	0.359	0.378	0.398	0.419	0.440	0.461	0.483	0.505	0.528
3.3	0.055	0.063	0.072	0.081	0.090	0.100	0.111	0.122	0.133	0.145	0.158	0.171	0.185	0.199	0.214	0.229	0.245	0.261	0.278	0.295	0.313	0.332	0.351	0.370	0.390	0.411	0.432	0.453	0.475	0.498	0.521	0.545
3.4	0.057	0.065	0.074	0.083	0.093	0.103	0.114	0.125	0.137	0.150	0.163	0.176	0.190	0.205	0.220	0.236	0.252	0.269	0.286	0.304	0.323	0.341	0.361	0.381	0.402	0.423	0.444	0.467	0.489	0.513	0.537	0.561
3.5	0.059	0.067	0.076	0.085	0.095	0.106	0.117	0.129	0.141	0.154	0.167	0.181	0.196	0.211	0.226	0.243	0.259	0.277	0.295	0.313	0.332	0.351	0.371	0.392	0.413	0.435	0.457	0.480	0.504	0.528	0.552	0.577
3.6	0.060	0.069	0.078	0.088	0.098	0.109	0.120	0.133	0.145	0.158	0.172	0.186	0.201	0.217	0.233	0.249	0.267	0.284	0.303	0.322	0.341	0.361	0.382	0.403	0.425	0.447	0.470	0.494	0.518	0.542	0.568	0.593
3.7	0.062	0.071	0.080	0.090	0.101	0.112	0.124	0.136	0.149	0.163	0.177	0.192	0.207	0.223	0.239	0.256	0.274	0.292	0.311	0.331	0.351	0.371	0.392	0.414	0.437	0.460	0.483	0.507	0.532	0.557	0.583	0.610
3.8	0.064	0.073	0.082	0.093	0.103	0.115	0.127	0.140	0.153	0.167	0.182	0.197	0.212	0.229	0.246	0.263	0.281	0.300	0.319	0.339	0.360	0.381	0.403	0.425	0.448	0.472	0.496	0.521	0.546	0.572	0.599	0.626
3.9	0.065	0.075	0.084	0.095	0.106	0.118	0.130	0.143	0.157	0.171	0.186	0.202	0.218	0.235	0.252	0.270	0.289	0.308	0.328	0.348	0.369	0.391	0.413	0.436	0.460	0.484	0.509	0.534	0.560	0.587	0.614	0.642
4.0	0.067	0.076	0.087	0.097	0.109	0.121	0.134	0.147	0.161	0.176	0.191	0.207	0.223	0.241	0.258	0.277	0.296	0.316	0.336	0.357	0.379	0.401	0.424	0.447	0.471	0.496	0.522	0.548	0.575	0.602	0.630	0.659
4.1	0.069	0.078	0.089	0.100	0.112	0.124	0.137	0.151	0.165	0.180	0.196	0.212	0.229	0.246	0.265	0.284	0.303	0.323	0.344	0.366	0.388	0.411	0.434	0.458	0.483	0.509	0.535	0.561	0.589	0.617	0.645	0.675
4.2	0.070	0.080	0.091	0.102	0.114	0.127	0.140	0.154	0.169	0.184	0.200	0.217	0.234	0.252	0.271	0.290	0.311	0.331	0.353	0.375	0.397	0.421	0.445	0.469	0.495	0.521	0.548	0.575	0.603	0.632	0.661	0.691
4.3	0.072	0.082	0.093	0.105	0.117	0.130	0.144	0.158	0.173	0.189	0.205	0.222	0.240	0.258	0.278	0.297	0.318	0.339	0.361	0.383	0.407	0.431	0.455	0.480	0.506	0.533	0.560	0.588	0.617	0.646	0.677	0.707
4.4	0.073	0.084	0.095	0.107	0.120	0.133	0.147	0.162	0.177	0.193	0.210	0.227	0.245	0.264	0.284	0.304	0.325	0.347	0.369	0.392	0.416	0.440	0.466	0.491	0.518	0.545	0.573	0.602	0.631	0.661	0.692	0.724
4.5	0.075	0.086	0.097	0.109	0.122	0.136	0.150	0.165	0.181	0.197	0.215	0.232	0.251	0.270	0.290	0.311	0.332	0.355	0.377	0.401	0.425	0.450	0.476	0.503	0.530	0.558	0.586	0.615	0.645	0.676	0.708	0.740
4.6	0.077	0.088	0.099	0.112	0.125	0.139	0.153	0.169	0.185	0.202	0.219	0.237	0.256	0.276	0.297	0.318	0.340	0.362	0.386	0.410	0.435	0.460	0.487	0.514	0.541	0.570	0.599	0.629	0.660	0.691	0.723	0.756
4.7	0.078	0.090	0.102	0.114	0.128	0.142	0.157	0.172	0.189	0.206	0.224	0.243	0.262	0.282	0.303	0.325	0.347	0.370	0.394	0.419	0.444	0.470	0.497	0.525	0.553	0.582	0.612	0.643	0.674	0.706	0.739	0.772
4.8	0.080	0.092	0.104	0.117	0.130	0.145	0.160	0.176	0.193	0.210	0.229	0.248	0.268	0.288	0.309	0.331	0.354	0.378	0.402	0.427	0.453	0.480	0.507	0.536	0.565	0.594	0.625	0.656	0.688	0.721	0.754	0.789
4.9	0.082	0.093	0.106	0.119	0.133	0.148	0.163	0.180	0.197	0.215	0.233	0.253	0.273	0.294	0.316	0.338	0.362	0.386	0.411	0.436	0.463	0.490	0.518	0.547	0.576	0.607	0.638	0.670	0.702	0.736	0.770	0.805
5.0	0.083	0.095	0.108	0.121	0.136	0.151	0.167	0.183	0.201	0.219	0.238	0.258	0.279	0.300	0.322	0.345	0.369	0.394	0.419	0.445	0.472	0.500	0.528	0.558	0.588	0.619	0.651	0.683	0.716	0.751	0.785	0.821
5.1	0.085	0.097	0.110	0.124	0.138	0.154	0.170	0.187	0.205	0.223	0.243	0.263	0.284	0.306	0.329	0.352	0.376	0.401	0.427	0.454	0.481	0.510	0.539	0.569	0.600	0.631	0.663	0.697	0.731	0.765	0.801	0.837
5.2	0.087	0.099	0.112	0.126	0.141	0.157	0.173	0.191	0.209	0.228	0.248	0.268	0.290	0.312	0.335	0.359	0.384	0.409	0.436	0.463	0.491	0.520	0.549	0.580	0.611	0.643	0.676	0.710	0.745	0.780	0.817	0.854
5.3	0.088	0.101	0.114	0.129	0.144	0.160	0.177	0.194	0.213	0.232	0.252	0.273	0.295	0.318	0.341	0.366	0.391	0.417	0.444	0.472	0.500	0.530	0.560	0.591	0.623	0.656	0.689	0.724	0.759	0.795	0.832	0.870
5.4	0.090	0.103	0.117	0.131	0.146	0.163	0.180	0.198	0.217	0.236	0.257	0.278	0.301	0.324	0.348	0.373	0.398	0.425	0.452	0.480	0.509	0.539	0.570	0.602	0.634	0.668	0.702	0.737	0.773	0.810	0.848	0.886
5.5	0.092	0.105	0.119	0.133	0.149	0.166	0.183	0.201	0.221	0.241	0.262	0.283	0.306	0.330	0.354	0.379	0.405	0.433	0.460	0.489	0.519	0.549	0.581	0.613	0.646	0.680	0.715	0.751	0.787	0.825	0.863	0.902
5.6	0.093	0.107	0.121	0.136	0.152	0.169	0.186	0.205	0.225	0.245	0.266	0.289	0.312	0.336	0.360	0.386	0.413	0.440	0.469	0.498	0.528	0.559	0.591	0.624	0.658	0.692	0.728	0.764	0.802	0.840	0.879	0.919
5.7	0.095	0.108	0.123	0.138	0.155	0.172	0.190	0.209	0.229	0.249	0.271	0.294	0.317	0.342	0.367	0.393	0.420	0.448	0.477	0.507	0.538	0.569	0.602	0.635	0.669	0.705	0.741	0.778	0.816	0.855	0.894	0.935
5.8	0.097	0.110	0.125	0.141	0.157	0.175	0.193	0.212	0.233	0.254	0.276	0.299	0.323	0.347	0.373	0.400	0.427	0.456	0.485	0.516	0.547	0.579	0.612	0.646	0.681	0.717	0.754	0.791	0.830	0.869	0.910	0.951
5.9	0.098	0.112	0.127	0.143	0.160	0.178	0.196	0.216	0.237	0.258	0.281	0.304	0.328	0.353	0.380	0.407	0.435	0.464	0.494	0.524	0.556	0.589	0.623	0.657	0.693	0.729	0.767	0.805	0.844	0.884	0.925	0.967
6.0	0.100	0.114	0.129	0.146	0.163	0.181	0.200	0.220	0.241	0.262	0.285	0.309	0.334	0.359	0.386	0.414	0.442	0.471	0.502	0.533	0.566	0.599	0.633	0.668	0.704	0.741	0.779	0.818	0.858	0.899	0.941	0.984

Note: The volume is located at the intersect of the small-end diameter (top of the table) and the log length (left side of the table).

Table A.1.J. New Zealand 3-D volume chart (m^3).

Small-end diameter in cm	0.8 cm of taper per metre of log length												1.2 cm of taper per metre of log length											
	Log length in metres												Log length in metres											
	1.0	2.0	3.0	4.0	5.0	6.0	7.0	8.0	9.0	10.0	11.0	12.0	1.0	2.0	3.0	4.0	5.0	6.0	7.0	8.0	9.0	10.0	11.0	12.0
14	0.017	0.035	0.056	0.079	0.104	0.131	0.161	0.192	0.225	0.260	0.296	0.335	0.017	0.037	0.061	0.087	0.116	0.148	0.183	0.221	0.262	0.305	0.351	0.400
15	0.019	0.040	0.063	0.089	0.117	0.146	0.178	0.212	0.248	0.286	0.325	0.367	0.019	0.042	0.068	0.097	0.129	0.164	0.202	0.242	0.286	0.332	0.382	0.434
16	0.021	0.045	0.071	0.099	0.130	0.162	0.197	0.233	0.272	0.313	0.356	0.401	0.022	0.047	0.076	0.107	0.142	0.180	0.221	0.265	0.312	0.362	0.414	0.470
17	0.024	0.050	0.079	0.110	0.144	0.179	0.217	0.256	0.298	0.342	0.389	0.437	0.025	0.053	0.084	0.119	0.156	0.197	0.242	0.289	0.339	0.392	0.449	0.508
18	0.027	0.056	0.088	0.122	0.158	0.197	0.238	0.281	0.326	0.373	0.423	0.475	0.027	0.058	0.093	0.130	0.172	0.216	0.263	0.314	0.368	0.425	0.485	0.548
19	0.030	0.062	0.097	0.134	0.174	0.216	0.260	0.306	0.355	0.406	0.459	0.515	0.030	0.064	0.102	0.143	0.187	0.235	0.286	0.341	0.398	0.459	0.523	0.590
20	0.033	0.068	0.106	0.147	0.190	0.235	0.283	0.333	0.386	0.441	0.498	0.557	0.033	0.071	0.112	0.156	0.204	0.255	0.310	0.369	0.430	0.495	0.563	0.635
21	0.036	0.075	0.116	0.160	0.207	0.256	0.308	0.361	0.418	0.477	0.538	0.601	0.037	0.077	0.122	0.170	0.222	0.277	0.336	0.398	0.464	0.533	0.605	0.681
22	0.040	0.082	0.127	0.175	0.225	0.278	0.333	0.391	0.451	0.514	0.580	0.647	0.040	0.084	0.133	0.184	0.240	0.299	0.362	0.429	0.499	0.572	0.649	0.730
23	0.043	0.089	0.138	0.189	0.244	0.300	0.360	0.422	0.487	0.554	0.623	0.696	0.044	0.092	0.144	0.200	0.259	0.323	0.390	0.461	0.535	0.613	0.695	0.781
24	0.047	0.097	0.149	0.205	0.263	0.324	0.388	0.454	0.523	0.595	0.669	0.746	0.048	0.099	0.155	0.215	0.279	0.347	0.419	0.494	0.573	0.656	0.743	0.833
25	0.051	0.105	0.161	0.221	0.284	0.349	0.417	0.488	0.561	0.638	0.717	0.798	0.051	0.107	0.167	0.232	0.300	0.372	0.449	0.529	0.613	0.701	0.793	0.889
26	0.055	0.113	0.174	0.238	0.305	0.375	0.447	0.523	0.601	0.682	0.766	0.853	0.056	0.116	0.180	0.249	0.322	0.399	0.480	0.565	0.654	0.747	0.845	0.946
27	0.059	0.121	0.187	0.255	0.327	0.401	0.479	0.559	0.642	0.728	0.817	0.909	0.060	0.124	0.193	0.267	0.344	0.426	0.512	0.603	0.697	0.796	0.898	1.005
28	0.063	0.130	0.200	0.273	0.350	0.429	0.511	0.597	0.685	0.776	0.871	0.968	0.064	0.133	0.207	0.285	0.368	0.455	0.546	0.642	0.741	0.846	0.954	1.066
29	0.068	0.139	0.214	0.292	0.373	0.458	0.545	0.636	0.729	0.826	0.926	1.028	0.069	0.142	0.221	0.304	0.392	0.484	0.581	0.682	0.787	0.897	1.012	1.130
30	0.073	0.149	0.228	0.311	0.398	0.487	0.580	0.676	0.775	0.877	0.983	1.091	0.073	0.152	0.236	0.324	0.417	0.514	0.617	0.724	0.835	0.951	1.071	1.196
31	0.077	0.159	0.243	0.331	0.423	0.518	0.616	0.718	0.822	0.930	1.042	1.156	0.078	0.162	0.251	0.344	0.443	0.546	0.654	0.767	0.884	1.006	1.133	1.264
32	0.082	0.169	0.259	0.352	0.449	0.550	0.653	0.761	0.871	0.985	1.102	1.223	0.083	0.172	0.266	0.365	0.469	0.579	0.692	0.811	0.935	1.063	1.196	1.334
33	0.088	0.179	0.275	0.374	0.476	0.582	0.692	0.805	0.922	1.042	1.165	1.292	0.089	0.183	0.282	0.387	0.497	0.612	0.732	0.857	0.987	1.122	1.262	1.407
34	0.093	0.190	0.291	0.396	0.504	0.616	0.732	0.851	0.974	1.100	1.230	1.363	0.094	0.194	0.299	0.409	0.525	0.647	0.773	0.905	1.041	1.183	1.330	1.482
35	0.098	0.201	0.308	0.418	0.533	0.651	0.772	0.898	1.027	1.160	1.297	1.437	0.099	0.205	0.316	0.433	0.555	0.682	0.815	0.953	1.097	1.245	1.399	1.558
36	0.104	0.213	0.325	0.442	0.562	0.686	0.814	0.946	1.082	1.222	1.365	1.512	0.105	0.216	0.333	0.456	0.585	0.719	0.858	1.004	1.154	1.310	1.471	1.638
37	0.110	0.224	0.343	0.465	0.592	0.723	0.858	0.996	1.139	1.285	1.436	1.590	0.111	0.228	0.352	0.481	0.616	0.757	0.903	1.055	1.213	1.376	1.545	1.719
38	0.116	0.236	0.361	0.490	0.623	0.761	0.902	1.048	1.197	1.351	1.508	1.670	0.117	0.240	0.370	0.506	0.648	0.795	0.949	1.108	1.273	1.444	1.621	1.803
39	0.122	0.249	0.380	0.515	0.655	0.799	0.948	1.100	1.257	1.418	1.583	1.752	0.123	0.253	0.389	0.532	0.680	0.835	0.996	1.163	1.335	1.514	1.699	1.889
40	0.128	0.261	0.399	0.541	0.688	0.839	0.995	1.154	1.318	1.487	1.659	1.836	0.129	0.266	0.409	0.558	0.714	0.876	1.044	1.219	1.399	1.586	1.778	1.977
41	0.135	0.274	0.419	0.568	0.722	0.880	1.043	1.210	1.381	1.557	1.737	1.922	0.136	0.279	0.429	0.585	0.748	0.918	1.094	1.276	1.465	1.659	1.860	2.068
42	0.141	0.288	0.439	0.595	0.756	0.922	1.092	1.267	1.446	1.630	1.818	2.010	0.143	0.292	0.449	0.613	0.784	0.961	1.144	1.335	1.532	1.735	1.945	2.161
43	0.148	0.301	0.460	0.623	0.791	0.964	1.142	1.325	1.512	1.704	1.900	2.101	0.149	0.306	0.470	0.642	0.820	1.005	1.197	1.395	1.600	1.812	2.031	2.256
44	0.155	0.315	0.481	0.652	0.828	1.008	1.194	1.384	1.580	1.780	1.985	2.194	0.156	0.320	0.492	0.671	0.857	1.050	1.250	1.457	1.671	1.892	2.119	2.353
45	0.162	0.330	0.503	0.681	0.865	1.053	1.247	1.445	1.649	1.857	2.071	2.289	0.163	0.335	0.514	0.701	0.895	1.096	1.304	1.520	1.743	1.973	2.210	2.453
46	0.169	0.344	0.525	0.711	0.902	1.099	1.301	1.508	1.720	1.937	2.159	2.386	0.171	0.350	0.536	0.731	0.933	1.143	1.360	1.585	1.817	2.056	2.302	2.556

Small-end diameter in cm																								
14	0.018	0.039	0.065	0.094	0.128	0.165	0.205	0.249	0.297	0.349	0.404	0.463	0.018	0.041	0.069	0.102	0.139	0.180	0.226	0.277	0.332	0.391	0.455	0.523
15	0.020	0.044	0.072	0.105	0.141	0.180	0.224	0.272	0.323	0.378	0.436	0.498	0.020	0.046	0.077	0.112	0.152	0.197	0.246	0.300	0.358	0.421	0.489	0.561
16	0.022	0.049	0.080	0.115	0.154	0.197	0.244	0.295	0.350	0.408	0.470	0.536	0.023	0.051	0.085	0.123	0.166	0.214	0.267	0.324	0.386	0.453	0.525	0.601
17	0.025	0.055	0.089	0.127	0.169	0.215	0.265	0.320	0.378	0.440	0.506	0.576	0.026	0.057	0.093	0.135	0.181	0.232	0.289	0.350	0.416	0.487	0.562	0.643
18	0.028	0.061	0.097	0.139	0.184	0.234	0.288	0.346	0.408	0.474	0.544	0.619	0.028	0.063	0.102	0.147	0.197	0.252	0.312	0.377	0.447	0.522	0.602	0.687
19	0.031	0.067	0.107	0.152	0.201	0.254	0.312	0.374	0.440	0.510	0.584	0.663	0.031	0.069	0.112	0.160	0.213	0.272	0.336	0.406	0.480	0.559	0.644	0.733
20	0.034	0.073	0.117	0.165	0.218	0.275	0.337	0.403	0.473	0.547	0.626	0.710	0.035	0.075	0.122	0.174	0.231	0.294	0.362	0.436	0.514	0.598	0.688	0.782
21	0.037	0.080	0.127	0.179	0.236	0.297	0.363	0.433	0.508	0.587	0.670	0.758	0.038	0.082	0.132	0.188	0.249	0.316	0.389	0.467	0.550	0.639	0.733	0.833
22	0.041	0.087	0.138	0.194	0.254	0.320	0.390	0.465	0.544	0.628	0.716	0.809	0.041	0.089	0.143	0.203	0.268	0.340	0.417	0.500	0.588	0.682	0.781	0.886
23	0.044	0.094	0.149	0.209	0.274	0.344	0.418	0.498	0.582	0.671	0.764	0.862	0.045	0.097	0.155	0.219	0.289	0.364	0.446	0.534	0.627	0.726	0.831	0.942
24	0.048	0.102	0.161	0.225	0.295	0.369	0.448	0.532	0.621	0.715	0.814	0.918	0.049	0.104	0.167	0.235	0.309	0.390	0.477	0.570	0.668	0.773	0.883	1.000
25	0.052	0.110	0.173	0.242	0.316	0.395	0.479	0.568	0.663	0.762	0.866	0.975	0.053	0.113	0.179	0.252	0.331	0.417	0.509	0.607	0.711	0.821	0.937	1.060
26	0.056	0.118	0.186	0.259	0.338	0.422	0.511	0.606	0.705	0.810	0.920	1.035	0.057	0.121	0.192	0.270	0.354	0.445	0.542	0.645	0.755	0.871	0.993	1.122
27	0.061	0.127	0.199	0.277	0.361	0.450	0.545	0.644	0.750	0.860	0.976	1.097	0.061	0.130	0.206	0.288	0.377	0.473	0.576	0.685	0.801	0.923	1.052	1.187
28	0.065	0.136	0.213	0.296	0.385	0.479	0.579	0.685	0.796	0.912	1.034	1.161	0.066	0.139	0.219	0.307	0.402	0.503	0.612	0.727	0.848	0.977	1.112	1.254
29	0.070	0.145	0.228	0.316	0.410	0.509	0.615	0.726	0.843	0.966	1.094	1.228	0.070	0.148	0.234	0.327	0.427	0.534	0.648	0.770	0.898	1.033	1.174	1.323
30	0.074	0.155	0.242	0.336	0.435	0.541	0.652	0.769	0.893	1.022	1.156	1.297	0.075	0.158	0.249	0.347	0.453	0.566	0.686	0.814	0.949	1.090	1.239	1.395
31	0.079	0.165	0.258	0.357	0.462	0.573	0.690	0.814	0.944	1.079	1.221	1.368	0.080	0.168	0.264	0.368	0.480	0.599	0.726	0.860	1.001	1.150	1.306	1.469
32	0.084	0.175	0.273	0.378	0.489	0.606	0.730	0.860	0.996	1.138	1.287	1.441	0.085	0.179	0.280	0.390	0.508	0.633	0.766	0.907	1.055	1.211	1.375	1.545
33	0.089	0.186	0.290	0.400	0.517	0.641	0.771	0.907	1.050	1.200	1.355	1.517	0.090	0.189	0.297	0.413	0.537	0.669	0.808	0.956	1.112	1.275	1.446	1.624
34	0.095	0.197	0.307	0.423	0.546	0.676	0.813	0.956	1.106	1.263	1.426	1.595	0.096	0.200	0.314	0.436	0.566	0.705	0.852	1.006	1.169	1.340	1.519	1.706
35	0.100	0.208	0.324	0.446	0.576	0.713	0.856	1.007	1.164	1.328	1.498	1.676	0.101	0.212	0.331	0.460	0.597	0.742	0.896	1.058	1.229	1.407	1.594	1.789
36	0.106	0.220	0.342	0.471	0.607	0.750	0.901	1.058	1.223	1.395	1.573	1.758	0.107	0.224	0.350	0.484	0.628	0.781	0.942	1.112	1.290	1.477	1.672	1.875
37	0.112	0.232	0.360	0.495	0.638	0.789	0.947	1.112	1.284	1.463	1.650	1.843	0.113	0.236	0.368	0.510	0.660	0.820	0.989	1.167	1.353	1.548	1.752	1.964
38	0.118	0.244	0.379	0.521	0.671	0.829	0.994	1.166	1.347	1.534	1.729	1.931	0.119	0.248	0.387	0.536	0.694	0.861	1.037	1.223	1.418	1.621	1.834	2.055
39	0.124	0.257	0.398	0.547	0.704	0.869	1.042	1.223	1.411	1.607	1.810	2.021	0.125	0.261	0.407	0.562	0.728	0.903	1.087	1.281	1.484	1.697	1.918	2.149
40	0.130	0.270	0.418	0.574	0.739	0.911	1.092	1.280	1.477	1.681	1.893	2.113	0.132	0.274	0.427	0.590	0.763	0.946	1.138	1.340	1.552	1.774	2.005	2.245
41	0.137	0.283	0.438	0.602	0.774	0.954	1.143	1.340	1.545	1.758	1.979	2.208	0.138	0.287	0.447	0.618	0.799	0.990	1.191	1.402	1.622	1.853	2.094	2.344
42	0.144	0.297	0.459	0.630	0.810	0.998	1.195	1.400	1.614	1.836	2.066	2.305	0.145	0.301	0.468	0.647	0.835	1.035	1.244	1.464	1.694	1.935	2.185	2.445
43	0.151	0.311	0.480	0.659	0.847	1.043	1.249	1.463	1.685	1.917	2.156	2.405	0.152	0.315	0.490	0.676	0.873	1.081	1.299	1.528	1.768	2.018	2.278	2.549
44	0.157	0.325	0.502	0.689	0.885	1.090	1.304	1.527	1.758	1.999	2.249	2.507	0.159	0.330	0.512	0.706	0.912	1.128	1.356	1.594	1.843	2.103	2.374	2.656
45	0.165	0.340	0.525	0.719	0.923	1.137	1.360	1.592	1.833	2.083	2.343	2.611	0.166	0.344	0.535	0.737	0.951	1.177	1.414	1.662	1.921	2.191	2.472	2.765
46	0.172	0.355	0.547	0.750	0.963	1.185	1.417	1.659	1.910	2.170	2.440	2.718	0.173	0.359	0.558	0.769	0.992	1.226	1.473	1.731	2.000	2.281	2.573	2.877

Note: This table is included for comparative purposes only as there are too many combinations of length, diameter and taper to include in this publication. To use this table, locate the volume at the intersect of the appropriate diameter and length in one of the four taper categories.

Table A.1.K. Hoppus cubic feet and cubic metre volume chart.

Imperial measure (ft^3)												Metric measure (m^3)											
Inches		Length of log in feet										Centimetres		Length of log in metres									
Dia.	**Girth**	**1**	**2**	**3**	**4**	**5**	**6**	**7**	**8**	**9**	**10**	**Dia.**	**Girth**	**1.0**	**2.0**	**3.0**	**4.0**	**5.0**	**6.0**	**7.0**	**8.0**	**9.0**	**10.0**
4.8	**15**	0.10	0.20	0.29	0.39	0.49	0.59	0.68	0.78	0.88	0.98	**19.1**	**60**	0.023	0.045	0.068	0.090	0.113	0.135	0.158	0.180	0.203	0.225
5.1	**16**	0.11	0.22	0.33	0.44	0.56	0.67	0.78	0.89	1.00	1.11	**19.7**	**62**	0.024	0.048	0.072	0.096	0.120	0.144	0.168	0.192	0.216	0.240
5.4	**17**	0.13	0.25	0.38	0.50	0.63	0.75	0.88	1.00	1.13	1.25	**20.4**	**64**	0.026	0.051	0.077	0.102	0.128	0.154	0.179	0.205	0.230	0.256
5.7	**18**	0.14	0.28	0.42	0.56	0.70	0.84	0.98	1.13	1.27	1.41	**21.0**	**66**	0.027	0.054	0.082	0.109	0.136	0.163	0.191	0.218	0.245	0.272
6.0	**19**	0.16	0.31	0.47	0.63	0.78	0.94	1.10	1.25	1.41	1.57	**21.6**	**68**	0.029	0.058	0.087	0.116	0.145	0.173	0.202	0.231	0.260	0.289
6.4	**20**	0.17	0.35	0.52	0.69	0.87	1.04	1.22	1.39	1.56	1.74	**22.3**	**70**	0.031	0.061	0.092	0.123	0.153	0.184	0.214	0.245	0.276	0.306
6.7	**21**	0.19	0.38	0.57	0.77	0.96	1.15	1.34	1.53	1.72	1.91	**22.9**	**72**	0.032	0.065	0.097	0.130	0.162	0.194	0.227	0.259	0.292	0.324
7.0	**22**	0.21	0.42	0.63	0.84	1.05	1.26	1.47	1.68	1.89	2.10	**23.6**	**74**	0.034	0.068	0.103	0.137	0.171	0.205	0.240	0.274	0.308	0.342
7.3	**23**	0.23	0.46	0.69	0.92	1.15	1.38	1.61	1.84	2.07	2.30	**24.2**	**76**	0.036	0.072	0.108	0.144	0.181	0.217	0.253	0.289	0.325	0.361
7.6	**24**	0.25	0.50	0.75	1.00	1.25	1.50	1.75	2.00	2.25	2.50	**24.8**	**78**	0.038	0.076	0.114	0.152	0.190	0.228	0.266	0.304	0.342	0.380
8.0	**25**	0.27	0.54	0.81	1.09	1.36	1.63	1.90	2.17	2.44	2.71	**25.5**	**80**	0.040	0.080	0.120	0.160	0.200	0.240	0.280	0.320	0.360	0.400
8.3	**26**	0.29	0.59	0.88	1.17	1.47	1.76	2.05	2.35	2.64	2.93	**26.1**	**82**	0.042	0.084	0.126	0.168	0.210	0.252	0.294	0.336	0.378	0.420
8.6	**27**	0.32	0.63	0.95	1.27	1.58	1.90	2.21	2.53	2.85	3.16	**26.7**	**84**	0.044	0.088	0.132	0.176	0.221	0.265	0.309	0.353	0.397	0.441
8.9	**28**	0.34	0.68	1.02	1.36	1.70	2.04	2.38	2.72	3.06	3.40	**27.4**	**86**	0.046	0.092	0.139	0.185	0.231	0.277	0.324	0.370	0.416	0.462
9.2	**29**	0.37	0.73	1.10	1.46	1.83	2.19	2.56	2.92	3.29	3.65	**28.0**	**88**	0.048	0.097	0.145	0.194	0.242	0.290	0.339	0.387	0.436	0.484
9.5	**30**	0.39	0.78	1.17	1.56	1.95	2.34	2.73	3.13	3.52	3.91	**28.6**	**90**	0.051	0.101	0.152	0.203	0.253	0.304	0.354	0.405	0.456	0.506
9.9	**31**	0.42	0.83	1.25	1.67	2.09	2.50	2.92	3.34	3.75	4.17	**29.3**	**92**	0.053	0.106	0.159	0.212	0.265	0.317	0.370	0.423	0.476	0.529
10.2	**32**	0.44	0.89	1.33	1.78	2.22	2.67	3.11	3.56	4.00	4.44	**29.9**	**94**	0.055	0.110	0.166	0.221	0.276	0.331	0.387	0.442	0.497	0.552
10.5	**33**	0.47	0.95	1.42	1.89	2.36	2.84	3.31	3.78	4.25	4.73	**30.6**	**96**	0.058	0.115	0.173	0.230	0.288	0.346	0.403	0.461	0.518	0.576
10.8	**34**	0.50	1.00	1.51	2.01	2.51	3.01	3.51	4.01	4.52	5.02	**31.2**	**98**	0.060	0.120	0.180	0.240	0.300	0.360	0.420	0.480	0.540	0.600
11.1	**35**	0.53	1.06	1.60	2.13	2.66	3.19	3.72	4.25	4.79	5.32	**31.8**	**100**	0.063	0.125	0.188	0.250	0.313	0.375	0.438	0.500	0.563	0.625
11.5	**36**	0.56	1.13	1.69	2.25	2.81	3.38	3.94	4.50	5.06	5.63	**32.5**	**102**	0.065	0.130	0.195	0.260	0.325	0.390	0.455	0.520	0.585	0.650
11.8	**37**	0.59	1.19	1.78	2.38	2.97	3.57	4.16	4.75	5.35	5.94	**33.1**	**104**	0.068	0.135	0.203	0.270	0.338	0.406	0.473	0.541	0.608	0.676
12.1	**38**	0.63	1.25	1.88	2.51	3.13	3.76	4.39	5.01	5.64	6.27	**33.7**	**106**	0.070	0.140	0.211	0.281	0.351	0.421	0.492	0.562	0.632	0.702
12.4	**39**	0.66	1.32	1.98	2.64	3.30	3.96	4.62	5.28	5.94	6.60	**34.4**	**108**	0.073	0.146	0.219	0.292	0.365	0.437	0.510	0.583	0.656	0.729
12.7	**40**	0.69	1.39	2.08	2.78	3.47	4.17	4.86	5.56	6.25	6.94	**35.0**	**110**	0.076	0.151	0.227	0.303	0.378	0.454	0.529	0.605	0.681	0.756
13.1	**41**	0.73	1.46	2.19	2.92	3.65	4.38	5.11	5.84	6.57	7.30	**35.7**	**112**	0.078	0.157	0.235	0.314	0.392	0.470	0.549	0.627	0.706	0.784
13.4	**42**	0.77	1.53	2.30	3.06	3.83	4.59	5.36	6.13	6.89	7.66	**36.3**	**114**	0.081	0.162	0.244	0.325	0.406	0.487	0.569	0.650	0.731	0.812
13.7	**43**	0.80	1.61	2.41	3.21	4.01	4.82	5.62	6.42	7.22	8.03	**36.9**	**116**	0.084	0.168	0.252	0.336	0.421	0.505	0.589	0.673	0.757	0.841
14.0	**44**	0.84	1.68	2.52	3.36	4.20	5.04	5.88	6.72	7.56	8.40	**37.6**	**118**	0.087	0.174	0.261	0.348	0.435	0.522	0.609	0.696	0.783	0.870
14.3	**45**	0.88	1.76	2.64	3.52	4.39	5.27	6.15	7.03	7.91	8.79	**38.2**	**120**	0.090	0.180	0.270	0.360	0.450	0.540	0.630	0.720	0.810	0.900
14.6	**46**	0.92	1.84	2.76	3.67	4.59	5.51	6.43	7.35	8.27	9.18	**38.8**	**122**	0.093	0.186	0.279	0.372	0.465	0.558	0.651	0.744	0.837	0.930
15.0	**47**	0.96	1.92	2.88	3.84	4.79	5.75	6.71	7.67	8.63	9.59	**39.5**	**124**	0.096	0.192	0.288	0.384	0.481	0.577	0.673	0.769	0.865	0.961
15.3	**48**	1.00	2.00	3.00	4.00	5.00	6.00	7.00	8.00	9.00	10.00	**40.1**	**126**	0.099	0.198	0.298	0.397	0.496	0.595	0.695	0.794	0.893	0.992

15.6	**49**	1.04	2.08	3.13	4.17	5.21	6.25	7.29	8.34	9.38	10.42	**40.7**	**128**	0.102	0.205	0.307	0.410	0.512	0.614	0.717	0.819	0.922	1.024
15.9	**50**	1.09	2.17	3.26	4.34	5.43	6.51	7.60	8.68	9.77	10.85	**41.4**	**130**	0.106	0.211	0.317	0.423	0.528	0.634	0.739	0.845	0.951	1.056
16.2	**51**	1.13	2.26	3.39	4.52	5.64	6.77	7.90	9.03	10.16	11.29	**42.0**	**132**	0.109	0.218	0.327	0.436	0.545	0.653	0.762	0.871	0.980	1.089
16.6	**52**	1.17	2.35	3.52	4.69	5.87	7.04	8.22	9.39	10.56	11.74	**42.7**	**134**	0.112	0.224	0.337	0.449	0.561	0.673	0.786	0.898	1.010	1.122
16.9	**53**	1.22	2.44	3.66	4.88	6.10	7.32	8.53	9.75	10.97	12.19	**43.3**	**136**	0.116	0.231	0.347	0.462	0.578	0.694	0.809	0.925	1.040	1.156
17.2	**54**	1.27	2.53	3.80	5.06	6.33	7.59	8.86	10.13	11.39	12.66	**43.9**	**138**	0.119	0.238	0.357	0.476	0.595	0.714	0.833	0.952	1.071	1.190
17.5	**55**	1.31	2.63	3.94	5.25	6.56	7.88	9.19	10.50	11.82	13.13	**44.6**	**140**	0.123	0.245	0.368	0.490	0.613	0.735	0.858	0.980	1.103	1.225
17.8	**56**	1.36	2.72	4.08	5.44	6.81	8.17	9.53	10.89	12.25	13.61	**45.2**	**142**	0.126	0.252	0.378	0.504	0.630	0.756	0.882	1.008	1.134	1.260
18.1	**57**	1.41	2.82	4.23	5.64	7.05	8.46	9.87	11.28	12.69	14.10	**45.8**	**144**	0.130	0.259	0.389	0.518	0.648	0.778	0.907	1.037	1.166	1.296
18.5	**58**	1.46	2.92	4.38	5.84	7.30	8.76	10.22	11.68	13.14	14.60	**46.5**	**146**	0.133	0.266	0.400	0.533	0.666	0.799	0.933	1.066	1.199	1.332
18.8	**59**	1.51	3.02	4.53	6.04	7.55	9.07	10.58	12.09	13.60	15.11	**47.1**	**148**	0.137	0.274	0.411	0.548	0.685	0.821	0.958	1.095	1.232	1.369
19.1	**60**	1.56	3.13	4.69	6.25	7.81	9.38	10.94	12.50	14.06	15.63	**47.7**	**150**	0.141	0.281	0.422	0.563	0.703	0.844	0.984	1.125	1.266	1.406
19.4	**61**	1.62	3.23	4.85	6.46	8.08	9.69	11.31	12.92	14.54	16.15	**48.4**	**152**	0.144	0.289	0.433	0.578	0.722	0.866	1.011	1.155	1.300	1.444
19.7	**62**	1.67	3.34	5.01	6.67	8.34	10.01	11.68	13.35	15.02	16.68	**49.0**	**154**	0.148	0.296	0.445	0.593	0.741	0.889	1.038	1.186	1.334	1.482
20.1	**63**	1.72	3.45	5.17	6.89	8.61	10.34	12.06	13.78	15.50	17.23	**49.7**	**156**	0.152	0.304	0.456	0.608	0.761	0.913	1.065	1.217	1.369	1.521
20.4	**64**	1.78	3.56	5.33	7.11	8.89	10.67	12.44	14.22	16.00	17.78	**50.3**	**158**	0.156	0.312	0.468	0.624	0.780	0.936	1.092	1.248	1.404	1.560
20.7	**65**	1.83	3.67	5.50	7.34	9.17	11.00	12.84	14.67	16.50	18.34	**50.9**	**160**	0.160	0.320	0.480	0.640	0.800	0.960	1.120	1.280	1.440	1.600
21.0	**66**	1.89	3.78	5.67	7.56	9.45	11.34	13.23	15.13	17.02	18.91	**51.6**	**162**	0.164	0.328	0.492	0.656	0.820	0.984	1.148	1.312	1.476	1.640
21.3	**67**	1.95	3.90	5.85	7.79	9.74	11.69	13.64	15.59	17.54	19.48	**52.2**	**164**	0.168	0.336	0.504	0.672	0.841	1.009	1.177	1.345	1.513	1.681
21.6	**68**	2.01	4.01	6.02	8.03	10.03	12.04	14.05	16.06	18.06	20.07	**52.8**	**166**	0.172	0.344	0.517	0.689	0.861	1.033	1.206	1.378	1.550	1.722
22.0	**69**	2.07	4.13	6.20	8.27	10.33	12.40	14.46	16.53	18.60	20.66	**53.5**	**168**	0.176	0.353	0.529	0.706	0.882	1.058	1.235	1.411	1.588	1.764
22.3	**70**	2.13	4.25	6.38	8.51	10.63	12.76	14.89	17.01	19.14	21.27	**54.1**	**170**	0.181	0.361	0.542	0.723	0.903	1.084	1.264	1.445	1.626	1.806
22.6	**71**	2.19	4.38	6.56	8.75	10.94	13.13	15.32	17.50	19.69	21.88	**54.7**	**172**	0.185	0.370	0.555	0.740	0.925	1.109	1.294	1.479	1.664	1.849
22.9	**72**	2.25	4.50	6.75	9.00	11.25	13.50	15.75	18.00	20.25	22.50	**55.4**	**174**	0.189	0.378	0.568	0.757	0.946	1.135	1.325	1.514	1.703	1.892
23.2	**73**	2.31	4.63	6.94	9.25	11.56	13.88	16.19	18.50	20.82	23.13	**56.0**	**176**	0.194	0.387	0.581	0.774	0.968	1.162	1.355	1.549	1.742	1.936
23.6	**74**	2.38	4.75	7.13	9.51	11.88	14.26	16.64	19.01	21.39	23.77	**56.7**	**178**	0.198	0.396	0.594	0.792	0.990	1.188	1.386	1.584	1.782	1.980
23.9	**75**	2.44	4.88	7.32	9.77	12.21	14.65	17.09	19.53	21.97	24.41	**57.3**	**180**	0.203	0.405	0.608	0.810	1.013	1.215	1.418	1.620	1.823	2.025
24.2	**76**	2.51	5.01	7.52	10.03	12.53	15.04	17.55	20.06	22.56	25.07	**57.9**	**182**	0.207	0.414	0.621	0.828	1.035	1.242	1.449	1.656	1.863	2.070
24.5	**77**	2.57	5.15	7.72	10.29	12.87	15.44	18.01	20.59	23.16	25.73	**58.6**	**184**	0.212	0.423	0.635	0.846	1.058	1.270	1.481	1.693	1.904	2.116
24.8	**78**	2.64	5.28	7.92	10.56	13.20	15.84	18.48	21.13	23.77	26.41	**59.2**	**186**	0.216	0.432	0.649	0.865	1.081	1.297	1.514	1.730	1.946	2.162
25.1	**79**	2.71	5.42	8.13	10.84	13.54	16.25	18.96	21.67	24.38	27.09	**59.8**	**188**	0.221	0.442	0.663	0.884	1.105	1.325	1.546	1.767	1.988	2.209
25.5	**80**	2.78	5.56	8.33	11.11	13.89	16.67	19.44	22.22	25.00	27.78	**60.5**	**190**	0.226	0.451	0.677	0.903	1.128	1.354	1.579	1.805	2.031	2.256
25.8	**81**	2.85	5.70	8.54	11.39	14.24	17.09	19.93	22.78	25.63	28.48	**61.1**	**192**	0.230	0.461	0.691	0.922	1.152	1.382	1.613	1.843	2.074	2.304
26.1	**82**	2.92	5.84	8.76	11.67	14.59	17.51	20.43	23.35	26.27	29.18	**61.8**	**194**	0.235	0.470	0.706	0.941	1.176	1.411	1.647	1.882	2.117	2.352
26.4	**83**	2.99	5.98	8.97	11.96	14.95	17.94	20.93	23.92	26.91	29.90	**62.4**	**196**	0.240	0.480	0.720	0.960	1.201	1.441	1.681	1.921	2.161	2.401
26.7	**84**	3.06	6.13	9.19	12.25	15.31	18.38	21.44	24.50	27.56	30.63	**63.0**	**198**	0.245	0.490	0.735	0.980	1.225	1.470	1.715	1.960	2.205	2.450
27.1	**85**	3.14	6.27	9.41	12.54	15.68	18.82	21.95	25.09	28.22	31.36	**63.7**	**200**	0.250	0.500	0.750	1.000	1.250	1.500	1.750	2.000	2.250	2.500

Note: To use this table find the volume at the intersect of the mid-girth or diameter and log length. For lengths that are longer than 10′ use 10 × the volume, e.g. for a 32′ log use 10 × the volume of 3′ + the volume of 2′. For lengths which are not on the full metre, use a percentage of the full metre lengths, e.g. for a 9.8 m log use the volume of a 9 m log + one tenth the volume of an 8 m log.

Table A.1.L. JAS cubic volume (m^3).

Length of log in m	Small-end diameter in cm																										
	10	11	12	13	14	16	18	20	22	24	26	28	30	32	34	36	38	40	42	44	46	48	50	52	54	56	58	60
2.0	0.020	0.024	0.029	0.034	0.039	0.051	0.065	0.080	0.097	0.115	0.135	0.157	0.180	0.205	0.231	0.259	0.289	0.320	0.353	0.387	0.423	0.461	0.500	0.541	0.583	0.627	0.673	0.720
2.2	0.022	0.027	0.032	0.037	0.043	0.056	0.071	0.088	0.106	0.127	0.149	0.172	0.198	0.225	0.254	0.285	0.318	0.352	0.388	0.426	0.466	0.507	0.550	0.595	0.642	0.690	0.740	0.792
2.4	0.024	0.029	0.035	0.041	0.047	0.061	0.078	0.096	0.116	0.138	0.162	0.188	0.216	0.246	0.277	0.311	0.347	0.384	0.423	0.465	0.508	0.553	0.600	0.649	0.700	0.753	0.807	0.864
2.6	0.026	0.031	0.037	0.044	0.051	0.067	0.084	0.104	0.126	0.150	0.176	0.204	0.234	0.266	0.301	0.337	0.375	0.416	0.459	0.503	0.550	0.599	0.650	0.703	0.758	0.815	0.875	0.936
2.8	0.028	0.034	0.040	0.047	0.055	0.072	0.091	0.112	0.136	0.161	0.189	0.220	0.252	0.287	0.324	0.363	0.404	0.448	0.494	0.542	0.592	0.645	0.700	0.757	0.816	0.878	0.942	1.008
3.0	0.030	0.036	0.043	0.051	0.059	0.077	0.097	0.120	0.145	0.173	0.203	0.235	0.270	0.307	0.347	0.389	0.433	0.480	0.529	0.581	0.635	0.691	0.750	0.811	0.875	0.941	1.009	1.080
3.2	0.032	0.039	0.046	0.054	0.063	0.082	0.104	0.128	0.155	0.184	0.216	0.251	0.288	0.328	0.370	0.415	0.462	0.512	0.564	0.620	0.677	0.737	0.800	0.865	0.933	1.004	1.076	1.152
3.4	0.034	0.041	0.049	0.057	0.067	0.087	0.110	0.136	0.165	0.196	0.230	0.267	0.306	0.348	0.393	0.441	0.491	0.544	0.600	0.658	0.719	0.783	0.850	0.919	0.991	1.066	1.144	1.224
3.6	0.036	0.044	0.052	0.061	0.071	0.092	0.117	0.144	0.174	0.207	0.243	0.282	0.324	0.369	0.416	0.467	0.520	0.576	0.635	0.697	0.762	0.829	0.900	0.973	1.050	1.129	1.211	1.296
3.8	0.038	0.046	0.055	0.064	0.074	0.097	0.123	0.152	0.184	0.219	0.257	0.298	0.342	0.389	0.439	0.492	0.549	0.608	0.670	0.736	0.804	0.876	0.950	1.028	1.108	1.192	1.278	1.368
4.0	0.040	0.048	0.058	0.068	0.078	0.102	0.130	0.160	0.194	0.230	0.270	0.314	0.360	0.410	0.462	0.518	0.578	0.640	0.706	0.774	0.846	0.922	1.000	1.082	1.166	1.254	1.346	1.440
4.2	0.042	0.051	0.060	0.071	0.082	0.108	0.136	0.168	0.203	0.242	0.284	0.329	0.378	0.430	0.486	0.544	0.606	0.672	0.741	0.813	0.889	0.968	1.050	1.136	1.225	1.317	1.413	1.512
4.4	0.044	0.053	0.063	0.074	0.086	0.113	0.143	0.176	0.213	0.253	0.297	0.345	0.396	0.451	0.509	0.570	0.635	0.704	0.776	0.852	0.931	1.014	1.100	1.190	1.283	1.380	1.480	1.584
4.6	0.046	0.056	0.066	0.078	0.090	0.118	0.149	0.184	0.223	0.265	0.311	0.361	0.414	0.471	0.532	0.596	0.664	0.736	0.811	0.891	0.973	1.060	1.150	1.244	1.341	1.443	1.547	1.656
4.8	0.048	0.058	0.069	0.081	0.094	0.123	0.156	0.192	0.232	0.276	0.324	0.376	0.432	0.492	0.555	0.622	0.693	0.768	0.847	0.929	1.016	1.106	1.200	1.298	1.400	1.505	1.615	1.728
5.0	0.050	0.061	0.072	0.085	0.098	0.128	0.162	0.200	0.242	0.288	0.338	0.392	0.450	0.512	0.578	0.648	0.722	0.800	0.882	0.968	1.058	1.152	1.250	1.352	1.458	1.568	1.682	1.800
5.2	0.052	0.063	0.075	0.088	0.102	0.133	0.168	0.208	0.252	0.300	0.352	0.408	0.468	0.532	0.601	0.674	0.751	0.832	0.917	1.007	1.100	1.198	1.300	1.406	1.516	1.631	1.749	1.872
5.4	0.054	0.065	0.078	0.091	0.106	0.138	0.175	0.216	0.261	0.311	0.365	0.423	0.486	0.553	0.624	0.700	0.780	0.864	0.953	1.045	1.143	1.244	1.350	1.460	1.575	1.693	1.817	1.944
5.6	0.056	0.068	0.081	0.095	0.110	0.143	0.181	0.224	0.271	0.323	0.379	0.439	0.504	0.573	0.647	0.726	0.809	0.896	0.988	1.084	1.185	1.290	1.400	1.514	1.633	1.756	1.884	2.016
5.8	0.058	0.070	0.084	0.098	0.114	0.148	0.188	0.232	0.281	0.334	0.392	0.455	0.522	0.594	0.670	0.752	0.838	0.928	1.023	1.123	1.227	1.336	1.450	1.568	1.691	1.819	1.951	2.088
6.0	0.073	0.086	0.101	0.118	0.135	0.173	0.217	0.265	0.317	0.375	0.437	0.505	0.577	0.653	0.735	0.821	0.913	1.009	1.109	1.215	1.325	1.441	1.561	1.685	1.815	1.949	2.089	2.233
6.2	0.075	0.089	0.105	0.122	0.140	0.179	0.224	0.273	0.328	0.388	0.452	0.521	0.596	0.675	0.760	0.849	0.943	1.042	1.146	1.256	1.370	1.489	1.613	1.742	1.876	2.014	2.158	2.307
6.4	0.077	0.092	0.108	0.125	0.144	0.185	0.231	0.282	0.339	0.400	0.467	0.538	0.615	0.697	0.784	0.876	0.973	1.076	1.183	1.296	1.414	1.537	1.665	1.798	1.936	2.079	2.228	2.381
6.6	0.080	0.095	0.112	0.129	0.149	0.191	0.238	0.291	0.349	0.413	0.481	0.555	0.634	0.719	0.809	0.904	1.004	1.109	1.220	1.337	1.458	1.585	1.717	1.854	1.997	2.144	2.297	2.456
6.8	0.082	0.098	0.115	0.133	0.153	0.197	0.245	0.300	0.360	0.425	0.496	0.572	0.653	0.741	0.833	0.931	1.034	1.143	1.257	1.377	1.502	1.633	1.769	1.910	2.057	2.209	2.367	2.530
7.0	0.093	0.109	0.128	0.147	0.168	0.214	0.266	0.324	0.387	0.455	0.529	0.609	0.695	0.786	0.882	0.984	1.092	1.206	1.325	1.449	1.579	1.715	1.857	2.004	2.156	2.314	2.478	2.648
7.2	0.095	0.113	0.131	0.151	0.173	0.221	0.274	0.333	0.398	0.468	0.545	0.627	0.714	0.808	0.907	1.013	1.123	1.240	1.362	1.491	1.625	1.764	1.910	2.061	2.218	2.381	2.549	2.723
7.4	0.098	0.116	0.135	0.156	0.178	0.227	0.281	0.342	0.409	0.481	0.560	0.644	0.734	0.830	0.933	1.041	1.155	1.274	1.400	1.532	1.670	1.813	1.963	2.118	2.279	2.447	2.620	2.799

Length of log in m																												
7.6	0.101	0.119	0.139	0.160	0.183	0.233	0.289	0.351	0.420	0.494	0.575	0.661	0.754	0.853	0.958	1.069	1.186	1.309	1.438	1.573	1.715	1.862	2.016	2.175	2.341	2.513	2.691	2.875
7.8	0.103	0.122	0.142	0.164	0.187	0.239	0.297	0.361	0.431	0.507	0.590	0.679	0.774	0.875	0.983	1.097	1.217	1.343	1.476	1.615	1.760	1.911	2.069	2.233	2.403	2.579	2.761	2.950
8.0	0.115	0.135	0.157	0.180	0.205	0.259	0.320	0.387	0.461	0.541	0.627	0.720	0.819	0.925	1.037	1.155	1.280	1.411	1.549	1.693	1.843	2.000	2.163	2.333	2.509	2.691	2.880	3.075
8.2	0.118	0.139	0.161	0.185	0.210	0.266	0.328	0.397	0.472	0.554	0.643	0.738	0.840	0.948	1.063	1.184	1.312	1.446	1.588	1.735	1.889	2.050	2.217	2.391	2.572	2.758	2.952	3.152
8.4	0.121	0.142	0.165	0.189	0.215	0.272	0.336	0.407	0.484	0.568	0.659	0.756	0.860	0.971	1.089	1.213	1.344	1.482	1.626	1.777	1.935	2.100	2.271	2.449	2.634	2.826	3.024	3.229
8.6	0.124	0.145	0.169	0.194	0.220	0.279	0.344	0.416	0.495	0.581	0.674	0.774	0.881	0.994	1.115	1.242	1.376	1.517	1.665	1.820	1.981	2.150	2.325	2.508	2.697	2.893	3.096	3.306
8.8	0.127	0.149	0.172	0.198	0.225	0.285	0.352	0.426	0.507	0.595	0.690	0.792	0.901	1.017	1.140	1.271	1.408	1.552	1.704	1.862	2.028	2.200	2.380	2.566	2.760	2.960	3.168	3.383
9.0	0.141	0.164	0.189	0.216	0.245	0.308	0.378	0.456	0.540	0.632	0.731	0.837	0.951	1.071	1.199	1.334	1.476	1.626	1.782	1.946	2.117	2.295	2.481	2.673	2.873	3.080	3.294	3.516
9.2	0.144	0.168	0.193	0.221	0.250	0.315	0.387	0.466	0.552	0.646	0.747	0.856	0.972	1.095	1.226	1.364	1.509	1.662	1.822	1.989	2.164	2.346	2.536	2.733	2.937	3.148	3.367	3.594
9.4	0.147	0.171	0.198	0.226	0.256	0.322	0.395	0.476	0.564	0.660	0.764	0.874	0.993	1.119	1.252	1.393	1.542	1.698	1.861	2.033	2.211	2.397	2.591	2.792	3.001	3.217	3.441	3.672
9.6	0.150	0.175	0.202	0.231	0.261	0.329	0.403	0.486	0.576	0.674	0.780	0.893	1.014	1.143	1.279	1.423	1.575	1.734	1.901	2.076	2.258	2.448	2.646	2.851	3.065	3.285	3.514	3.750
9.8	0.153	0.179	0.206	0.235	0.267	0.335	0.412	0.496	0.588	0.688	0.796	0.912	1.035	1.166	1.306	1.453	1.607	1.770	1.941	2.119	2.305	2.499	2.701	2.911	3.128	3.354	3.587	3.828
10.0	0.169	0.196	0.225	0.256	0.289	0.361	0.441	0.529	0.625	0.729	0.841	0.961	1.089	1.225	1.369	1.521	1.681	1.849	2.025	2.209	2.401	2.601	2.809	3.025	3.249	3.481	3.721	3.969
10.2	0.172	0.200	0.230	0.261	0.295	0.368	0.450	0.540	0.638	0.744	0.858	0.980	1.111	1.250	1.396	1.551	1.715	1.886	2.066	2.253	2.449	2.653	2.865	3.086	3.314	3.551	3.795	4.048
10.4	0.176	0.204	0.234	0.266	0.301	0.375	0.459	0.550	0.650	0.758	0.875	0.999	1.133	1.274	1.424	1.582	1.748	1.923	2.106	2.297	2.497	2.705	2.921	3.146	3.379	3.620	3.870	4.128
10.6	0.179	0.208	0.239	0.271	0.306	0.383	0.467	0.561	0.663	0.773	0.891	1.019	1.154	1.299	1.451	1.612	1.782	1.960	2.147	2.342	2.545	2.757	2.978	3.207	3.444	3.690	3.944	4.207
10.8	0.183	0.212	0.243	0.276	0.312	0.390	0.476	0.571	0.675	0.787	0.908	1.038	1.176	1.323	1.479	1.643	1.815	1.997	2.187	2.386	2.593	2.809	3.034	3.267	3.509	3.759	4.019	4.287
11.0	0.200	0.231	0.264	0.299	0.337	0.418	0.508	0.607	0.715	0.832	0.957	1.091	1.234	1.386	1.547	1.716	1.894	2.081	2.277	2.482	2.695	2.917	3.148	3.388	3.637	3.894	4.160	4.435
11.2	0.204	0.235	0.269	0.305	0.343	0.426	0.518	0.619	0.728	0.847	0.975	1.111	1.257	1.411	1.575	1.747	1.929	2.119	2.319	2.527	2.744	2.971	3.206	3.450	3.703	3.965	4.236	4.516
11.4	0.208	0.240	0.274	0.310	0.349	0.433	0.527	0.630	0.741	0.862	0.992	1.131	1.279	1.437	1.603	1.779	1.963	2.157	2.360	2.572	2.793	3.024	3.263	3.511	3.769	4.036	4.312	4.597
11.6	0.211	0.244	0.279	0.316	0.355	0.441	0.536	0.641	0.754	0.877	1.009	1.151	1.302	1.462	1.631	1.810	1.998	2.195	2.401	2.617	2.842	3.077	3.320	3.573	3.835	4.107	4.387	4.677
11.8	0.215	0.248	0.283	0.321	0.361	0.449	0.545	0.652	0.767	0.892	1.027	1.171	1.324	1.487	1.659	1.841	2.032	2.233	2.443	2.662	2.891	3.130	3.377	3.635	3.901	4.177	4.463	4.758
12.0	0.235	0.270	0.307	0.347	0.389	0.480	0.581	0.691	0.811	0.941	1.080	1.229	1.387	1.555	1.733	1.920	2.117	2.323	2.539	2.765	3.000	3.245	3.499	3.763	4.037	4.320	4.613	4.915

Note: The volume is located at the intersect of the small-end diameter (top of the table) and the log length (left side of the table).

Table A.1.M. Summarized volumes of control group of logs used for modeling of conversions used in Section 2.5 (m^3 for cubic measure; mbf for product output rules).

	2.5 to 4.6 m 8 to 15 ft.		4.7 to 6.4 m 16 to 21′		6.5 to 9.5 m 22 to 31′		9.6 to 12.5 m 32 to 41′		Total all lengths	
	Gross	Net	Gross	Net	Gross	Net	Gross	Net	Gross	Net
Small-end diameter 11.43–19.05 cm (4.5–7.49″)										
USFS Cubic	0.531	0.501	0.658	0.620	1.833	1.716	6.646	6.442	9.668	9.279
BC Firmwood	0.564	0.556	0.712	0.712	1.898	1.898	7.224	7.183	10.398	10.349
Alberta Cubic	0.552	0.517	0.685	0.650	1.861	1.761	7.076	6.866	10.174	9.793
Ontario Cubic	0.537	0.532	0.704	0.704	1.828	1.828	6.903	6.903	9.972	9.968
Swedish Cubic*	0.474	–	0.499	–	1.551	–	5.683	–	8.207	–
Russian Standard*	0.569	–	0.697	–	1.857	–	7.137	–	10.259	–
Cubage au Réel*	0.528	–	0.662	–	1.791	–	6.795	–	9.775	–
New Zealand 3-D*	0.643	–	0.740	–	1.935	–	7.081	–	10.399	–
Brereton (PNG)	0.532	–	0.638	–	1.713	–	6.368	–	9.251	–
Hoppus	0.445	–	0.547	–	1.454	–	5.574	–	8.019	–
JAS Scale*	0.512	–	0.515	–	1.584	–	5.925	–	8.536	–
Scribner Short R**	0.090	0.090	0.120	0.110	0.290	0.260	1.110	1.060	1.610	1.520
Scribner Short NR**	0.090	0.090	0.120	0.090	0.270	0.250	1.070	1.000	1.550	1.430
Scribner Long Log**	0.090	0.080	0.090	0.080	0.220	0.200	0.760	0.760	1.160	1.120
Doyle**	0.022	0.020	0.017	0.016	0.151	0.141	0.439	0.424	0.628	0.601
International ¼″**	0.075	0.070	0.090	0.080	0.340	0.325	1.225	1.175	1.730	1.650
Small-end diameter 19.06–29.19 cm (7.5–11.49″)										
USFS Cubic	1.855	1.795	3.008	2.817	4.680	4.474	9.882	9.410	19.425	18.496
BC Firmwood	2.049	2.049	3.218	3.195	5.005	4.911	10.626	10.548	20.898	20.703
Alberta Cubic	1.993	1.929	3.100	2.903	4.875	4.656	10.303	9.836	20.270	19.324
Ontario Cubic	2.003	2.003	3.129	3.101	4.838	4.635	10.268	10.167	20.237	19.906
Swedish Cubic*	1.758	–	2.734	–	4.311	–	8.903	–	17.706	–
Russian Standard*	1.960	–	3.178	–	4.930	–	10.395	–	20.463	–
Cubage au Réel*	1.943	–	3.116	–	4.767	–	10.419	–	20.245	–
New Zealand 3-D*	2.093	–	3.305	–	4.889	–	10.061	–	20.348	–
Brereton (PNG)	1.902	–	3.041	–	4.626	–	9.754	–	19.322	–
Hoppus	1.559	–	2.522	–	3.863	–	8.301	–	16.245	–
JAS Scale*	1.909	–	2.863	–	4.822	–	9.621	–	19.214	–
Scribner Short R**	0.330	0.320	0.500	0.440	0.850	0.780	1.780	1.630	3.460	3.170
Scribner Short NR**	0.320	0.310	0.470	0.420	0.830	0.780	1.780	1.650	3.400	3.160
Scribner Long Log**	0.260	0.260	0.450	0.410	0.670	0.650	1.310	1.250	2.690	2.570
Doyle**	0.257	0.251	0.358	0.335	0.706	0.675	1.390	1.320	2.710	2.580
International ¼″**	0.395	0.380	0.645	0.595	1.025	0.965	2.190	2.065	4.255	4.005
Small-end diameter 29.2–39.35 cm (11.5–15.49″)										
USFS Cubic	2.473	2.438	4.214	3.811	5.194	4.856	30.738	29.163	42.619	40.269
BC Firmwood	2.622	2.622	4.418	4.169	5.367	5.367	32.663	31.550	45.071	43.708
Alberta Cubic	2.561	2.513	4.263	3.890	5.309	5.070	31.724	30.128	43.856	41.601
Ontario Cubic	2.615	2.615	4.264	3.998	5.328	5.328	32.073	30.613	44.280	42.554
Swedish Cubic*	2.402	–	3.873	–	4.944	–	29.024	–	40.243	–
Russian Standard*	2.593	–	4.354	–	5.347	–	32.248	–	44.542	–
Cubage au Réel*	2.504	–	4.227	–	5.272	–	31.662	–	43.665	–
New Zealand 3-D*	2.657	–	4.388	–	5.550	–	31.403	–	43.999	–
Brereton (PNG)	2.521	–	4.172	–	5.170	–	30.931	–	42.793	–
Hoppus	2.044	–	3.425	–	4.218	–	25.594	–	35.280	–

continued

Table A.1.M. *continued*

	2.5 to 4.6 m 8 to 15′		4.7 to 6.4 m 16 to 21′		6.5 to 9.5 m 22 to 31′		9.6 to 12.5 m 32 to 41′		Total all lengths	
	Gross	Net	Gross	Net	Gross	Net	Gross	Net	Gross	Net
Small-end diameter 29.2–39.35 cm (11.5–15.49″)										
JAS Scale*	2.605	–	4.309	–	5.806	–	32.658	–	45.377	–
Scribner Short R**	0.530	0.520	0.840	0.760	1.120	1.060	6.620	6.140	9.110	8.480
Scribner Short NR**	0.530	0.520	0.840	0.760	1.120	1.060	6.620	6.140	9.110	8.480
Scribner Long Log**	0.490	0.480	0.830	0.750	0.860	0.830	5.000	4.540	7.180	6.600
Doyle**	0.493	0.485	0.771	0.690	0.970	0.907	6.241	5.772	8.475	7.854
International ¼″**	0.605	0.595	0.995	0.890	1.290	1.200	7.665	7.150	10.555	9.835
Small-end diameter >39.36 cm (15.5″)										
USFS Cubic	7.463	6.691	18.500	16.973	32.415	30.531	38.192	36.486	96.570	90.681
BC Firmwood	8.036	7.714	19.195	17.827	33.266	32.031	39.423	38.184	99.920	95.756
Alberta Cubic	7.860	7.057	18.491	16.962	33.074	31.415	38.335	36.505	97.759	91.940
Ontario Cubic	8.057	7.667	18.884	17.289	32.975	31.575	39.514	38.149	99.430	94.680
Swedish Cubic*	7.752	–	18.064	–	32.054	–	36.727	–	94.597	–
Russian Standard*	7.887	–	18.776	–	33.260	–	38.833	–	98.756	–
Cubage au Réel*	7.727	–	18.427	–	32.589	–	38.927	–	97.671	–
New Zealand 3-D*	8.517	–	20.248	–	35.281	–	40.011	–	104.057	–
Brereton (PNG)	7.826	–	18.504	–	32.673	–	38.278	–	97.282	–
Hoppus	6.211	–	14.783	–	26.260	–	31.119	–	78.373	–
JAS Scale*	8.725	–	20.395	–	36.441	–	42.971	–	108.532	–
Scribner Short R**	2.010	1.750	4.900	4.430	8.800	8.160	9.730	9.240	25.440	23.580
Scribner Short NR**	2.010	1.750	4.900	4.430	8.790	8.150	9.730	9.240	25.430	23.570
Scribner Long Log**	1.970	1.700	4.620	4.150	7.740	7.210	7.940	7.520	22.270	20.580
Doyle**	2.122	1.860	4.940	4.473	8.882	8.266	9.529	8.963	25.473	23.562
International ¼″**	2.080	1.820	5.200	4.695	9.000	8.355	10.470	9.900	26.750	24.770
Total all diameters										
USFS Cubic	12.321	11.426	26.381	24.222	44.121	41.577	85.458	81.501	168.282	158.725
BC Firmwood	13.272	12.942	27.542	25.902	45.537	44.208	89.936	87.464	176.287	170.516
Alberta Cubic	12.966	12.016	26.538	24.405	45.118	42.901	87.437	83.335	172.060	162.657
Ontario Cubic	13.212	12.817	26.982	25.092	44.969	43.367	88.757	85.832	173.920	167.108
Swedish Cubic*	12.385	–	25.170	–	42.860	–	80.337	–	160.753	–
Russian Standard*	13.009	–	27.005	–	45.394	–	88.612	–	174.021	–
Cubage au Réel*	12.701	–	26.432	–	44.419	–	87.803	–	171.356	–
New Zealand 3-D*	13.909	–	28.682	–	47.655	–	88.556	–	178.802	–
Brereton (PNG)	12.782	–	26.354	–	44.182	–	85.331	–	168.648	–
Hoppus	10.259	–	21.276	–	35.795	–	70.588	–	137.918	–
JAS Scale*	13.751	–	28.082	–	48.652	–	91.175	–	181.660	–
Scribner Short R**	2.960	2.680	6.360	5.740	11.060	10.260	19.240	18.070	39.620	36.750
Scribner Short NR**	2.950	2.670	6.330	5.700	11.010	10.240	19.200	18.030	39.490	36.640
Scribner Long Log**	2.810	2.520	5.990	5.390	9.490	8.890	15.010	14.070	33.300	30.870
Doyle**	2.893	2.616	6.085	5.513	10.710	9.989	17.599	16.478	37.286	34.596
International ¼″**	3.155	2.865	6.930	6.260	11.655	10.845	21.550	20.290	43.290	40.260

*Swedish Cubic, Russian Standard, Cubage au Réel, New Zealand 3-D, Brereton, Hoppus, and JAS Scale include only gross scale.

**Product output rules are reflected in units of 1000 board feet (mbf). Scribner Short R = revised Scribner used in California, Oregon, Washington and Alaska, Scribner NR = non-revised Scribner used elsewhere.

Table A.1.N. Long Log Scribner volume per Short Log Scribner volume index (1.00 = 100% of Scribner Short Log).

Small-end diameter in inches as measured using Scribner Short Log procedures	Log length in feet																		
	8	10	12	14	16	17	18	20	22	24	25	26	28	30	32	34	36	38	40
4	0.50	0.50	0.50	1.00	1.00	1.00	1.00	1.00	0.50	0.75	0.75	0.75	0.75	0.50	0.50	0.50	0.50	0.67	0.83
5	1.00	1.00	1.00	1.00	0.75	0.75	0.75	0.75	0.75	1.25	0.83	0.83	0.83	0.83	0.50	0.60	0.60	0.70	0.70
6	1.00	1.00	1.00	0.75	1.00	1.00	1.00	1.00	0.83	1.00	1.00	1.00	0.75	0.70	0.80	0.90	1.00	0.83	0.83
7	1.00	1.00	0.75	1.00	0.83	0.83	0.83	0.83	1.17	0.88	0.88	0.88	1.00	0.90	0.79	0.79	0.86	0.81	0.81
8	1.00	0.75	1.00	1.00	1.00	1.00	1.00	0.88	0.80	0.80	0.90	0.90	1.00	0.92	0.72	0.72	0.78	0.75	0.73
9	0.75	1.00	0.83	0.83	0.88	0.88	0.88	0.90	0.75	0.71	0.79	0.79	0.86	0.72	0.73	0.71	0.75	0.79	0.81
10	0.83	0.83	0.88	0.88	0.83	0.83	0.83	0.86	0.93	0.94	0.83	0.83	0.94	0.82	0.75	0.77	0.80	0.78	0.79
11	1.00	0.88	1.00	0.90	0.93	0.93	0.88	0.94	0.85	0.95	0.95	0.91	0.92	0.92	0.76	0.78	0.79	0.78	0.83
12	0.88	0.90	0.83	0.86	0.94	0.94	0.94	0.90	0.83	0.85	0.79	0.86	0.87	0.82	0.79	0.76	0.77	0.78	0.79
13	0.90	0.92	0.93	0.94	0.90	0.90	0.91	0.92	0.80	0.84	0.84	0.85	0.86	0.87	0.73	0.73	0.74	0.72	0.73
14	0.92	0.93	0.89	0.90	0.95	0.92	0.92	0.93	0.81	0.80	0.79	0.83	0.84	0.81	0.78	0.78	0.77	0.76	0.78
15	0.93	0.89	0.91	0.92	0.89	0.90	0.91	0.89	0.86	0.83	0.83	0.84	0.87	0.86	0.80	0.77	0.78	0.78	0.79
16	0.94	0.95	0.96	0.93	0.94	0.94	0.94	0.95	0.88	0.87	0.87	0.88	0.88	0.89	0.81	0.80	0.81	0.80	0.81
17	0.94	0.92	0.93	0.94	0.94	0.93	0.93	0.93	0.84	0.87	0.87	0.85	0.86	0.88	0.82	0.81	0.81	0.80	0.81
18	0.91	0.96	0.94	0.92	0.93	0.93	0.94	0.93	0.87	0.88	0.89	0.88	0.86	0.87	0.82	0.81	0.82	0.81	0.80
19	0.96	0.93	0.94	0.95	0.94	0.96	0.94	0.95	0.86	0.87	0.85	0.88	0.88	0.87	0.84	0.83	0.84	0.83	0.83
20	0.93	0.94	0.93	0.94	0.93	0.92	0.94	0.93	0.89	0.89	0.88	0.88	0.89	0.90	0.85	0.83	0.85	0.84	0.84
21	0.97	0.95	0.96	0.94	0.97	0.97	0.96	0.96	0.91	0.92	0.91	0.90	0.91	0.91	0.86	0.86	0.86	0.85	0.86

Small-end diameter in inches as measured using Scribner Short Log procedures																			
22	0.94	0.95	0.96	0.97	0.95	0.96	0.95	0.95	0.90	0.91	0.88	0.89	0.90	0.90	0.88	0.87	0.86	0.86	0.87
23	0.95	0.94	0.95	0.94	0.93	0.94	0.95	0.95	0.91	0.91	0.91	0.91	0.91	0.92	0.85	0.84	0.85	0.85	0.86
24	0.98	0.98	0.97	0.97	0.98	0.97	0.97	0.97	0.91	0.91	0.91	0.91	0.91	0.91	0.87	0.86	0.87	0.86	0.87
25	0.93	0.93	0.94	0.94	0.93	0.94	0.93	0.94	0.89	0.92	0.90	0.90	0.90	0.90	0.86	0.85	0.85	0.85	0.86
26	0.96	0.97	0.96	0.95	0.96	0.96	0.96	0.96	0.92	0.92	0.93	0.92	0.91	0.91	0.89	0.89	0.89	0.88	0.89
27	0.96	0.96	0.95	0.96	0.95	0.96	0.95	0.96	0.92	0.92	0.93	0.92	0.92	0.93	0.91	0.90	0.90	0.90	0.91
28	0.97	0.97	0.97	0.97	0.97	0.97	0.98	0.97	0.95	0.94	0.95	0.95	0.95	0.95	0.91	0.91	0.91	0.91	0.91
29	0.98	0.97	0.98	0.98	0.98	0.98	0.98	0.98	0.94	0.94	0.94	0.94	0.95	0.94	0.90	0.90	0.91	0.90	0.90
30	0.95	0.96	0.97	0.96	0.96	0.96	0.96	0.96	0.93	0.93	0.93	0.93	0.93	0.93	0.90	0.90	0.91	0.91	0.91
31	0.96	0.97	0.96	0.96	0.96	0.97	0.96	0.96	0.95	0.95	0.95	0.95	0.95	0.94	0.92	0.92	0.92	0.91	0.91
32	0.99	0.98	0.98	0.98	0.98	0.98	0.98	0.98	0.95	0.95	0.95	0.95	0.95	0.95	0.94	0.94	0.94	0.94	0.94
33	0.97	0.97	0.97	0.96	0.97	0.97	0.97	0.97	0.96	0.96	0.96	0.96	0.96	0.96	0.92	0.92	0.92	0.92	0.92
34	0.99	0.99	0.99	0.99	0.99	0.99	0.99	0.99	0.94	0.94	0.94	0.94	0.94	0.94	0.92	0.92	0.92	0.92	0.92
35	0.95	0.95	0.95	0.95	0.95	0.96	0.96	0.96	0.93	0.93	0.93	0.93	0.93	0.93	0.88	0.87	0.88	0.88	0.88
36	0.98	0.97	0.98	0.98	0.98	0.97	0.97	0.97	0.91	0.92	0.92	0.92	0.92	0.92	0.90	0.90	0.90	0.90	0.91
37	0.95	0.95	0.95	0.95	0.95	0.95	0.95	0.95	0.93	0.93	0.93	0.93	0.93	0.93	0.91	0.91	0.91	0.90	0.91
38	0.98	0.98	0.98	0.98	0.98	0.98	0.98	0.98	0.96	0.96	0.96	0.96	0.96	0.96	0.93	0.92	0.93	0.92	0.93
39	0.97	0.98	0.98	0.97	0.98	0.97	0.98	0.98	0.94	0.94	0.94	0.94	0.94	0.94	0.92	0.91	0.91	0.91	0.91
40	0.97	0.97	0.97	0.97	0.97	0.96	0.97	0.97	0.94	0.94	0.94	0.94	0.94	0.94	0.92	0.91	0.91	0.91	0.91

Note: Assumptions are that ½ of the diameters in each diameter class using short log rounding conventions will fall into the next lower inch diameter using long log rounding conventions. Taper is assumed to be Region 6 East-side standard taper (1″/segment for logs 21–31′; 2″/segment for logs 32–40′).

Table A.1.O. Washington and Oregon mill survey Scribner to BC cubic metre index by length and small-end diameter class (mbf per m^3 BC Firmwood).

Scribner Long Log to BC Firmwood

Average S.E.D.			Average log length									
			8–14′ 2.5–4.6 m		16–20′ 4.7–6.4 m		22–31′ 6.5–9.5 m		32–40′ 9.6–12.5 m		Total All lengths	
cm	Inches	Diameter class	Gross	Net	Gross	Net	Gross	Net	Gross	Net	Gross	Net
10.2	4	<5″	0.224	0.149	0.106	0.106	0.097	0.097	0.094	0.094	0.102	0.099
18.7	7.3	5–10″	0.131	0.132	0.138	0.126	0.123	0.118	0.121	0.118	0.124	0.120
36.5	14.4	11–20″	0.201	0.194	0.199	0.194	0.176	0.170	0.166	0.159	0.174	0.167
67.2	26.5	>20″	0.262	0.230	0.254	0.245	0.246	0.239	0.214	0.208	0.239	0.229

Scribner Short Log to BC Firmwood

Average S.E.D.			Average log length									
			8–14′ 2.5–4.6 m		16–20′ 4.7–6.4 m		22–31′ 6.5–9.5 m		32–40′ 9.6–12.5 m		Total All lengths	
cm	Inches	Diameter class	Gross	Net	Gross	Net	Gross	Net	Gross	Net	Gross	Net
10.2	4	<5″	0.257	0.172	0.111	0.111	0.135	0.135	0.132	0.132	0.135	0.132
18.7	7.4	5–10″	0.162	0.163	0.157	0.139	0.162	0.149	0.159	0.151	0.160	0.150
37.1	14.6	11–20″	0.213	0.204	0.210	0.206	0.221	0.209	0.213	0.206	0.214	0.206
68.1	26.8	>20″	0.264	0.235	0.263	0.255	0.274	0.265	0.257	0.252	0.265	0.256

Note: This table includes the volume relationship from the logs in Table A.1.M, sorted by the diameters as determined by the nominal Scribner diameters based on the respective rule, e.g. the 5–10″ group of Scribner Long Log contains the volume index for diameters which would be classified as being 5–10″ via Scribner Long Log and thus are not comparable to the 5–10″ group as classified via Scribner Short Log.

Appendix 2
Physical Properties and Weight-to-Volume Data

Table A.2.A	Weight-to-volume data for common timber species of the world in the log form	227
Table A.2.B	Bone-dry weight and volume conversions for selected tree species of the world	243

Table A.2.A. Weight-to-volume data for common timber species of the world in the log form.

Common name	Latin name	Size class	Source	kg/m³	lbs/ft³	Specific gravity (green m³)	Bark % whole log weight (bark & wood)	Bark % log wood volume
		North American conifers						
Pacific silver fir	*Abies amabilis*	ns	13	999.4	62.4	0.40	–	–
Balsam fir	*Abies balsamea*	ns	7	864.1	54.0	0.33	10.7	15.0
White fir	*Abies concolor*	ns	11	1034.1	64.6	0.37	–	–
Grand fir	*Abies grandis*	a	2	884.6	55.2	0.35	11.7	14.3
Grand fir	*Abies grandis*	b	2	854.0	53.3	0.35	11.7	14.3
Grand fir	*Abies grandis*	c	2	840.0	52.4	0.35	11.7	14.3
Grand fir	*Abies grandis*	d	2	817.6	51.0	0.35	11.7	14.3
Subalpine fir	*Abies lasiocarpa*	a	2	756.5	47.2	0.31	15.5	15.7
Subalpine fir	*Abies lasiocarpa*	b	2	749.9	46.8	0.31	15.5	15.7
Subalpine fir	*Abies lasiocarpa*	c, d	2	753.7	47.1	0.31	15.5	15.7
Noble fir	*Abies procera*	a	8	934.6	58.4	0.37	–	9.0
Noble fir	*Abies procera*	d	13	663.1	41.4	0.37	–	–
Port orford cedar	*Chamaecyparis lawsoniana*	ns	11	1007.5	62.9	0.39	–	–
Yellow cedar	*Chamaecyparis nootkatensis*	ns	12	730.0	45.6	0.42	–	–
Eastern red cedar	*Juniperus virginiana*	ns	7	765.6	47.8	0.44	10.0	12.0
Tamarack	*Larix laricina*	ns	7	951.7	59.4	0.49	7.4	13.0
Western larch	*Larix occidentalis*	a	2	921.4	57.5	0.48	10.1	19.5
Western larch	*Larix occidentalis*	b	2	898.7	56.1	0.48	10.1	19.5
Western larch	*Larix occidentalis*	c	2	882.4	55.1	0.48	10.1	19.5
Western larch	*Larix occidentalis*	d	2	867.0	54.1	0.48	10.1	19.5
Incense cedar	*Libocedrus decurrens*	ns	11	906.7	56.6	0.35	–	–
Engelmann spruce	*Picea engelmannii*	a	2	927.3	57.9	0.33	11.0	13.0
Engelmann spruce	*Picea engelmannii*	b	2	846.6	52.9	0.33	11.0	13.0
Engelmann spruce	*Picea engelmannii*	c	2	776.4	48.5	0.33	11.0	13.0
Engelmann spruce	*Picea engelmannii*	d	2	689.3	43.0	0.33	11.0	13.0
White spruce	*Picea glauca*	a	2	797.7	49.8	0.33	7.2	10.0

continued

Table A.2.A. *continued*

Common name	Latin name	Size class	Source	kg/m^3	lbs/ft^3	Specific gravity (green m^3)	Bark % whole log weight (bark & wood)	Bark % log wood volume
White spruce	*Picea glauca*	b	2	766.3	47.8	0.33	7.5	10.0
White spruce	*Picea glauca*	c	2	740.1	46.2	0.33	7.8	10.0
White spruce	*Picea glauca*	d	2	726.4	45.4	0.33	7.9	10.0
Black spruce	*Picea mariana*	ns	7	848.8	53.0	0.38	7.2	14.0
Sitka spruce	*Picea sitchensis*	a	8	925.9	57.8	0.37	–	10.0
Sitka spruce	*Picea sitchensis*	d	13	701.5	43.8	0.37	–	–
Jack pine	*Pinus banksiana*	ns	7	861.7	53.8	0.40	12.6	17.0
Lodgepole pine	*Pinus contorta*	a	2	905.2	56.5	0.38	5.4	6.0
Lodgepole pine	*Pinus contorta*	b	2	796.0	49.7	0.38	5.4	6.0
Lodgepole pine	*Pinus contorta*	c	2	749.2	46.8	0.38	5.4	6.0
Lodgepole pine	*Pinus contorta*	d	2	744.9	46.5	0.38	5.4	6.0
Shortleaf pine	*Pinus echinata*	ns	7	1025.0	64.0	0.47	9.4	15.0
Slash pine	*Pinus elliottii*	ns	9	1039.5	64.9	0.54	9.2	15.0
Limber pine	*Pinus flexilis*	a, b, c, d	2	930.2	58.1	0.37	12.1	–
Sugar pine	*Pinus lambertiana*	ns	11	1009.7	63.0	0.34	–	–
Western white pine	*Pinus monticola*	a, b, c, d	2	842.5	52.6	0.35	12.4	–
Longleaf pine	*Pinus palustris*	ns	9	1140.3	71.2	0.54	8.4	15.0
Ponderosa pine	*Pinus ponderosa*	a	2	1015.6	63.4	0.38	11.7	20.3
Ponderosa pine	*Pinus ponderosa*	b	2	971.6	60.7	0.38	11.7	20.3
Ponderosa pine	*Pinus ponderosa*	c	2	933.8	58.3	0.38	11.7	20.3
Ponderosa pine	*Pinus ponderosa*	d	2	896.4	56.0	0.38	11.7	20.3
Monterey pine	*Pinus radiata*	ns	9	1042.8	65.1	0.40	–	–
Red pine	*Pinus resinosa*	ns	7	859.1	53.6	0.41	8.7	16.0
Eastern white pine	*Pinus strobus*	ns	7	839.9	52.4	0.34	18.0	16.0
Loblolly pine	*Pinus taeda*	a, b, c, d	2	1026.6	64.1	0.47	9.4	15.0
Douglas fir	*Pseudotsuga menziesii*	a	2	950.8	59.4	0.45	13.7	20.1
Douglas fir	*Pseudotsuga menziesii*	b	2	909.1	56.8	0.45	13.7	20.1
Douglas fir	*Pseudotsuga menziesii*	c	2	871.8	54.4	0.45	13.7	20.1

Douglas fir	*Pseudotsuga menziesii*	d	2	836.1	52.2	0.45	13.7	20.1
Coastal redwood	*Sequoia sempervirens*	ns	11	949.8	59.3	0.36	–	–
Bald cypress	*Taxodium distichum*	ns	7	1027.4	64.2	0.42	11.1	13.0
Northern white cedar	*Thuja occidentalis*	ns	7	649.6	40.6	0.29	11.2	12.0
Western red cedar	*Thuja plicata*	a, b, c, d	2	621.1	38.8	0.31	11.8	12.6
Eastern hemlock	*Tsuga canadensis*	ns	7	979.1	61.1	0.38	18.2	21.0
Western hemlock	*Tsuga heterophylla*	a, b, c, d	2	942.8	58.9	0.42	9.2	10.3
Mountain hemlock	*Tsuga mertensiana*	a	2	1073.0	67.0	0.42	15.0	25.1
Mountain hemlock	*Tsuga mertensiana*	b	2	1031.1	64.4	0.42	15.0	25.1
Mountain hemlock	*Tsuga mertensiana*	c	2	997.4	62.3	0.42	15.0	25.1
Mountain hemlock	*Tsuga mertensiana*	d	2	956.7	59.7	0.42	15.0	25.1
		North American hardwoods						
Red maple	*Acer rubrum*	ns	10	993.0	62.0	0.54	11.0	13.0
Soft maple	*Acer saccharinum*	ns	7	993.0	62.0	0.49	11.2	12.0
Hard maple	*Acer saccharum*	ns	7	1121.1	70.0	0.56	10.5	12.0
Yellow birch	*Betula alleghaniensis*	ns	7	1089.1	68.0	0.55	11.1	12.0
Paper birch	*Betula papyrifera*	ns	7	1009.0	63.0	0.48	14.7	16.0
Pecan hickory	*Carya illinoensis*	ns	7	1057.0	66.0	0.60	11.6	13.0
Hickory	*Carya* spp.	ns	10	1201.2	75.0	0.64	17.0	17.0
Hackberry	*Celtis* spp.	ns	7	961.0	60.0	0.49	13.3	15.0
American beech	*Fagus grandifolia*	ns	7	1025.0	64.0	0.56	6.6	7.0
White ash	*Fraxinus americana*	ns	7	961.0	60.0	0.54	9.1	16.0
Black ash	*Fraxinus nigra*	ns	7	912.9	57.0	0.45	12.3	14.0
Green ash	*Fraxinus pennsylvania*	ns	10	880.9	55.0	0.53	11.8	13.0
Butternut	*Juglans cinerea*	ns	7	832.8	52.0	0.36	14.4	15.0
Black walnut	*Juglans nigra*	ns	7	993.0	62.0	0.51	7.5	15.0
Sweetgum	*Liquidambar styraciflua*	ns	10	1121.1	70.0	0.48	12.0	11.0
Yellow poplar	*Liriodendron tulipifera*	ns	10	1025.0	64.0	0.44	12.4	15.0
Water tupelo	*Nyssa aquatica*	ns	10	944.9	59.0	0.45	11.0	11.0
Black gum	*Nyssa sylvatica*	ns	10	1041.0	65.0	0.48	11.0	12.0
Sycamore	*Platanus occidentalis*	ns	10	1089.1	68.0	0.45	4.0	4.6

continued

Table A.2.A. *continued*

Common name	Latin name	Size class	Source	kg/m^3	lbs/ft^3	Specific gravity (green m^3)	Bark % whole log weight (bark & wood)	Bark % log wood volume
Balsam poplar	*Populas balsamifera*	ns	7	993.0	62.0	0.37	19.7	18.0
Cottonwood	*Populus* spp.	ns	7	944.9	59.0	0.37	15.0	15.0
Aspen	*Populus tremuloides*	ns	7	944.9	59.0	0.39	18.9	18.0
Black cherry	*Prunus serotina*	ns	7	864.9	54.0	0.47	9.4	10.0
White oak	*Quercus alba*	ns	10	1185.2	74.0	0.64	10.0	11.0
Scarlet oak	*Quercus coccinea*	ns	10	1217.2	76.0	0.57	13.0	15.2
Laurel oak	*Quercus laurifolia*	ns	10	1249.2	78.0	0.59	12.8	13.0
Water oak	*Quercus nigra*	ns	10	1249.2	78.0	0.57	12.8	13.0
Chestnut oak	*Quercus prinus*	ns	10	1153.1	72.0	0.62	19.0	24.1
Post oak	*Quercus stellata*	ns	10	1185.2	74.0	0.64	16.0	19.4
Southern red oak	*Quercus falcata*	ns	10	1249.2	78.0	0.57	17.0	20.0
Black willow	*Salix nigra*	ns	7	896.9	56.0	0.34	16.9	16.0
American basswood	*Tilia americana*	ns	7	784.8	49.0	0.32	18.3	16.0
Elm	*Ulmus* spp.	ns	10	1089.1	68.0	0.46	10.1	14.0
Central and South American conifers and hardwoods								
Amburana	*Amburana cearensis*	ns	1	900.0	56.2	0.55	–	–
Espave	*Anacardium* spp.	ns	1	800.0	50.0	0.41	–	–
Angelin, red cabbage tree	*Andira inermis, A. coriacea*	ns	1	1075.0	67.1	0.64	–	–
Parana pine (conifer)	*Araucaria angustifolia*	ns	1	700.0	43.7	0.46	–	–
Chilean pine (conifer)	*Araucaria araucana*	ns	1	700.0	43.7	–	–	–
Amirillo	*Aspidosperma* spp.	ns	1	800.0	50.0	–	–	–
Araracanea	*Aspidosperma* spp.	ns	1	1000.0	62.4	–	–	–
Goncalo Alves	*Astronium* spp.	ns	1	1230.0	76.8	0.80	–	–
Cow-tree	*Brosimum* spp. *(utile group)*	ns	1	750.0	46.8	0.44	–	–
Maracaibo lignum-vitae	*Bulnesia arborea*	ns	1	1250.0	78.0	1.00	–	–
Coffeewood	*Caesalpinia* spp.	ns	1	1200.0	74.9	1.05	–	–
Abarco, Jequitibe	*Cariniana pyriformis, C. brasiliensis*	ns	1	800.0	50.0	0.46	–	–

Andiroba	*Carapa guianensis*	ns	1	900.0	56.2	0.56	–	–
Cedro-rana	*Cedrelinga catenaeformis*	ns	1	950.0	59.3	0.53	–	–
Fromager	*Ceiba pentandra, C. samarina*	ns	1	550.0	34.3	0.25	–	–
Copaiba	*Copaifera* spp.	ns	1	1000.0	62.4	–	–	–
Violet wood	*Dalbergia cearensis*	ns	1	1200.0	74.9	–	–	–
Brasilian tulipwood	*Dalbergia decipularis*	ns	1	1100.0	68.7	–	–	–
Jacaranda	*Dalbergia nigra*	ns	1	1025.0	64.0	0.67	–	–
Cocobolo	*Dalbergia retusa*	ns	1	1300.0	81.2	0.89	–	–
Jacaranda	*Dalbergia spruceana*	ns	1	1100.0	68.7	–	–	–
Honduras rosewood	*Dalbergia stevensonii*	ns	1	1100.0	68.7	0.82	–	–
Angelique	*Dicorynia guianensis*	ns	1	1000.0	62.4	0.60	–	–
Wallaba	*Eperua* spp.	ns	1	1100.0	68.7	0.78	–	–
Alerce (conifer)	*Fitzroya cupressoides*	ns	1	750.0	46.8	0.38	–	–
Kabukalli	*Goupia glabra*	ns	1	1050.0	65.6	0.72	–	–
Lignum vitae	*Guaiacum* spp.	ns	1	1450.0	90.5	1.09	–	–
Pilon	*Hyeronima* spp.	ns	1	1050.0	65.6	0.64	–	–
Hura	*Hura crepitans*	ns	1	625.0	39.0	0.36	–	–
Algarrobo	*Hymenaea courbaril*	ns	1	1100.0	68.7	0.77	–	–
Caviuna	*Machaerium* spp.	ns	1	1100.0	68.7	0.70	–	–
Balata	*Manilkara bidentata*	ns	1	1250.0	78.0	0.85	–	–
Coigüe	*Nothofagus dombeyi*	ns	1	1000.0	62.4	0.49	–	–
Rauli	*Nothofagus procera*	ns	1	1000.0	62.4	0.49	–	–
Demerara greenheart	*Ocotea rodiaei*	ns	1	1300.0	81.2	0.83	–	–
Determa	*Ocotea rubra*	ns	1	825.0	51.5	0.56	–	–
Peroba de campos	*Paratecoma peroba*	ns	1	1025.0	64.0	0.60	–	–
Amaranth	*Peltogyne* spp.	ns	1	1100.0	68.7	–	–	–
Ocote pine	*Pinus oocarpa, pinus patula*	ns	1	900.0	56.2	0.55	–	–
Snakewood	*Piratinera guianensis*	ns	1	1350.0	84.3	0.96	–	–
Macawood	*Platymiscium* spp.	ns	1	1150.0	71.8	0.84	–	–
Manio (conifer)	*Podocarpus* spp.	ns	1	1000.0	62.4	0.46	–	–
Grignon fou	*Qualea* spp.	ns	1	825.0	51.5	0.55	–	–
Quebracho	*Schinopsis* spp.	ns	1	1350.0	84.3	1.00	–	–

continued

Table A.2.A. *continued*

Common name	Latin name	Size class	Source	kg/m^3	lbs/ft^3	Specific gravity (green m^3)	Bark % whole log weight (bark & wood)	Bark % log wood volume
Mahogany	*Swietenia macrophylla*	ns	1	600.0	37.5	0.45	–	–
Ipe, Lapacho	*Tabebuia* spp.	ns	1	1250.0	78.0	0.92	–	–
Baboen	*Virola* spp.	ns	1	775.0	48.4	0.44	–	–
Kwari	*Vochysia* spp.	ns	1	825.0	51.5	0.40	–	–
Acapu	*Vouacapoua americana*	ns	1	1100.0	68.7	0.79	–	–
West Indian satinwood	*Zanthoxylum flavum*	ns	1	1100.0	68.7	0.73	–	–
	Asian, Australian, and Oceania conifers and hardwoods							
Mulga	*Acacia aneura*	ns	3	1330.0	83.0	1.02	–	–
Gidgee	*Acacia cambagei*	ns	3	1265.0	79.0	0.97	–	–
Miniritchie	*Acacia monticola*	ns	3	1275.0	79.6	1.05	–	–
Haldu	*Adina* spp.	ns	1	1000.0	62.4	0.59	–	–
Kauri pine (conifer)	*Agathis australis, Agathis* spp.	ns	1	840.0	52.4	0.44	–	–
Pulai	*Alstonia* spp.	ns	1	800.0	50.0	0.37	–	–
Amoora	*Amoora* spp.	ns	1	775.0	48.4	–	–	–
Mersawa	*Anisoptera* spp.	ns	1	1000.0	62.4	0.51	–	–
Hoop-pine, Klinki pine (conifer)	*Araucaria* spp.	ns	1	880.0	54.9	0.42	–	–
Chengal	*Balanocarpus* spp.	ns	1	1200.0	74.9	0.76	–	–
White cypress pine (conifer)	*Callitris glauca*	ns	3	770.0	48.1	0.58	–	–
Bintangor	*Calophyllum* spp.	ns	1	800.0	50.0	–	–	–
Australian black 'oak'	*Casuarina* spp.	ns	3	1320.0	82.4	1.09	–	–
Deodar cedar (conifer)	*Cedrus deodara*	ns	1	750.0	46.8	–	–	–
East Indian satinwood	*Chloroxylon swietenia*	ns	1	1050.0	65.6	0.80	–	–
Medang	*Cinnamomum* spp.	ns	1	850.0	53.1	–	–	–
Sugi (conifer)	*Cryptomeria japonica*	ns	1	600.0	37.5	–	–	–
San-mu (conifer)	*Cunninghamia lanceolata*	ns	1	600.0	37.5	–	–	–
Silver pine (conifer)	*Dacrydium colensoi*	ns	5	864.9	54.0	0.54	–	–
Rimu (conifer)	*Dacrydium cupressinum*	ns	5	961.0	60.0	0.49	–	–

Sempilor (conifer)	*Dacrydium* spp.	ns	1	850.0	53.1	–	–	–
Medang	*Dehaasia* spp.	ns	1	850.0	53.1	–	–	–
Asian ebony	*Diospyros* spp.	ns	1	1350.0	84.3	–	–	–
Kayu malam African ebony	*Diospyros* spp.	ns	1	1325.0	82.7	–	–	–
Apitong	*Dipterocarpus* spp.	ns	1	1025.0	64.0	0.59	–	–
New Guinea wood	*Dracontomelum* spp.	ns	1	900.0	56.2	0.50	–	–
Duabanga, Magas	*Duabanga* spp.	ns	1	650.0	40.6	0.33	–	–
Corkwood	*Duboisia myoporoides*	ns	3	1225.0	76.5	0.80	–	–
Hinau	*Elaeocarpus denatus*	ns	5	1121.1	70.0	0.57	–	–
Brown mallet	*Eucalyptus astringens*	ns	3	1120.0	69.9	0.77	–	–
Dundas 'mahogany'	*Eucalyptus brockwayi*	ns	3	1260.0	78.7	–	–	–
Red gum	*Eucalyptus camaldulendsis*	ns	1	1150.0	71.8	–	–	–
Spotted gum	*Eucalyptus citriodora*	ns	3	1150.0	71.8	0.79	–	–
Sugar gum	*Eucalyptus cladocalyx*	ns	3	1105.0	69.0	0.75	–	–
Cleland's blackbutt	*Eucalyptus clelandii*	ns	3	1215.0	75.9	0.97	–	–
Deglupta	*Eucalyptus deglupta*	ns	1	1000.0	62.4	0.55	–	–
Karri	*Eucalyptus diversicolor*	ns	1	1150.0	71.8	0.70	–	–
Merrit	*Eucalyptus flocktoniae*	ns	3	1145.0	71.5	–	–	–
Tasmanian 'oak'	*Eucalyptus gigantea*	ns	1	1050.0	65.6	–	–	–
Blue gum	*Eucalyptus globulus*	ns	1	1100.0	68.7	–	–	–
Goldfields blackbutt	*Eucalyptus lesouefii*	ns	3	1180.0	73.7	0.88	–	–
Red morrel	*Eucalyptus longicornis*	ns	3	1240.0	77.4	0.95	–	–
Jarrah	*Eucalyptus marginata*	ns	1	1150.0	71.8	0.67	–	–
Yellow stringy bark	*Eucalyptus muelleriana*	ns	3	1100.0	68.7	0.69	–	–
Blackbutt	*Eucalyptus pilularis*	ns	1	1150.0	71.8	–	–	–
Salt gum	*Eucalyptus salicola*	ns	3	1215.0	75.9	0.94	–	–
Salmon gum	*Eucalyptus salmonophloia*	ns	3	1160.0	72.4	0.87	–	–
Gimlet	*Eucalyptus salubris*	ns	3	1230.0	76.8	0.94	–	–
Forest red gum	*Eucalyptus tereticornis*	ns	1	1150.0	71.8	–	–	–
Wandoo	*Eucalyptus wandoo*	ns	3	1280.0	79.9	0.92	–	–
Belian	*Eusideroxylon zwageri*	ns	1	1250.0	78.0	0.89	–	–
Japanese ash	*Fraxinus mandschurica*	ns	1	900.0	56.2	–	–	–

continued

Table A.2.A. *continued*

Common name	Latin name	Size class	Source	kg/m^3	lbs/ft^3	Specific gravity (green m^3)	Bark % whole log weight (bark & wood)	Bark % log wood volume
Scaly ash	*Ganophyllum falcatum*	ns	1	1100.0	68.7	–	–	–
Beefwood	*Grevillea striata*	ns	3	1230.0	76.8	0.82	–	–
Huynh, heritiera	*Heritiera* spp.	ns	1	930.0	58.1	0.56	–	–
Malas	*Homalium tomentosum*	ns	1	1000.0	62.4	–	–	–
Japanese walnut	*Juglans sieboldiana*	ns	1	700.0	43.7	–	–	–
Sen	*Kalopanaxpictus nakai*	ns	1	800.0	50.0	–	–	–
Bungur	*Lagerstroemina* spp.	ns	1	900.0	56.2	0.55	–	–
Medang	*Litsea* spp.	ns	1	850.0	53.1	0.40	–	–
Machang	*Mangifera* spp.	ns	1	850.0	53.1	0.51	–	–
Thinwin	*Millettia pendula*	ns	1	1300.0	81.2	–	–	–
Red beech	*Nothofagus fusca*	ns	5	977.0	61.0	0.57	–	–
Silver beech	*Nothofagus menziesii*	ns	5	880.9	55.0	0.45	–	–
Mountain beech	*Nothofagus solandri*	ns	5	961.0	60.0	0.52	–	–
New Guinea 'beech'	*Nothofagus* spp.	ns	1	1050.0	65.6	–	–	–
Hard beech	*Nothofagus truncata*	ns	5	1105.1	69.0	0.63	–	–
Nyatoh, bitis	*Palaquium* spp., *Payena* spp.	ns	1	1075.0	67.1	0.55	–	–
Bagtikan	*Parashorea* spp.	ns	1	1100.0	68.7	0.44	–	–
Thitka	*Pentacme* spp.	ns	1	800.0	50.0	0.56	–	–
White lauan	*Pentacme contorta*	ns	1	950.0	59.3	0.43	–	–
Tropical Asian pine (conifer)	*Pinus kesyia*	ns	1	875.0	54.6	0.47	–	–
Merkus pine (conifer)	*Pinus merkusii*	ns	1	875.0	54.6	0.57	–	–
Chir pine (conifer)	*Pinus roxburghii*	ns	1	800.0	50.0	–	–	–
Red planchonella	*Planchnella torricellensis*	ns	1	875.0	54.6	–	–	–
Kahikatea (conifer)	*Podocarpus dacrydioides*	ns	5	816.8	51.0	0.39	–	–
Miro (conifer)	*Podocarpus ferrugineus*	ns	5	864.9	54.0	0.51	–	–
Matai (conifer)	*Podocarpus spicaus*	ns	5	1089.1	68.0	0.54	–	–
Podocarp (conifer)	*Podocarpus* spp.	ns	1	850.0	53.1	–	–	–
Podo totara (conifer)	*Podocarpus* spp.	ns	5	944.9	59.0	–	–	–

Kasai	*Pometia* spp.	ns	1	1000.0	62.4	0.54	–	–
Amboyna burl	*Pterocarpus indicus, P. vidalianus*	ns	1	800.0	50.0	–	–	–
Amberoi	*Pterocymbium beccarii*	ns	1	625.0	39.0	–	–	–
Bayur	*Pterospermum* spp.	ns	1	750.0	46.8	–	–	–
Japanese oak	*Quercus grosseserrata*	ns	1	900.0	56.2	–	–	–
Alan	*Shorea albida*	ns	1	900.0	56.2	–	–	–
Almon, white lauan	*Shorea almon*	ns	1	725.0	45.3	–	–	–
Balau	*Shorea* spp. *(Balau group)*	ns	1	1200.0	74.9	0.70	–	–
White meranti	*Shorea* spp. *(white meranti)*	ns	1	725.0	45.3	0.47	–	–
Yellow meranti	*Shorea* spp. *(yellow meranti)*	ns	1	900.0	56.2	0.46	–	–
Red meranti	*Shorea* spp. Subgen. *Rubroshorea*	ns	1	800.0	50.0	–	–	–
Teak	*Tectona grandis*	ns	1	850.0	53.1	0.57	–	–
Brown terminalia	*Terminalia brassii*	ns	1	675.0	42.1	–	–	–
Indian almond wood	*Terminalia* spp.	ns	1	825.0	51.5	0.51	–	–
White bombay	*Terminalia procera*	ns	1	800.0	50.0	–	–	–
Red cedar	*Toona sureni, T. calantas*	ns	1	750.0	46.8	0.42	–	–
Brush box	*Tristania* spp.	ns	1	1200.0	74.9	0.80	–	–
Pyinkado	*Xylia xylocarpa*	ns	1	1200.0	74.9	0.81	–	–
Mangrove cedar	*Xylocarpus* spp.	ns	1	950.0	59.3	–	–	–
		African conifers and hardwoods						
Adina	*Adina microcephala*	ns	1	1200.0	74.9	–	–	–
Afzelia	*Afzelia cuanzensis*	ns	1	1250.0	78.0	0.67	–	–
Doussié	*Afzelia* spp.	ns	1	1150.0	71.8	0.67	–	–
Albizzia	*Albizia* spp.	ns	1	975.0	60.9	0.51	–	–
Alstonia	*Alstonia congensis*	ns	1	600.0	37.5	0.33	–	–
Lati	*Amphimas* spp.	ns	1	950.0	59.3	–	–	–
Mecrusse	*Androstachys johnsonii*	ns	1	1200.0	74.9	0.72	–	–
Aningeria	*Aningeria* spp.	ns	1	900.0	56.2	0.44	–	–
Bodioa	*Anopyxis klaineana*	ns	1	1025.0	64.0	–	–	–
Ako	*Antiaris* spp.	ns	1	800.0	50.0	0.38	–	–
Onzabili	*Antrocaryon* spp.	ns	1	850.0	53.1	–	–	–

continued

Table A.2.A. *continued*

Common name	Latin name	Size class	Source	kg/m^3	lbs/ft^3	Specific gravity (green m^3)	Bark % whole log weight (bark & wood)	Bark % log wood volume
Gaboon (okoumé)	*Aucoumea klaineana*	ns	1	600.0	37.5	0.37	–	–
Mukulungu	*Autranella congolensis*	ns	1	1100.0	68.7	0.78	–	–
Rhodesian-teak	*Baikiaea plurijuga*	ns	1	1150.0	71.8	0.73	–	–
Dimpampi	*Baillonella toxisperma*	ns	1	1050.0	65.6	0.71	–	–
Kanda	*Beilschmieda* spp.	ns	1	1000.0	62.4	–	–	–
Berlinia	*Berlinia* spp.	ns	1	900.0	56.2	0.58	–	–
East African bombax	*Bombax rhodognaphalon*	ns	1	900.0	56.2	0.40	–	–
Bombax	*Bombax* spp.	ns	1	800.0	50.0	0.40	–	–
Muhuhu	*Brachylaena hutchinsii*	ns	1	1200.0	74.9	0.75	–	–
Okwen	*Brachystegia* spp.	ns	1	950.0	59.3	0.51	–	–
South African boxwood	*Bruxus macowani*	ns	1	1100.0	68.7	–	–	–
Burkea	*Burkea africana*	ns	1	1150.0	71.8	0.70	–	–
Miama	*Calpocalyx heitzii*	ns	1	1100.0	68.7	–	–	–
Abel	*Canarium schweinfurthii*	ns	1	800.0	50.0	0.40	–	–
Pillarwood	*Cassipourea malosana*	ns	1	1000.0	62.4	0.59	–	–
Ceiba	*Ceiba pentandra*	ns	1	900.0	56.2	0.26	–	–
African celtis	*Celtis* spp.	ns	1	950.0	59.3	0.59	–	–
Esa	*Celtis* spp.	ns	1	950.0	59.3	0.59	–	–
Ohia	*Celtis* spp.	ns	1	950.0	59.3	0.59	–	–
Mtambara	*Cephalosphaera usambarensis*	ns	1	850.0	53.1	0.48	–	–
Kambala	*Chlorophora excelsa* and *C. regia*	ns	1	1075.0	67.1	0.55	–	–
Sobu	*Cleistopholis paterns, C. glauca*	ns	1	550.0	34.3	–	–	–
Ekoune	*Coelocaryon preussii*	ns	1	800.0	50.0	–	–	–
Abale	*Combretodenodron africanum*	ns	1	1100.0	68.7	0.70	–	–
Ghéombi	*Copaifera letesturi*	ns	1	900.0	56.2	–	–	–
N'téné	*Copaifera religosa*	ns	1	800.0	50.0	–	–	–
Etimoé	*Copaifera salikounda*	ns	1	950.0	59.3	–	–	–
West African cordia	*Cordia millenii* and *C. platythyrsa*	ns	1	750.0	46.8	0.34	–	–

Cordyla	*Cordyla africana*	ns	1	1100.0	68.7	–	–	–
Tsanya	*Corynanthe paniculata*	ns	1	950.0	59.3	–	–	–
Musine	*Croton megalocarpus*	ns	1	950.0	59.3	0.57	–	–
Tani	*Cryptosepalum staudtii*	ns	1	1050.0	65.6	–	–	–
Okan	*Cylicodiscus gabunensi*	ns	1	1050.0	65.6	0.80	–	–
Muhimbi	*Cynometra alexandri*	ns	1	1000.0	62.4	0.74	–	–
Apomé	*Cynometra ananta*	ns	1	1200.0	74.9	–	–	–
Nganga	*Cynometra hankei*	ns	1	1150.0	71.8	–	–	–
Safukala	*Dacryodes pubescens, D. heterotricha*	ns	1	975.0	60.9	0.61	–	–
Adjouaba	*Dacryodes* spp.	ns	1	850.0	53.1	0.61	–	–
African blackwood	*Dalbergia melanoxylon*	ns	1	1400.0	87.4	1.08	–	–
Daniellia	*Daniellia ogea* and *D. thurifera*	ns	1	900.0	56.2	0.40	–	–
Alep	*Desbordesia glaucescens*	ns	1	1250.0	78.0	–	–	–
Mambodé	*Detarium* spp.	ns	1	925.0	57.8	–	–	–
Eyoum	*Dialium* spp.	ns	1	1050.0	65.6	0.71	–	–
Sapo	*Didelotia brevipaniculata*	ns	1	950.0	59.3	0.53	–	–
African ebony	*Diospyros crassiflora*	ns	1	1325.0	82.7	0.82	–	–
Ayan	*Distemonanthus benthamianus*	ns	1	900.0	56.2	0.58	–	–
Yungu	*Drypetes gossweileri*	ns	1	950.0	59.3	–	–	–
Moambe jaune	*Enantia chlorantha*	ns	1	900.0	56.2	–	–	–
Gedu nohor	*Entandrophragma angolense*	ns	1	850.0	53.1	0.45	–	–
Kosipo	*Entandrophragma candollei*	ns	1	900.0	56.2	–	–	–
Aboudikro	*Entandrophragma cylindricum*	ns	1	900.0	56.2	0.60	–	–
Mutaco	*Entandrophragma ekebergioides*	ns	1	850.0	53.1	–	–	–
Sipo	*Entandrophragma utile*	ns	1	775.0	48.4	0.53	–	–
Missanda	*Erythrophleum ivorense*	ns	1	1150.0	71.8	0.72	–	–
Landa	*Erythroxylum manii*	ns	1	850.0	53.1	0.50	–	–
Olon	*Fagara heitzii*	ns	1	750.0	46.8	–	–	–
East African satinwood	*Fagara macrophylla*	ns	1	950.0	59.3	–	–	–
Mafu	*Fagaropsis angolensis*	ns	1	950.0	59.3	0.56	–	–
Mutando	*Funtumia* spp.	ns	1	850.0	53.1	–	–	–
Longui	*Gambeya* spp.	ns	1	1050.0	65.6	0.63	–	–

continued

Table A.2.A. *continued*

Common name	Latin name	Size class	Source	kg/m^3	lbs/ft^3	Specific gravity (green m^3)	Bark % whole log weight (bark & wood)	Bark % log wood volume
Limbali	*Gilbertiodendron* spp.	ns	1	1050.0	65.6	–	–	–
Kamassi	*Gonioma kamassi*	ns	1	1100.0	68.7	0.75	–	–
Agba	*Gossweilerodendron balsamiferum*	ns	1	800.0	50.0	0.40	–	–
Scented Guarea	*Guarea cedrata*	ns	1	900.0	56.2	0.48	–	–
Black Guarea	*Guarea thompsonii*	ns	1	975.0	60.9	0.48	–	–
Benge	*Guibourtia arnoldiana*	ns	1	1050.0	65.6	0.64	–	–
Rhodesian copalwood	*Guibourtia coleosperma*	ns	1	1150.0	71.8	–	–	–
Amazakoue	*Guibourtia ehie*	ns	1	1050.0	65.6	0.67	–	–
Bubinga	*Guibourtia* spp.	ns	1	1100.0	68.7	0.72	–	–
Abura	*Hallea ciliata* and *H. stipulosa*	ns	1	925.0	57.8	–	–	–
Idewa	*Haplormosia monophylla*	ns	1	1150.0	71.8	–	–	–
Kékélé	*Holoptelea grandis*	ns	1	900.0	56.2	–	–	–
Alumbi	*Julbernardia seratii*	ns	1	1150.0	71.8	–	–	–
African pencil cedar (conifer)	*Juniperus procera*	ns	1	750.0	46.8	0.48	–	–
Senegal mahogany	*Khaya grandifoliola*	ns	1	775.0	48.4	0.60	–	–
African mahogany	*Khaya ivorensis*	ns	1	700.0	43.7	0.45	–	–
East African mahogany	*Khaya nyasica*	ns	1	800.0	50.0	0.45	–	–
Kumbi	*Lannea welwitschii*	ns	1	750.0	46.8	–	–	–
Oguomo	*Lecomtedoxa klaineana*	ns	1	1150.0	71.8	–	–	–
Congotali	*Letestua durissima*	ns	1	1200.0	74.9	–	–	–
Azobe	*Lophira alata*	ns	1	1200.0	74.9	0.90	–	–
African walnut	*Lovoa trichilioides*	ns	1	800.0	50.0	0.45	–	–
Musizi	*Maesopsis eminii*	ns	1	700.0	43.7	0.35	–	–
Oboto	*Mammea africana*	ns	1	1100.0	68.7	0.62	–	–
Monghinza	*Manilkara* spp.	ns	1	1200.0	74.9	–	–	–
Aprono	*Mansonia altissima*	ns	1	925.0	57.8	0.54	–	–
Zebrano	*Microberlinia* spp.	ns	1	1150.0	71.8	0.70	–	–
Panga panga	*Millettia* spp.	ns	1	1150.0	71.8	0.71	–	–

Bahia	*Mitragyna ciliata*	ns	1	925.0	57.8	0.45	–	–
Andoung	*Monopetalanthus heitzii*	ns	1	900.0	56.2	0.39	–	–
Difou	*Morus mesozygia*	ns	1	1050.0	65.6	0.70	–	–
Umbrella tree	*Musanga cecropioides*	ns	1	750.0	46.8	0.23	–	–
Bilinga	*Nauclea diderrichii*	ns	1	1050.0	65.6	0.63	–	–
Danta	*Nesogordonia papaverifera*	ns	1	925.0	57.8	0.65	–	–
East African newtonia	*Newtonia buchananii*	ns	1	900.0	56.2	–	–	–
Ossimiale	*Newtonia leucocarpa*	ns	1	1000.0	62.4	–	–	–
East African camphorwood	*Ocotea usambarensis*	ns	1	875.0	54.6	0.51	–	–
Vésa'mbata	*Oldfieldia africana*	ns	1	1050.0	65.6	–	–	–
East African olive	*Olea hochstetteri*	ns	1	1200.0	74.9	0.72	–	–
Elgon olive	*Olea welwitschii*	ns	1	1050.0	65.6	–	–	–
Angueuk	*Ongokea gore*	ns	1	1050.0	65.6	0.72	–	–
Tola mafuta	*Oxystigma oxyphyllum*	ns	1	900.0	56.2	0.53	–	–
Awoura	*Paraberlinia bifoliolata*	ns	1	1000	62.4	–	–	–
Mubara	*Parinari excelsa*	ns	1	1000.0	62.4	0.69	–	–
Kiasose	*Pentadesma butyracea, P. lebruntii*	ns	1	1000.0	62.4	–	–	–
Aformosia	*Pericopsis elata*	ns	1	1125.0	70.2	0.57	–	–
Agboin	*Piptadeniastrum africanum*	ns	1	1000.0	62.4	0.56	–	–
Podo (conifer)	*Podocarpus gracilior, P. usambarensis*	ns	1	700.0	43.7	0.43	–	–
Ovoga	*Poga oleosa*	ns	1	950.0	59.3	0.36	–	–
Osanga	*Pteleopsis hylodendron*	ns	1	1000.0	62.4	–	–	–
Muninga	*Pterocarpus angolensis*	ns	1	1050.0	65.6	0.59	–	–
African padauk	*Pterocarpus soyauxii*	ns	1	950.0	59.3	0.61	–	–
Koto	*Pterygota* spp.	ns	1	900.0	56.2	0.52	–	–
Akomu	*Pycnanthus angolensis*	ns	1	800.0	50.0	0.44	–	–
Mueri	*Pygeum africanum*	ns	1	950.0	59.3	0.58	–	–
Mugaita	*Rapancea rhododendroides*	ns	1	950.0	59.3	–	–	–
Erimado	*Ricinodendron heudelotii*	ns	1	800.0	50.0	0.20	–	–
Erimado	*Ricinodendron rautanenii*	ns	1	750.0	46.8	–	–	–
Ozouga	*Sacoglottis gabonensis*	ns	1	1050.0	65.6	–	–	–
Odoko	*Scottellia coriaceae*	ns	1	950.0	59.3	0.56	–	–

continued

Table A.2.A. *continued*

Common name	Latin name	Size class	Source	kg/m^3	lbs/ft^3	Specific gravity (green m^3)	Bark % whole log weight (bark & wood)	Bark % log wood volume
Ossoko	*Seyphocephalium mannii*	ns	1	925.0	57.8	–	–	–
Tamboti	*Spirostachys africana*	ns	1	1150.0	71.8	–	–	–
Niove	*Staudtia stipitata*	ns	1	1050.0	65.6	0.75	–	–
Eyong	*Sterculia oblonga*	ns	1	1100.0	68.7	–	–	–
Brown stericula	*Sterculia rhinopetala*	ns	1	1000.0	62.4	0.64	–	–
Dina	*Swarztia fistuloides*	ns	1	1150.0	71.8	0.82	–	–
Pao rosa	*Swartzia madagascasiensis*	ns	1	1150.0	71.8	–	–	–
Ossol	*Symphonia globulifera*	ns	1	925.0	57.8	0.58	–	–
Niangon	*Tarrietia utilis* and *T.densiflora*	ns	1	950.0	59.3	–	–	–
Emeri	*Terminalia ivorensis*	ns	1	800.0	50.0	0.43	–	–
Akom	*Terminalia superba*	ns	1	762.5	47.6	0.49	–	–
Wamba	*Tessmannia* spp.	ns	1	1050.0	65.6	–	–	–
Izombe	*Testulea gabonensis*	ns	1	1000.0	62.4	0.60	–	–
Ekop	*Tetraberlinia tubmaniana*	ns	1	825.0	51.5	0.60	–	–
Baku	*Tieghemella* spp.	ns	1	937.5	58.5	0.55	–	–
Abachi	*Triplochiton scleroxylon*	ns	1	625.0	39.0	0.33	–	–
Avodire	*Turraeanthus africanus*	ns	1	800.0	50.0	0.51	–	–
Sugar-plum	*Uapaca* spp.	ns	1	1000.0	62.4	0.60	–	–
Vitex	*Vitex* spp.	ns	1	825.0	51.5	0.40	–	–
Kenyan greenheart	*Warburgia ugandensis*	ns	1	1150.0	71.8	–	–	–
Mlanje-cedar (conifer)	*Widdringtonia whytei*	ns	1	700.0	43.7	0.45	–	–
		European conifers						
Silver fir	*Abies alba*	ns	4	920.0	57.4	0.40	–	–
Common juniper	*Juniperus communis*	ns	4	1020.0	63.7	–	–	–
European larch	*Larix decidua*	ns	4	860.0	53.7	0.45	–	20.0
Siberian larch	*Larix sibiricus*	ns	6	885.0	55.3	0.46	–	–
Norway spruce	*Picea abies*	ns	4	860.0	53.7	0.38	–	9.0

Siberian yellow pine	*Pinus cembra*	ns	4	870.0	54.3	–	–	–
Aleppo pine	*Pinus halepensis*	ns	4	1000.0	62.4	–	–	–
Corsican pine	*Pinus laricio*	ns	4	900.0	56.2	–	–	21.0
Swiss mountain pine	*Pinus montana*	ns	4	800.0	50.0	–	–	–
Austrian pine	*Pinus nigra*	ns	4	900.0	56.2	–	–	–
Maritime pine	*Pinus pinaster*	ns	4	950.0	59.3	–	–	–
Umbrella pine	*Pinus pinea*	ns	4	850.0	53.1	–	–	–
Scots pine	*Pinus sylvestris*	ns	4	880.0	54.9	0.40	–	15.5
English yew	*Taxus baccata*	ns	4	1020.0	63.7		–	–
		European hardwoods						
Field maple	*Acer campestre*	ns	4	900.0	56.2	–	–	–
Neapolitan maple	*Acer opalus*	ns	4	870.0	54.3	–	–	–
Norway maple	*Acer platanoides*	ns	4	880.0	54.9	–	–	–
Sycamore	*Acer pseudoplatanus*	ns	4	820.0	51.2	–	–	–
Horse chestnut	*Aesculus hyppocastanum*	ns	4	900.0	56.2	–	–	–
Napoletano alder	*Alnus cordata*	ns	4	830.0	51.8	–	–	–
Black alder	*Alnus glutinosa*	ns	4	860.0	53.7	–	–	–
White alder	*Alnus incana*	ns	4	840.0	52.4	0.45	–	–
Strawberry tree	*Arbutus unedo*	ns	4	900.0	56.2	–	–	–
White birch	*Betula alba*	ns	4	950.0	59.3	–	–	–
European boxwood	*Buxus sempervirens*	ns	4	1070.0	66.8	–	–	–
European hornbeam	*Carpinus betulus*	ns	4	1000.0	62.4	–	–	–
Chestnut	*Castagna sativa*	ns	4	1000.0	62.4	–	–	–
Hackberry	*Celtis australis*	ns	4	960.0	59.9	–	–	–
Carob tree	*Ceratonia siliqua*	ns	4	820.0	51.2	–	–	–
Judas tree	*Cercis siliquastrum*	ns	4	880.0	54.9	–	–	–
Cornelian cherry	*Cornus mas*	ns	4	1020.0	63.7	–	–	–
Common hazel	*Corylus avellana*	ns	4	920.0	57.4	–	–	–
Heath	*Erica arborea*	ns	4	950.0	59.3	–	–	–
Beech	*Fagus sylvatica*	ns	4	1050.0	65.6	0.58	–	–
Common ash	*Fraxinus excelsior*	ns	4	960.0	59.9	–	–	–

continued

Table A.2.A. *continued*

Common name	Latin name	Size class	Source	kg/m^3	lbs/ft^3	Specific gravity (green m^3)	Bark % whole log weight (bark & wood)	Bark % log wood volume
Flowering ash	*Fraxinus ornus*	ns	4	820.0	51.2	–	–	–
Common walnut	*Juglans regia*	ns	4	930.0	58.1	–	–	–
Laburnum	*Laburnum anagyroides*	ns	4	1000.0	62.4	–	–	–
Laurel	*Lauras nobilis*	ns	4	900.0	56.2	–	–	–
Black mullberry	*Morus nigra*	ns	4	1000.0	62.4	–	–	–
Myrtle	*Myrtus communis*	ns	4	1000.0	62.4	–	–	–
Olive	*Olea europaea*	ns	4	1120.0	69.9	–	–	–
European hophornbean	*Ostrya carpinifolia*	ns	4	1050.0	65.6	–	–	–
Plane tree	*Platanus orientalis*	ns	4	960.0	59.9	–	–	–
White poplar	*Populas alba*	ns	4	800.0	50.0	–	–	–
Black poplar	*Populas negra*	ns	4	840.0	52.4	–	–	–
Quaking aspen	*Populas tremula*	ns	4	860.0	53.7	–	–	–
Wild cherry	*Prunus avium*	ns	4	850.0	53.1	–	–	–
Apple	*Pyrus malus*	ns	4	880.0	54.9	–	–	–
Austrian oak	*Quercus cerris*	ns	4	1100.0	68.7	–	–	–
Evergreen oak	*Quercus ilex*	ns	4	1100.0	68.7	–	–	–
Pedunculata oak	*Quercus robur*	ns	4	1080.0	67.4	–	–	–
Cork oak	*Quercus suber*	ns	4	1100.0	68.7	–	–	–
Black locust	*Robinia pseudoacacia*	ns	4	1050.0	65.6	–	–	–
White willow	*Salix alba*	ns	4	880.0	54.9	–	–	–
Mountain ash	*Sorbus aucuparia*	ns	4	1000.0	62.4	–	–	–
Summer lime	*Tilia plathyphylla*	ns	4	900.0	56.2	–	–	–
Mountain elm	*Ulmus montana*	ns	4	1000.0	62.4	–	–	–

Note: size classes are based on small end diameter of delivered log lengths: a = 4.5–7.49″ (11.4–19.05 cm); b = 7.5–11.49″ (19.06–29.19 cm); c = 11.5–15.49″ (29.2– 39.35 cm); d = 15.5″+ (39.36 cm+); ns = size not specified. Green weight to volume ratios for logs includes weight of bark, and volume of wood only. If weight referenced listed a range, e.g. 900-1000 kg/m^3, weight shown is the average of the range, e.g. 950 kg/m^3.

Source: **1**. International Technical Association of Tropical Timber, 1982; **2**. Author's own research; **3**. Forest Products Commission of Western Australia, 2004; **4**. Giordano, G., 1976; **5**. Entrican, Alex R., Hinds, H.V., Reid, J.S., 1957; **6**. University of Helsinki Finland, 2003; **7**. Smith, Brad W., 1985; **8**. Hamilton, G.J., 1975; **9**. Ellis, Elliot, 2001; **10**. Clark, Alexander III, Phillips, Douglas R., Frederick, Douglas J., 1986, 1985; **11**. Kingsbury J., 2004 (personal correspondence); **12**. British Columbia Ministry of Forestry, Harvest Billing System, 2004; **13**. Brereton B., 1929.

Table A.2.B. Bone-dry weight and volume conversions for selected tree species of the world.

Common name	Latin name	Size class	Green with bark kg/m^3	Green with bark lbs/ft^3	Specific gravity (green volume)	BD weight kg/m^3	BD weight lbs/ft^3	ft^3/BDU*	m^3/BDU*	ft^3/BDT*	m^3/BDMT*	Ton/BDU**	Tonne/BDU**	Ton/BDT** Tonne/BDMT
			North American conifers											
Pacific silver fir	*Abies amabilis*	ns	999.4	62.4	0.40	400.0	25.0	96.2	2.72	80.1	2.50	3.00	2.72	2.50
Balsam fir	*Abies balsamea*	ns	864.1	54.0	0.33	330.0	20.6	116.6	3.30	97.1	3.03	3.14	2.85	2.62
White fir	*Abies concolor*	ns	1034.1	64.6	0.37	370.0	23.1	104.0	2.94	86.6	2.70	3.36	3.05	2.80
Grand fir	*Abies grandis*	a	884.6	55.2	0.35	350.0	21.8	109.9	3.11	91.6	2.86	3.03	2.75	2.53
Grand fir	*Abies grandis*	b	854.0	53.3	0.35	350.0	21.8	109.9	3.11	91.6	2.86	2.93	2.66	2.44
Grand fir	*Abies grandis*	c	840.0	52.4	0.35	350.0	21.8	109.9	3.11	91.6	2.86	2.88	2.61	2.40
Grand fir	*Abies grandis*	d	817.6	51.0	0.35	350.0	21.8	109.9	3.11	91.6	2.86	2.80	2.54	2.34
Subalpine fir	*Abies lasiocarpa*	a	756.5	47.2	0.31	310.0	19.3	124.1	3.51	103.4	3.23	2.93	2.66	2.44
Subalpine fir	*Abies lasiocarpa*	b	749.9	46.8	0.31	310.0	19.3	124.1	3.51	103.4	3.23	2.90	2.63	2.42
Subalpine fir	*Abies lasiocarpa*	c, d	753.7	47.1	0.31	310.0	19.3	124.1	3.51	103.4	3.23	2.92	2.65	2.43
Noble fir	*Abies procera*	a	934.6	58.4	0.37	370.0	23.1	104.0	2.94	86.6	2.70	3.04	2.75	2.53
Noble fir	*Abies procera*	d	663.1	41.4	0.37	370.0	23.1	104.0	2.94	86.6	2.70	2.15	1.95	1.79
Port orford cedar	*Chamaecyparis lawsoniana*	ns	1007.5	62.9	0.39	390.0	24.3	98.6	2.79	82.2	2.56	3.10	2.81	2.58
Yellow cedar	*Chamaecyparis nootkatensis*	ns	730.0	45.6	0.42	420.0	26.2	91.6	2.59	76.3	2.38	2.09	1.89	1.74
Eastern red cedar	*Juniperus virginiana*	ns	765.6	47.8	0.44	440.0	27.5	87.4	2.47	72.8	2.27	2.09	1.89	1.74
Tamarack	*Larix laricina*	ns	951.7	59.4	0.49	490.0	30.6	78.5	2.22	65.4	2.04	2.33	2.12	1.94
Western larch	*Larix occidentalis*	a	921.4	57.5	0.48	480.0	30.0	80.1	2.27	66.8	2.08	2.30	2.09	1.92
Western larch	*Larix occidentalis*	b	898.7	56.1	0.48	480.0	30.0	80.1	2.27	66.8	2.08	2.25	2.04	1.87
Western larch	*Larix occidentalis*	c	882.4	55.1	0.48	480.0	30.0	80.1	2.27	66.8	2.08	2.21	2.00	1.84
Western larch	*Larix occidentalis*	d	867.0	54.1	0.48	480.0	30.0	80.1	2.27	66.8	2.08	2.17	1.97	1.81
Incense cedar	*Libocedrus decurrens*	ns	906.7	56.6	0.35	350.0	21.8	109.9	3.11	91.6	2.86	3.11	2.82	2.59
Engelmann spruce	*Picea engelmannii*	a	927.3	57.9	0.33	330.0	20.6	116.6	3.30	97.1	3.03	3.37	3.06	2.81
Engelmann spruce	*Picea engelmannii*	b	846.6	52.9	0.33	330.0	20.6	116.6	3.30	97.1	3.03	3.08	2.79	2.57

continued

Table A.2.B. *continued*

Common name	Latin name	Size class	Green with bark kg/m^3	Green with bark lbs/ft^3	Specific gravity (green volume)	BD weight kg/m^3	BD weight lbs/ft^3	ft^3/BDU*	m^3/BDU*	ft^3/BDT*	m^3/BDMT*	Ton/BDU**	Tonne/BDU**	Ton/BDT Tonne/BDMT**
Engelmann spruce	*Picea engelmannii*	c	776.4	48.5	0.33	330.0	20.6	116.6	3.30	97.1	3.03	2.83	2.56	2.35
Engelmann spruce	*Picea engelmannii*	d	689.3	43.0	0.33	330.0	20.6	116.6	3.30	97.1	3.03	2.51	2.27	2.09
White spruce	*Picea glauca*	a	797.7	49.8	0.33	330.0	20.6	116.6	3.30	97.1	3.03	2.90	2.63	2.42
White spruce	*Picea glauca*	b	766.3	47.8	0.33	330.0	20.6	116.6	3.30	97.1	3.03	2.79	2.53	2.32
White spruce	*Picea glauca*	c	740.1	46.2	0.33	330.0	20.6	116.6	3.30	97.1	3.03	2.69	2.44	2.24
White spruce	*Picea glauca*	d	726.4	45.4	0.33	330.0	20.6	116.6	3.30	97.1	3.03	2.64	2.40	2.20
Black spruce	*Picea mariana*	ns	848.8	53.0	0.38	380.0	23.7	101.2	2.86	84.3	2.63	2.68	2.43	2.24
Sitka spruce	*Picea sitchensis*	ns	840.0	52.4	0.37	370.0	23.1	104.0	2.94	86.6	2.70	2.73	2.47	2.27
Sitka spruce	*Picea sitchensis*	ns	701.5	43.8	0.37	370.0	23.1	104.0	2.94	86.6	2.70	2.28	2.06	1.90
Jack pine	*Pinus banksiana*	ns	861.7	53.8	0.40	400.0	25.0	96.2	2.72	80.1	2.50	2.59	2.35	2.16
Lodgepole pine	*Pinus contorta*	a	905.2	56.5	0.38	380.0	23.7	101.2	2.86	84.3	2.63	2.86	2.59	2.38
Lodgepole pine	*Pinus contorta*	b	796.0	49.7	0.38	380.0	23.7	101.2	2.86	84.3	2.63	2.52	2.28	2.10
Lodgepole pine	*Pinus contorta*	c	749.2	46.8	0.38	380.0	23.7	101.2	2.86	84.3	2.63	2.37	2.15	1.97
Lodgepole pine	*Pinus contorta*	d	744.9	46.5	0.38	380.0	23.7	101.2	2.86	84.3	2.63	2.35	2.14	1.96
Shortleaf pine	*Pinus echinata*	ns	1025.0	64.0	0.47	470.0	29.3	81.8	2.32	68.2	2.13	2.62	2.38	2.18
Slash pine	*Pinus elliottii*	ns	1039.5	64.9	0.54	540.0	33.7	71.2	2.02	59.4	1.85	2.31	2.10	1.93
Limber pine	*Pinus flexilis*	a, b, c, d	930.2	58.1	0.37	370.0	23.1	104.0	2.94	86.6	2.70	3.02	2.74	2.52
Sugar pine	*Pinus lambertiana*	ns	1009.7	63.0	0.34	340.0	21.2	113.1	3.20	94.3	2.94	3.56	3.23	2.97
Western white pine	*Pinus monticola*	a, b, c, d	842.5	52.6	0.35	350.0	21.8	109.9	3.11	91.6	2.86	2.89	2.62	2.41
Longleaf pine	*Pinus palustris*	ns	1140.3	71.2	0.54	540.0	33.7	71.2	2.02	59.4	1.85	2.54	2.30	2.11
Ponderosa pine	*Pinus ponderosa*	a	1015.6	63.4	0.38	380.0	23.7	101.2	2.86	84.3	2.63	3.21	2.91	2.67
Ponderosa pine	*Pinus ponderosa*	b	971.6	60.7	0.38	380.0	23.7	101.2	2.86	84.3	2.63	3.07	2.78	2.56
Ponderosa pine	*Pinus ponderosa*	c	933.8	58.3	0.38	380.0	23.7	101.2	2.86	84.3	2.63	2.95	2.68	2.46
Ponderosa pine	*Pinus ponderosa*	d	896.4	56.0	0.38	380.0	23.7	101.2	2.86	84.3	2.63	2.83	2.57	2.36
Monterey pine	*Pinus radiata*	ns	1042.8	65.1	0.40	400.0	25.0	96.2	2.72	80.1	2.50	3.13	2.84	2.61

Red pine	*Pinus resinosa*	ns	859.1	53.6	0.41	410.0	25.6	93.8	2.65	78.2	2.44	2.52	2.28	2.10
Eastern white pine	*Pinus strobus*	ns	839.9	52.4	0.34	340.0	21.2	113.1	3.20	94.3	2.94	2.97	2.69	2.47
Loblolly pine	*Pinus taeda*	a, b, c, d	1026.6	64.1	0.47	470.0	29.3	81.8	2.32	68.2	2.13	2.62	2.38	2.19
Douglas fir	*Pseudotsuga menziesii*	a	950.8	59.4	0.45	450.0	28.1	85.5	2.42	71.2	2.22	2.54	2.30	2.11
Douglas fir	*Pseudotsuga menziesii*	b	909.1	56.8	0.45	450.0	28.1	85.5	2.42	71.2	2.22	2.43	2.20	2.02
Douglas fir	*Pseudotsuga menziesii*	c	871.8	54.4	0.45	450.0	28.1	85.5	2.42	71.2	2.22	2.33	2.11	1.94
Douglas fir	*Pseudotsuga menziesii*	d	836.1	52.2	0.45	450.0	28.1	85.5	2.42	71.2	2.22	2.23	2.02	1.86
Coastal redwood	*Sequoia sempervirens*	ns	949.8	59.3	0.36	360.0	22.5	106.8	3.02	89.0	2.78	3.17	2.87	2.64
Bald cypress	*Taxodium distichum*	ns	1027.4	64.2	0.42	420.0	26.2	91.6	2.59	76.3	2.38	2.94	2.66	2.45
Northern white cedar	*Thuja occidentalis*	ns	649.6	40.6	0.29	290.0	18.1	132.6	3.75	110.5	3.45	2.69	2.44	2.24
Western red cedar	*Thuja plicata*	a, b, c, d	621.1	38.8	0.31	310.0	19.3	124.1	3.51	103.4	3.23	2.41	2.18	2.00
Eastern hemlock	*Tsuga canadensis*	ns	979.1	61.1	0.38	380.0	23.7	101.2	2.86	84.3	2.63	3.09	2.81	2.58
Western hemlock	*Tsuga heterophylla*	a, b, c, d	942.8	58.9	0.42	420.0	26.2	91.6	2.59	76.3	2.38	2.70	2.44	2.25
Mountain hemlock	*Tsuga mertensiana*	a	1073.0	67.0	0.42	420.0	26.2	91.6	2.59	76.3	2.38	3.07	2.78	2.56
Mountain hemlock	*Tsuga mertensiana*	b	1031.1	64.4	0.42	420.0	26.2	91.6	2.59	76.3	2.38	2.95	2.67	2.46
Mountain hemlock	*Tsuga mertensiana*	c	997.4	62.3	0.42	420.0	26.2	91.6	2.59	76.3	2.38	2.85	2.59	2.38
Mountain hemlock	*Tsuga mertensiana*	d	956.7	59.7	0.42	420.0	26.2	91.6	2.59	76.3	2.38	2.74	2.48	2.28
			North American hardwoods											
Red maple	*Acer rubrum*	ns	993.0	62.0	0.54	540.0	33.7	71.2	2.02	59.4	1.85	2.21	2.00	1.84
Soft maple	*Acer saccharinum*	ns	993.0	62.0	0.49	490.0	30.6	78.5	2.22	65.4	2.04	2.43	2.21	2.03
Hard maple	*Acer saccharum*	ns	1121.1	70.0	0.56	560.0	34.9	68.7	1.94	57.2	1.79	2.40	2.18	2.00
Yellow birch	*Betula alleghaniensis*	ns	1089.1	68.0	0.55	550.0	34.3	69.9	1.98	58.3	1.82	2.38	2.16	1.98
Paper birch	*Betula papyrifera*	ns	1009.0	63.0	0.48	480.0	30.0	80.1	2.27	66.8	2.08	2.52	2.29	2.10
Pecan hickory	*Carya illinoensis*	ns	1057.0	66.0	0.60	600.0	37.4	64.1	1.81	53.4	1.67	2.12	1.92	1.76
Hickory	*Carya* spp.	ns	1201.2	75.0	0.64	640.0	39.9	60.1	1.70	50.1	1.56	2.25	2.04	1.88
Hackberry	*Celtis* spp.	ns	961.0	60.0	0.49	490.0	30.6	78.5	2.22	65.4	2.04	2.35	2.14	1.96
American beech	*Fagus grandifolia*	ns	1025.0	64.0	0.56	560.0	34.9	68.7	1.94	57.2	1.79	2.20	1.99	1.83
White ash	*Fraxinus americana*	ns	961.0	60.0	0.54	540.0	33.7	71.2	2.02	59.4	1.85	2.14	1.94	1.78
Black ash	*Fraxinus nigra*	ns	912.9	57.0	0.45	450.0	28.1	85.5	2.42	71.2	2.22	2.44	2.21	2.03
Green ash	*Fraxinus pennsylvania*	ns	880.9	55.0	0.53	530.0	33.1	72.6	2.05	60.5	1.89	2.00	1.81	1.66
Butternut	*Juglans cinerea*	ns	832.8	52.0	0.36	360.0	22.5	106.8	3.02	89.0	2.78	2.78	2.52	2.31

continued

Table A.2.B. *continued*

Common name	Latin name	Size class	Green with bark kg/m^3	Green with bark lbs/ft^3	Specific gravity (green volume)	BD weight kg/m^3	BD weight lbs/ft^3	ft^3/BDU*	m^3/BDU*	ft^3/BDT*	m^3/BDMT*	Ton/BDU**	Tonne/BDU**	Ton/BDT** Tonne/BDMT
Black walnut	*Juglans nigra*	ns	993.0	62.0	0.51	510.0	31.8	75.4	2.13	62.8	1.96	2.34	2.12	1.95
Sweetgum	*Liquidambar styraciflua*	ns	1121.1	70.0	0.48	480.0	30.0	80.1	2.27	66.8	2.08	2.80	2.54	2.34
Yellow poplar	*Liriodendron tulipifera*	ns	1025.0	64.0	0.44	440.0	27.5	87.4	2.47	72.8	2.27	2.80	2.54	2.33
Water tupelo	*Nyssa aquatica*	ns	944.9	59.0	0.45	450.0	28.1	85.5	2.42	71.2	2.22	2.52	2.29	2.10
Black gum	*Nyssa sylvatica*	ns	1041.0	65.0	0.48	480.0	30.0	80.1	2.27	66.8	2.08	2.60	2.36	2.17
Sycamore	*Platanus occidentalis*	ns	1089.1	68.0	0.45	450.0	28.1	85.5	2.42	71.2	2.22	2.91	2.64	2.42
Balsam poplar	*Populas balsamifera*	ns	993.0	62.0	0.37	370.0	23.1	104.0	2.94	86.6	2.70	3.22	2.92	2.69
Cottonwood	*Populus* spp.	ns	944.9	59.0	0.37	370.0	23.1	104.0	2.94	86.6	2.70	3.07	2.78	2.56
Aspen	*Populus tremuloides*	ns	944.9	59.0	0.39	390.0	24.3	98.6	2.79	82.2	2.56	2.91	2.64	2.42
Black cherry	*Prunus serotina*	ns	864.9	54.0	0.47	470.0	29.3	81.8	2.32	68.2	2.13	2.21	2.00	1.84
White oak	*Quercus alba*	ns	1185.2	74.0	0.64	640.0	39.9	60.1	1.70	50.1	1.56	2.22	2.02	1.85
Scarlet oak	*Quercus coccinea*	ns	1217.2	76.0	0.57	570.0	35.6	67.5	1.91	56.2	1.75	2.56	2.33	2.14
Laurel oak	*Quercus laurifolia*	ns	1249.2	78.0	0.59	590.0	36.8	65.2	1.84	54.3	1.69	2.54	2.31	2.12
Water oak	*Quercus nigra*	ns	1249.2	78.0	0.57	570.0	35.6	67.5	1.91	56.2	1.75	2.63	2.39	2.19
Chestnut oak	*Quercus prinus*	ns	1153.1	72.0	0.62	620.0	38.7	62.0	1.76	51.7	1.61	2.23	2.03	1.86
Post oak	*Quercus stellata*	ns	1185.2	74.0	0.64	640.0	39.9	60.1	1.70	50.1	1.56	2.22	2.02	1.85
Southern red oak	*Quercus falcata*	ns	1249.2	78.0	0.57	570.0	35.6	67.5	1.91	56.2	1.75	2.63	2.39	2.19
Black willow	*Salix nigra*	ns	896.9	56.0	0.34	340.0	21.2	113.1	3.20	94.3	2.94	3.17	2.87	2.64
American basswood	*Tilia americana*	ns	784.8	49.0	0.32	320.0	20.0	120.2	3.40	100.2	3.13	2.94	2.67	2.45
Elm	*Ulmus* spp.	ns	1089.1	68.0	0.46	460.0	28.7	83.6	2.37	69.7	2.17	2.84	2.58	2.37
Central and South American conifers and hardwoods														
Amburana	*Amburana cearensis*	ns	900.0	56.2	0.55	550.0	34.3	69.9	1.98	58.3	1.82	1.97	1.78	1.64
Espave	*Anacardium* spp.	ns	800.0	50.0	0.41	410.0	25.6	93.8	2.65	78.2	2.44	2.35	2.13	1.95
Angelin, red cabbage tree	*Andira inermis, A. coriacea*	ns	1075.0	67.1	0.64	640.0	39.9	60.1	1.70	50.1	1.56	2.02	1.83	1.68

Parana-pine (conifer)	*Araucaria angustifolia*	ns	700.0	43.7	0.46	460.0	28.7	83.6	2.37	69.7	2.17	1.83	1.66	1.52
Goncalo Alves	*Astronium* spp.	ns	1230.0	76.8	0.80	800.0	49.9	48.1	1.36	40.1	1.25	1.85	1.67	1.54
Cow-tree	*Brosimum* spp. *(utile)*	ns	750.0	46.8	0.44	440.0	27.5	87.4	2.47	72.8	2.27	2.05	1.86	1.70
Maracaibo lignum-vitae	*Bulnesia arborea*	ns	1250.0	78.0	1.00	1000.0	62.4	38.5	1.09	32.1	1.00	1.50	1.36	1.25
Coffeewood	*Caesalpinia* spp.	ns	1200.0	74.9	1.05	1050.0	65.5	36.6	1.04	30.5	0.95	1.37	1.24	1.14
Abarco, Jequitibe	*Cariniana pyriformis*	ns	800.0	50.0	0.46	460.0	28.7	83.6	2.37	69.7	2.17	2.09	1.90	1.74
Andiroba	*Carapa guianensis*	ns	900.0	56.2	0.56	560.0	34.9	68.7	1.94	57.2	1.79	1.93	1.75	1.61
Cedro-rana	*Cedrelinga catenaeformis*	ns	950.0	59.3	0.53	530.0	33.1	72.6	2.05	60.5	1.89	2.15	1.95	1.79
Fromager	*Ceiba pentandra*	ns	550.0	34.3	0.25	250.0	15.6	153.8	4.35	128.2	4.00	2.64	2.39	2.20
Jacaranda	*Dalbergia nigra*	ns	1025.0	64.0	0.67	670.0	41.8	57.4	1.62	47.8	1.49	1.84	1.67	1.53
Cocobolo	*Dalbergia retusa*	ns	1300.0	81.2	0.89	890.0	55.5	43.2	1.22	36.0	1.12	1.75	1.59	1.46
Honduras rosewood	*Dalbergia stevensonii*	ns	1100.0	68.7	0.82	820.0	51.2	46.9	1.33	39.1	1.22	1.61	1.46	1.34
Angelique	*Dicorynia guianensis*	ns	1000.0	62.4	0.60	600.0	37.4	64.1	1.81	53.4	1.67	2.00	1.81	1.67
Wallaba	*Eperua* spp.	ns	1100.0	68.7	0.78	780.0	48.7	49.3	1.40	41.1	1.28	1.69	1.54	1.41
Alerce (conifer)	*Fitzroya cupressoides*	ns	750.0	46.8	0.38	380.0	23.7	101.2	2.86	84.3	2.63	2.37	2.15	1.97
Kabukalli	*Goupia glabra*	ns	1050.0	65.6	0.72	720.0	44.9	53.4	1.51	44.5	1.39	1.75	1.59	1.46
Lignum vitae	*Guaiacum* spp.	ns	1450.0	90.5	1.09	1090.0	68.0	35.3	1.00	29.4	0.92	1.60	1.45	1.33
Pilon	*Hyeronima* spp.	ns	1050.0	65.6	0.64	640.0	39.9	60.1	1.70	50.1	1.56	1.97	1.79	1.64
Hura	*Hura crepitans*	ns	625.0	39.0	0.36	360.0	22.5	106.8	3.02	89.0	2.78	2.08	1.89	1.74
Algarrobo	*Hymenaea courbaril*	ns	1100.0	68.7	0.77	770.0	48.0	50.0	1.41	41.6	1.30	1.72	1.56	1.43
Caviuna	*Machaerium* spp.	ns	1100.0	68.7	0.70	700.0	43.7	54.9	1.55	45.8	1.43	1.89	1.71	1.57
Balata	*Manilkara bidentata*	ns	1250.0	78.0	0.85	850.0	53.0	45.2	1.28	37.7	1.18	1.76	1.60	1.47
Coigüe	*Nothofagus dombeyi*	ns	1000.0	62.4	0.49	490.0	30.6	78.5	2.22	65.4	2.04	2.45	2.22	2.04
Rauli	*Nothofagus procera*	ns	1000.0	62.4	0.49	490.0	30.6	78.5	2.22	65.4	2.04	2.45	2.22	2.04
Demerara greenheart	*Ocotea rodiaei*	ns	1300.0	81.2	0.83	830.0	51.8	46.3	1.31	38.6	1.20	1.88	1.71	1.57
Determa	*Ocotea rubra*	ns	825.0	51.5	0.56	560.0	34.9	68.7	1.94	57.2	1.79	1.77	1.60	1.47
Peroba de campos	*Paratecoma peroba*	ns	1025.0	64.0	0.60	600.0	37.4	64.1	1.81	53.4	1.67	2.05	1.86	1.71
Ocote pine	*Pinus oocarpa, P. patula*	ns	900.0	56.2	0.55	550.0	34.3	69.9	1.98	58.3	1.82	1.97	1.78	1.64
Snakewood	*Piratinera guianensi*	ns	1350.0	84.3	0.96	960.0	59.9	40.1	1.13	33.4	1.04	1.69	1.53	1.41
Macawood	*Platymiscium* spp.	ns	1150.0	71.8	0.84	840.0	52.4	45.8	1.30	38.2	1.19	1.64	1.49	1.37
Manio (conifer)	*Podocarpus* spp.	ns	1000.0	62.4	0.46	460.0	28.7	83.6	2.37	69.7	2.17	2.61	2.37	2.17

continued

Table A.2.B. *continued*

Common name	Latin name	Size class	Green with bark kg/m^3	Green with bark lbs/ft^3	Specific gravity (green volume)	BD weight kg/m^3	BD weight lbs/ft^3	ft^3/BDU*	m^3/BDU*	ft^3/BDT*	m^3/BDMT*	Ton/BDU**	Tonne/BDU**	Ton/BDT** Tonne/BDMT
Grignon fou	*Qualea* spp.	ns	825.0	51.5	0.55	550.0	34.3	69.9	1.98	58.3	1.82	1.80	1.63	1.50
Quebracho	*Schinopsis* spp.	ns	1350.0	84.3	1.00	1000.0	62.4	38.5	1.09	32.1	1.00	1.62	1.47	1.35
Mahogany	*Swietenia macrophylla*	ns	600.0	37.5	0.45	450.0	28.1	85.5	2.42	71.2	2.22	1.60	1.45	1.34
White cedar	*Tabebuia* spp.	ns	1250.0	78.0	0.92	920.0	57.4	41.8	1.18	34.8	1.09	1.63	1.48	1.36
Baboen	*Virola* spp.	ns	775.0	48.4	0.44	440.0	27.5	87.4	2.47	72.8	2.27	2.12	1.92	1.76
Kwari	*Vochysia* spp.	ns	825.0	51.5	0.40	400.0	25.0	96.2	2.72	80.1	2.50	2.48	2.25	2.06
Acapu	*Vouacapoua americana*	ns	1100.0	68.7	0.79	790.0	49.3	48.7	1.38	40.6	1.27	1.67	1.52	1.39
West Indian satinwood	*Zanthoxylum flavum*	ns	1100.0	68.7	0.73	730.0	45.6	52.7	1.49	43.9	1.37	1.81	1.64	1.51
Asia, Australia and Oceania conifers and hardwoods														
Mulga	*Acacia aneura*	ns	1330.0	83.0	1.02	1020.0	63.6	37.7	1.07	31.4	0.98	1.56	1.42	1.30
Gidgee	*Acacia cambagei*	ns	1265.0	79.0	0.97	970.0	60.5	39.7	1.12	33.0	1.03	1.57	1.42	1.31
Miniritchie	*Acacia monticola*	ns	1275.0	79.6	1.05	1050.0	65.5	36.6	1.04	30.5	0.95	1.46	1.32	1.21
Haldu	*Adina* spp.	ns	1000.0	62.4	0.59	590.0	36.8	65.2	1.84	54.3	1.69	2.03	1.84	1.69
Kauri pine (conifer)	*Agathis australis, Agathis* spp.	ns	840.0	52.4	0.44	440.0	27.5	87.4	2.47	72.8	2.27	2.29	2.08	1.91
Pulai	*Alstonia* spp.	ns	800.0	50.0	0.37	370.0	23.1	104.0	2.94	86.6	2.70	2.60	2.36	2.17
Mersawa	*Anisoptera* spp.	ns	1000.0	62.4	0.51	510.0	31.8	75.4	2.13	62.8	1.96	2.35	2.13	1.96
Hoop-pine, Klinki pine (conifer)	*Araucaria* spp.	ns	880.0	54.9	0.42	420.0	26.2	91.6	2.59	76.3	2.38	2.51	2.28	2.09
Chengal	*Balanocarpus* spp.	ns	1200.0	74.9	0.76	760.0	47.4	50.6	1.43	42.2	1.32	1.90	1.72	1.58
White cypress pine (conifer)	*Callitris glauca*	ns	770.0	48.1	0.58	580.0	36.2	66.3	1.88	55.3	1.72	1.59	1.45	1.33
Australian black 'oak'	*Casuarina* spp.	ns	1320.0	82.4	1.09	1090.0	68.0	35.3	1.00	29.4	0.92	1.45	1.32	1.21
East Indian satinwood	*Chloroxylon swietenia*	ns	1050.0	65.6	0.80	800.0	49.9	48.1	1.36	40.1	1.25	1.58	1.43	1.31

Silver pine (conifer)	*Dacrydium colensoi*	ns	864.9	54.0	0.54	540.0	33.7	71.2	2.02	59.4	1.85	1.92	1.74	1.60
Rimu (conifer)	*Dacrydium cupressinum*	ns	961.0	60.0	0.49	490.0	30.6	78.5	2.22	65.4	2.04	2.35	2.14	1.96
Apitong	*Dipterocarpus* spp.	ns	1025.0	64.0	0.59	590.0	36.8	65.2	1.84	54.3	1.69	2.09	1.89	1.74
New Guinea wood	*Dracontomelum* spp.	ns	900.0	56.2	0.50	500.0	31.2	76.9	2.18	64.1	2.00	2.16	1.96	1.80
Duabanga, magas	*Duabanga* spp.	ns	650.0	40.6	0.33	330.0	20.6	116.6	3.30	97.1	3.03	2.37	2.15	1.97
Corkwood	*Duboisia myoporoides*	ns	1225.0	76.5	0.80	800.0	49.9	48.1	1.36	40.1	1.25	1.84	1.67	1.53
Hinau	*Elaeocarpus denatus*	ns	1121.1	70.0	0.57	570.0	35.6	67.5	1.91	56.2	1.75	2.36	2.14	1.97
Brown mallet	*Eucalyptus astringens*	ns	1120.0	69.9	0.77	770.0	48.0	50.0	1.41	41.6	1.30	1.75	1.58	1.45
Spotted gum	*Eucalyptus citriodora*	ns	1150.0	71.8	0.79	790.0	49.3	48.7	1.38	40.6	1.27	1.75	1.59	1.46
Sugar gum	*Eucalyptus cladocalyx*	ns	1105.0	69.0	0.75	750.0	46.8	51.3	1.45	42.7	1.33	1.77	1.60	1.47
Cleland's blackbutt	*Eucalyptus clelandii*	ns	1215.0	75.9	0.97	970.0	60.5	39.7	1.12	33.0	1.03	1.50	1.36	1.25
Deglupta	*Eucalyptus deglupta*	ns	1000.0	62.4	0.55	550.0	34.3	69.9	1.98	58.3	1.82	2.18	1.98	1.82
Karri	*Eucalyptus diversicolor*	ns	1150.0	71.8	0.70	700.0	43.7	54.9	1.55	45.8	1.43	1.97	1.79	1.64
Goldfields blackbutt	*Eucalyptus lesouefii*	ns	1180.0	73.7	0.88	880.0	54.9	43.7	1.24	36.4	1.14	1.61	1.46	1.34
Red morrel	*Eucalyptus longicornis*	ns	1240.0	77.4	0.95	950.0	59.3	40.5	1.15	33.7	1.05	1.57	1.42	1.31
Jarrah	*Eucalyptus marginata*	ns	1150.0	71.8	0.67	670.0	41.8	57.4	1.62	47.8	1.49	2.06	1.87	1.72
Yellow stringy bark	*Eucalyptus muelleriana*	ns	1100.0	68.7	0.69	690.0	43.1	55.7	1.58	46.5	1.45	1.91	1.74	1.60
Salt gum	*Eucalyptus salicola*	ns	1215.0	75.9	0.94	940.0	58.7	40.9	1.16	34.1	1.06	1.55	1.41	1.29
Salmon gum	*Eucalyptus salmonophloia*	ns	1160.0	72.4	0.87	870.0	54.3	44.2	1.25	36.8	1.15	1.60	1.45	1.33
Gimlet	*Eucalyptus salubris*	ns	1230.0	76.8	0.94	940.0	58.7	40.9	1.16	34.1	1.06	1.57	1.43	1.31
Wandoo	*Eucalyptus wandoo*	ns	1280.0	79.9	0.92	920.0	57.4	41.8	1.18	34.8	1.09	1.67	1.51	1.39
Belian	*Eusideroxylon zwageri*	ns	1250.0	78.0	0.89	890.0	55.5	43.2	1.22	36.0	1.12	1.69	1.53	1.40
Beefwood	*Grevillea striata*	ns	1230.0	76.8	0.82	820.0	51.2	46.9	1.33	39.1	1.22	1.80	1.63	1.50
Huynh, heritiera	*Heritiera* spp.	ns	930.0	58.1	0.56	560.0	34.9	68.7	1.94	57.2	1.79	2.00	1.81	1.66
Bungur	*Lagerstroemina* spp.	ns	900.0	56.2	0.55	550.0	34.3	69.9	1.98	58.3	1.82	1.97	1.78	1.64
Medang	*Litsea* spp.	ns	850.0	53.1	0.40	400.0	25.0	96.2	2.72	80.1	2.50	2.55	2.32	2.13
Machang	*Mangifera* spp.	ns	850.0	53.1	0.51	510.0	31.8	75.4	2.13	62.8	1.96	2.00	1.82	1.67
Red beech	*Nothofagus fusca*	ns	977.0	61.0	0.57	570.0	35.6	67.5	1.91	56.2	1.75	2.06	1.87	1.72
Silver beech	*Nothofagus menziesii*	ns	880.9	55.0	0.45	450.0	28.1	85.5	2.42	71.2	2.22	2.35	2.13	1.96
Mountain beech	*Nothofagus solandri*	ns	961.0	60.0	0.52	520.0	32.4	74.0	2.09	61.6	1.92	2.22	2.01	1.85
Hard beech	*Nothofagus truncata*	ns	1105.1	69.0	0.63	630.0	39.3	61.1	1.73	50.9	1.59	2.11	1.91	1.76
Nyatoh, bitis	*Palaquium* spp., *Payena* spp.	ns	1075.0	67.1	0.55	550.0	34.3	69.9	1.98	58.3	1.82	2.35	2.13	1.96

continued

Table A.2.B. *continued*

Common name	Latin name	Size class	Green with bark kg/m^3	Green with bark lbs/ft^3	Specific gravity (green volume)	BD weight kg/m^3	BD weight lbs/ft^3	ft^3/BDU*	m^3/BDU*	ft^3/BDT*	m^3/BDMT*	Ton/BDU**	Tonne/BDU**	Ton/BDT** Tonne/BDMT
Bagtikan	*Parashorea* spp.	ns	1100.0	68.7	0.44	440.0	27.5	87.4	2.47	72.8	2.27	3.00	2.72	2.50
Thitka	*Pentace* spp.	ns	800.0	50.0	0.56	560.0	34.9	68.7	1.94	57.2	1.79	1.72	1.56	1.43
White lauan	*Pentacme contorta*	ns	950.0	59.3	0.43	430.0	26.8	89.4	2.53	74.5	2.33	2.65	2.41	2.21
Tropical Asian pine (conifer)	*Pinus kesyia, Pinus merkusii*	ns	875.0	54.6	0.47	470.0	29.3	81.8	2.32	68.2	2.13	2.23	2.03	1.86
Merkus pine (conifer)	*Pinus merkusii*	ns	875.0	54.6	0.57	570.0	35.6	67.5	1.91	56.2	1.75	1.84	1.67	1.54
Kahikatea (conifer)	*Podocarpus dacrydioides*	ns	816.8	51.0	0.39	390.0	24.3	98.6	2.79	82.2	2.56	2.51	2.28	2.10
Miro (conifer)	*Podocarpus ferrugineus*	ns	864.9	54.0	0.51	510.0	31.8	75.4	2.13	62.8	1.96	2.04	1.85	1.70
Matai (conifer)	*Podocarpus spicaus*	ns	1089.1	68.0	0.54	540.0	33.7	71.2	2.02	59.4	1.85	2.42	2.20	2.02
Kasai	*Pometia* spp.	ns	1000.0	62.4	0.54	540.0	33.7	71.2	2.02	59.4	1.85	2.22	2.02	1.85
Balau	*Shorea* spp. (Balau group)	ns	1200.0	74.9	0.70	700.0	43.7	54.9	1.55	45.8	1.43	2.06	1.87	1.71
White meranti	*Shorea* spp. (white meranti)	ns	725.0	45.3	0.47	470.0	29.3	81.8	2.32	68.2	2.13	1.85	1.68	1.54
Yellow meranti	*Shorea* spp. (yellow meranti)	ns	900.0	56.2	0.46	460.0	28.7	83.6	2.37	69.7	2.17	2.35	2.13	1.96
Teak	*Tectona grandis*	ns	850.0	53.1	0.57	570.0	35.6	67.5	1.91	56.2	1.75	1.79	1.62	1.49
Indian almond wood	*Terminalia* spp.	ns	825.0	51.5	0.51	510.0	31.8	75.4	2.13	62.8	1.96	1.94	1.76	1.62
Red cedar	*Toona sureni, T. calantas*	ns	750.0	46.8	0.42	420.0	26.2	91.6	2.59	76.3	2.38	2.14	1.94	1.79
Brush box	*Tristania* spp.	ns	1200.0	74.9	0.80	800.0	49.9	48.1	1.36	40.1	1.25	1.80	1.63	1.50
Pyinkado	*Xylia xylocarpa*	ns	1200.0	74.9	0.81	810.0	50.5	47.5	1.34	39.6	1.23	1.78	1.61	1.48
		African conifers and hardwoods												
Afzelia	*Afzelia cuanzensis*	ns	1250.0	78.0	0.67	670.0	41.8	57.4	1.62	47.8	1.49	2.24	2.03	1.87
Doussié	*Afzelia* spp.	ns	1150.0	71.8	0.67	670.0	41.8	57.4	1.62	47.8	1.49	2.06	1.87	1.72
Albizzia	*Albizia* spp.	ns	975.0	60.9	0.51	510.0	31.8	75.4	2.13	62.8	1.96	2.30	2.08	1.91
Alstonia	*Alstonia congensis*	ns	600.0	37.5	0.33	330.0	20.6	116.6	3.30	97.1	3.03	2.19	1.98	1.82
Mecrusse	*Androstachys johnsonii*	ns	1200.0	74.9	0.72	720.0	44.9	53.4	1.51	44.5	1.39	2.00	1.81	1.67

Aningeria	*Aningeria* spp.	ns	900.0	56.2	0.44	440.0	27.5	87.4	2.47	72.8	2.27	2.46	2.23	2.05
Ako	*Antiaris* spp.	ns	800.0	50.0	0.38	380.0	23.7	101.2	2.86	84.3	2.63	2.53	2.30	2.11
Gaboon (okoumé)	*Aucoumea klaineana*	ns	600.0	37.5	0.37	370.0	23.1	104.0	2.94	86.6	2.70	1.95	1.77	1.62
Mukulungu	*Autranella congolensis*	ns	1100.0	68.7	0.78	780.0	48.7	49.3	1.40	41.1	1.28	1.69	1.54	1.41
Rhodesian-teak	*Baikiaea plurijuga*	ns	1150.0	71.8	0.73	730.0	45.6	52.7	1.49	43.9	1.37	1.89	1.72	1.58
Dimpampi	*Baillonella toxisperma*	ns	1050.0	65.6	0.71	710.0	44.3	54.2	1.53	45.1	1.41	1.78	1.61	1.48
Berlinia	*Berlinia* spp.	ns	900.0	56.2	0.58	580.0	36.2	66.3	1.88	55.3	1.72	1.86	1.69	1.55
East African bombax	*Bombax rhodognaphalon*	ns	900.0	56.2	0.40	400.0	25.0	96.2	2.72	80.1	2.50	2.70	2.45	2.25
Bombax	*Bombax* spp.	ns	800.0	50.0	0.40	400.0	25.0	96.2	2.72	80.1	2.50	2.40	2.18	2.00
Muhuhu	*Brachylaena hutchinsii*	ns	1200.0	74.9	0.75	750.0	46.8	51.3	1.45	42.7	1.33	1.92	1.74	1.60
Okwen	*Brachystegia* spp.	ns	950.0	59.3	0.51	510.0	31.8	75.4	2.13	62.8	1.96	2.24	2.03	1.86
Burkea	*Burkea africana*	ns	1150.0	71.8	0.70	700.0	43.7	54.9	1.55	45.8	1.43	1.97	1.79	1.64
Abel	*Canarium schweinfurthii*	ns	800.0	50.0	0.40	400.0	25.0	96.2	2.72	80.1	2.50	2.40	2.18	2.00
Pillarwood	*Cassipourea malosana*	ns	1000.0	62.4	0.59	590.0	36.8	65.2	1.84	54.3	1.69	2.03	1.84	1.69
Ceiba	*Ceiba pentandra*	ns	900.0	56.2	0.26	260.0	16.2	147.9	4.19	123.3	3.85	4.16	3.77	3.46
African celtis	*Celtis* spp.	ns	950.0	59.3	0.59	590.0	36.8	65.2	1.84	54.3	1.69	1.93	1.75	1.61
Esa	*Celtis* spp.	ns	950.0	59.3	0.59	590.0	36.8	65.2	1.84	54.3	1.69	1.93	1.75	1.61
Ohia	*Celtis* spp.	ns	950.0	59.3	0.59	590.0	36.8	65.2	1.84	54.3	1.69	1.93	1.75	1.61
Mtambara	*Cephalosphaera usambarensis*	ns	850.0	53.1	0.48	480.0	30.0	80.1	2.27	66.8	2.08	2.13	1.93	1.77
Kambala	*Chlorophora excelsa* and *C. regia*	ns	1075.0	67.1	0.55	550.0	34.3	69.9	1.98	58.3	1.82	2.35	2.13	1.96
Abale	*Combretodenodron africanum*	ns	1100.0	68.7	0.70	700.0	43.7	54.9	1.55	45.8	1.43	1.89	1.71	1.57
West African cordia	*Cordia millenii, C. platythyrsa*	ns	750.0	46.8	0.34	340.0	21.2	113.1	3.20	94.3	2.94	2.65	2.40	2.21
Musine	*Croton megalocarpus*	ns	950.0	59.3	0.57	570.0	35.6	67.5	1.91	56.2	1.75	2.00	1.81	1.67
Okan	*Cylicodiscus gabunensi*	ns	1050.0	65.6	0.80	800.0	49.9	48.1	1.36	40.1	1.25	1.58	1.43	1.31
Muhimbi	*Cynometra alexandri*	ns	1000.0	62.4	0.74	740.0	46.2	52.0	1.47	43.3	1.35	1.62	1.47	1.35
Safukala	*Dacryodes pubescens*	ns	975.0	60.9	0.61	610.0	38.1	63.1	1.78	52.5	1.64	1.92	1.74	1.60
Adjouaba	*Dacryodes* spp.	ns	850.0	53.1	0.61	610.0	38.1	63.1	1.78	52.5	1.64	1.67	1.52	1.40
African blackwood	*Dalbergia melanoxylon*	ns	1400.0	87.4	1.08	1080.0	67.4	35.6	1.01	29.7	0.93	1.56	1.41	1.30
Daniellia	*Daniellia ogea* and *D. thurifera*	ns	900.0	56.2	0.40	400.0	25.0	96.2	2.72	80.1	2.50	2.70	2.45	2.25
Eyoum	*Dialium* spp.	ns	1050.0	65.6	0.71	710.0	44.3	54.2	1.53	45.1	1.41	1.78	1.61	1.48
Sapo	*Didelotia brevipaniculata*	ns	950.0	59.3	0.53	530.0	33.1	72.6	2.05	60.5	1.89	2.15	1.95	1.79

continued

Table A.2.B. *continued*

Common name	Latin name	Size class	Green with bark kg/m^3	Green with bark lbs/ft^3	Specific gravity (green volume)	BD weight kg/m^3	BD weight lbs/ft^3	ft^3/BDU*	m^3/BDU*	ft^3/BDT*	m^3/BDMT*	Ton/BDU**	Tonne/BDU**	Ton/BDT** Tonne/BDMT
African ebony	*Diospyros crassiflora*	ns	1325.0	82.7	0.82	820.0	51.2	46.9	1.33	39.1	1.22	1.94	1.76	1.62
Ayan	*Distemonanthus benthamianus*	ns	900.0	56.2	0.58	580.0	36.2	66.3	1.88	55.3	1.72	1.86	1.69	1.55
Gedu nohor	*Entandrophragma angolense*	ns	850.0	53.1	0.45	450.0	28.1	85.5	2.42	71.2	2.22	2.27	2.06	1.89
Aboudikro	*Entandrophragma cylindricum*	ns	900.0	56.2	0.60	600.0	37.4	64.1	1.81	53.4	1.67	1.80	1.63	1.50
Sipo	*Entandrophragma utile*	ns	775.0	48.4	0.53	530.0	33.1	72.6	2.05	60.5	1.89	1.76	1.59	1.46
Missanda	*Erythrophleum ivorense*	ns	1150.0	71.8	0.72	720.0	44.9	53.4	1.51	44.5	1.39	1.92	1.74	1.60
Landa	*Erythroxylum manii*	ns	850.0	53.1	0.50	500.0	31.2	76.9	2.18	64.1	2.00	2.04	1.85	1.70
Mafu	*Fagaropsis angolensis*	ns	950.0	59.3	0.56	560.0	34.9	68.7	1.94	57.2	1.79	2.04	1.85	1.70
Longui	*Gambeya* spp.	ns	1050.0	65.6	0.63	630.0	39.3	61.1	1.73	50.9	1.59	2.00	1.82	1.67
Kamassi	*Gonioma kamassi*	ns	1100.0	68.7	0.75	750.0	46.8	51.3	1.45	42.7	1.33	1.76	1.60	1.47
Agba	*Gossweilerodendron* spp.	ns	800.0	50.0	0.40	400.0	25.0	96.2	2.72	80.1	2.50	2.40	2.18	2.00
Scented Guarea	*Guarea cedrata*	ns	900.0	56.2	0.48	480.0	30.0	80.1	2.27	66.8	2.08	2.25	2.04	1.88
Black Guarea	*Guarea thompsonii*	ns	975.0	60.9	0.48	480.0	30.0	80.1	2.27	66.8	2.08	2.44	2.21	2.03
Benge	*Guibourtia arnoldiana*	ns	1050.0	65.6	0.64	640.0	39.9	60.1	1.70	50.1	1.56	1.97	1.79	1.64
Amazakoue	*Guibourtia ehie*	ns	1050.0	65.6	0.67	670.0	41.8	57.4	1.62	47.8	1.49	1.88	1.71	1.57
Bubinga	*Guibourtia* spp.	ns	1100.0	68.7	0.72	720.0	44.9	53.4	1.51	44.5	1.39	1.83	1.66	1.53
African pencil cedar (conifer)	*Juniperus procera*	ns	750.0	46.8	0.48	480.0	30.0	80.1	2.27	66.8	2.08	1.88	1.70	1.56
Senegal mahogany	*Khaya grandifoliola*	ns	775.0	48.4	0.60	600.0	37.4	64.1	1.81	53.4	1.67	1.55	1.41	1.29
African mahogany	*Khaya ivorensis*	ns	700.0	43.7	0.45	450.0	28.1	85.5	2.42	71.2	2.22	1.87	1.69	1.56
East African Mahogany	*Khaya nyasica*	ns	800.0	50.0	0.45	450.0	28.1	85.5	2.42	71.2	2.22	2.14	1.94	1.78
Azobe	*Lophira alata*	ns	1200.0	74.9	0.90	900.0	56.2	42.7	1.21	35.6	1.11	1.60	1.45	1.33
African walnut	*Lovoa trichilioides*	ns	800.0	50.0	0.45	450.0	28.1	85.5	2.42	71.2	2.22	2.14	1.94	1.78
Musizi	*Maesopsis eminii*	ns	700.0	43.7	0.35	350.0	21.8	109.9	3.11	91.6	2.86	2.40	2.18	2.00
Oboto	*Mammea africana*	ns	1100.0	68.7	0.62	620.0	38.7	62.0	1.76	51.7	1.61	2.13	1.93	1.78

Aprono	*Mansonia altissima*	ns	925.0	57.8	0.54	540.0	33.7	71.2	2.02	59.4	1.85	2.06	1.87	1.72
Zebrano	*Microberlinia* spp.	ns	1150.0	71.8	0.70	700.0	43.7	54.9	1.55	45.8	1.43	1.97	1.79	1.64
Panga panga	*Millettia* spp.	ns	1150.0	71.8	0.71	710.0	44.3	54.2	1.53	45.1	1.41	1.94	1.76	1.62
Bahia	*Mitragyna ciliata*	ns	925.0	57.8	0.45	450.0	28.1	85.5	2.42	71.2	2.22	2.47	2.24	2.06
Andoung	*Monopetalanthus heitzii*	ns	900.0	56.2	0.39	390.0	24.3	98.6	2.79	82.2	2.56	2.77	2.51	2.31
Difou	*Morus mesozygia*	ns	1050.0	65.6	0.70	700.0	43.7	54.9	1.55	45.8	1.43	1.80	1.63	1.50
Umbrella tree	*Musanga cecropioides*	ns	750.0	46.8	0.23	230.0	14.4	167.2	4.73	139.4	4.35	3.91	3.55	3.26
Bilinga	*Nauclea diderrichii*	ns	1050.0	65.6	0.63	630.0	39.3	61.1	1.73	50.9	1.59	2.00	1.82	1.67
Danta	*Nesogordonia papaverifera*	ns	925.0	57.8	0.65	650.0	40.6	59.2	1.67	49.3	1.54	1.71	1.55	1.43
E. African camphorwood	*Ocotea usambarensis*	ns	875.0	54.6	0.51	510.0	31.8	75.4	2.13	62.8	1.96	2.06	1.87	1.72
East African olive	*Olea hochstetteri*	ns	1200.0	74.9	0.72	720.0	44.9	53.4	1.51	44.5	1.39	2.00	1.81	1.67
Angueuk	*Ongokea gore*	ns	1050.0	65.6	0.72	720.0	44.9	53.4	1.51	44.5	1.39	1.75	1.59	1.46
Tola mafuta	*Oxystigma oxyphyllum*	ns	900.0	56.2	0.53	530.0	33.1	72.6	2.05	60.5	1.89	2.04	1.85	1.70
Mubara	*Parinari excelsa*	ns	1000.0	62.4	0.69	690.0	43.1	55.7	1.58	46.5	1.45	1.74	1.58	1.45
Aformosia	*Pericopsis elata*	ns	1125.0	70.2	0.57	570.0	35.6	67.5	1.91	56.2	1.75	2.37	2.15	1.97
Agboin	*Piptadeniastrum africanum*	ns	1000.0	62.4	0.56	560.0	34.9	68.7	1.94	57.2	1.79	2.14	1.94	1.79
Podo (conifer)	*Podocarpus gracilior*	ns	700.0	43.7	0.43	430.0	26.8	89.4	2.53	74.5	2.33	1.95	1.77	1.63
Ovoga	*Poga oleosa*	ns	950.0	59.3	0.36	360.0	22.5	106.8	3.02	89.0	2.78	3.17	2.87	2.64
Muninga	*Pterocarpus angolensis*	ns	1050.0	65.6	0.59	590.0	36.8	65.2	1.84	54.3	1.69	2.14	1.94	1.78
African padauk	*Pterocarpus soyauxii*	ns	950.0	59.3	0.61	610.0	38.1	63.1	1.78	52.5	1.64	1.87	1.70	1.56
Koto	*Pterygota* spp.	ns	900.0	56.2	0.52	520.0	32.4	74.0	2.09	61.6	1.92	2.08	1.89	1.73
Akomu	*Pycnanthus angolensis*	ns	800.0	50.0	0.44	440.0	27.5	87.4	2.47	72.8	2.27	2.19	1.98	1.82
Mueri	*Pygeum africanum*	ns	950.0	59.3	0.58	580.0	36.2	66.3	1.88	55.3	1.72	1.97	1.78	1.64
Essessang	*Ricinodendron heudelotii*	ns	800.0	50.0	0.20	200.0	12.5	192.3	5.44	160.3	5.00	4.81	4.36	4.01
Odoko	*Scottellia coriaceae*	ns	950.0	59.3	0.56	560.0	34.9	68.7	1.94	57.2	1.79	2.04	1.85	1.70
Niove	*Staudtia stipitata*	ns	1050.0	65.6	0.75	750.0	46.8	51.3	1.45	42.7	1.33	1.68	1.53	1.40
Brown stericula	*Sterculia rhinopetala*	ns	1000.0	62.4	0.64	640.0	39.9	60.1	1.70	50.1	1.56	1.88	1.70	1.56
Dina	*Swarztia fistuloides*	ns	1150.0	71.8	0.82	820.0	51.2	46.9	1.33	39.1	1.22	1.68	1.53	1.40
Ossol	*Symphonia globulifera*	ns	925.0	57.8	0.58	580.0	36.2	66.3	1.88	55.3	1.72	1.92	1.74	1.60
Emeri	*Terminalia ivorensis*	ns	800.0	50.0	0.43	430.0	26.8	89.4	2.53	74.5	2.33	2.24	2.03	1.86
Akom	*Terminalia superba*	ns	762.5	47.6	0.49	490.0	30.6	78.5	2.22	65.4	2.04	1.87	1.69	1.56
Izombe	*Testulea gabonensis*	ns	1000.0	62.4	0.60	600.0	37.4	64.1	1.81	53.4	1.67	2.00	1.81	1.67

continued

Table A.2.B. *continued*

Common name	Latin name	Size class	Green with bark kg/m³	Green with bark lbs/ft³	Specific gravity (green volume)	BD weight kg/m³	BD weight lbs/ft³	ft³/BDU*	m³/BDU*	ft³/BDT*	m³/BDMT*	Ton/BDU**	Tonne/BDU**	Ton/BDT** Tonne/BDMT
Ekop	*Tetraberlinia tubmaniana*	ns	825.0	51.5	0.60	600.0	37.4	64.1	1.81	53.4	1.67	1.65	1.50	1.38
Baku	*Tieghemella* spp.	ns	937.5	58.5	0.55	550.0	34.3	69.9	1.98	58.3	1.82	2.05	1.86	1.70
Abachi	*Triplochiton scleroxylon*	ns	625.0	39.0	0.33	330.0	20.6	116.6	3.30	97.1	3.03	2.27	2.06	1.89
Avodire	*Turraeanthus africanus*	ns	800.0	50.0	0.51	510.0	31.8	75.4	2.13	62.8	1.96	1.89	1.71	1.57
Sugar-plum	*Uapaca* spp.	ns	1000.0	62.4	0.60	600.0	37.4	64.1	1.81	53.4	1.67	2.00	1.81	1.67
Vitex	*Vitex* spp.	ns	825.0	51.5	0.40	400.0	25.0	96.2	2.72	80.1	2.50	2.48	2.25	2.06
Mlanje-cedar (conifer)	*Widdringtonia whytei*	ns	700.0	43.7	0.45	450.0	28.1	85.5	2.42	71.2	2.22	1.87	1.69	1.56
				European conifers										
Siberian larch	*Larix sibiricus*	ns	885.0	55.3	0.46	460.0	28.7	83.6	2.37	69.7	2.17	2.31	2.10	1.93
Norway spruce	*Picea abies*	ns	860.0	53.7	0.38	380.0	23.7	101.2	2.86	84.3	2.63	2.72	2.46	2.26
Scots pine	*Pinus sylvestris*	ns	880.0	54.9	0.40	400.0	25.0	96.2	2.72	80.1	2.50	2.64	2.40	2.20

Note: Green weight-to-volume ratios for logs includes weight of bark, and volume of wood only. Bone dry weight-to-volume ratios includes weight and volume of wood only.

*Volume (ft, m³) is green log volume.

**Ton and Tonne are for green logs weighed with bark.

Source: Calculated by the author.

Appendix 3
Glossary

Arbor: A rotating shaft on which cutting tools such as saws are fastened.

Basal area: Cross-sectional area of the tree at breast height and reflected in squared area (ft^2, m^2, etc.)

Basic density: The weight of a m^3 of wood (volume normally measured in the green state), when bone dry, and often reflected in kg/m^3.

Bleached paper: Having the lignin removed through a chemical bleaching, giving it a whiter colour.

Block (log): A short processed log length generally less than 10′ in nominal length (3.05 m), such as a stud or peeler block.

Boards: Vernacular for lumber which is normally less than 2″ in nominal thickness and used in application where appearance is the primary concern.

Board foot: In principle, represents a piece of wood which is nominally 1′ wide, 1′ long, and 1′ thick, but in actuality may represent volumes and/or dimensions that are significantly different depending on the product being reflected, e.g. logs, lumber (type of lumber, size of lumber).

Bone dry: Wood fibre having all of the moisture removed (0% moisture).

British thermal unit (BTU): Amount of heat needed to raise 1 pound of water 1° Fahrenheit at its maximum density.

Broadleaved: Trees that are botanically classified as Angiospermae. Also called hardwoods.

Buck: The process of cutting a tree stem into shorter segments via crosscutting.

Bunk: A platform or rack, generally with stakes, either on a truck or trailer, or in a stationary location, used to store, transport or contain logs temporarily.

Butt: The large-end of a log or tree. Often referring to the buttressed and flared end which occurs just above the ground level.

Butt cut: The crosscut of the log which is next to the stump (closest to the ground). A butt cut log normally has a buttressed and irregular shape on the butt-end, and thus often gives a non-representative diameter.

Buttress: Flared area of a tree located in the lower part, near the ground, with the biological purpose of aiding in the support of the tree.

Calliper: A tool used to measure the diameter of a log, tree or other object by measuring the distance between two arms, which clamp on the outside of the log or object.

Cambium layer: A narrow area of cells which lie between the inner bark and sapwood of a tree, and is responsible for each years growth of wood and bark.

Cant: A semi-processed piece of roundwood with at least one (generally 2 or 4) flat face (either sawn or chipped) on it. Typically, a cant is further processed into lumber or veneer.

Canter: A milling machine that utilizes chipping heads, saws, or both, to convert a log into a cant.

Char: Wood fibre which has been partly burned, forming a crumbly carbonized area of the wood that is black and lacks strength.

Check: A split in the wood fibre that occurs across the annual rings of a tree. Checks can occur radially from the outside of the log toward the heart, or they may occur entirely inside the log.

Chemi-mechanical process: A process used to make wood pulp (for paper) which utilizes chemical treatment followed by mechanical processing to separate fibres.

Chipper: A piece of equipment (stationary or portable) which generally has a large rotating head, that makes chips from larger pieces of wood.

Chips: Wood particles larger and thicker than sawdust or shavings, which are produced via a chipping head from larger pieces of wood or roundwood.

Chop saw: A large stationary saw, generally located at the infeed of a sawmill, used to crosscut (buck) logs into shorter segments.

Clipper: A milling machine used to either cut veneer into preset sizes or to remove defective areas of veneer.

Compression wood: The wood that resides on the lower side of a leaning tree trunk. It is characterized by: wide annual rings which lack strength, and often possess excessive lengthwise shrinkage.

Conic: Having a cone-like shape with even taper throughout its length.

Conifer: A tree botanically classified as Gymnospermae. Also called softwood.

Cord: A unit of measure of stacked wood (wood, bark and void space) occupying a space of 128 ft^3, e.g. a stack of wood $4' \times 4' \times 8'$ is one cord.

Core (peeler): The residual core of a peeler after it has been processed into veneer via rotary peeling. Generally, the peeler core is slightly larger than the chucks used to spin the log.

Crook: An abrupt bend or curve in the lineal direction of a log or tree (also called pistol butt or kink).

Cubic foot: $1' \times 1' \times 1'$ (or the equivalent).

Cubic metre: 1 m × 1 m × 1 m (or the equivalent).

Cull: Worthless, of no value, rejected.

Cunit (ccf): 100 ft^3 of solid wood.

Cuttings: A portion of a board or flitch that is obtained via ripping and crosscutting, which fits the requirements needed for the manufacture of particular products. Normally used in grading hardwood lumber and softwood shop lumber.

Cut-to-length system (CTL): A piece of equipment used for timber harvesting which cuts the tree down, and processes the stem into logs by removing the limbs and cutting the tree stem into measured log lengths.

Diameter at breast height (DBH): The diameter of a tree measured at 4.5′ (1.3 m in metric countries) above the ground level.

Debarker: A machine (generally stationary) used to remove the bark from logs.

Decimal C: Refers to the Scribner log rule, where the volume is reflected in units of 10 bf, e.g. 50 bf is reflected as 5 decimal C.

De-limbing knives: Knives that are normally mounted on timber harvesting equipment for removing the limbs of a log.

Diagram rule: A log rule which is based on diagrams of a log with lumber placed hypothetically within the log to maximize recovery, and based on preset assumptions. The diagrammed lumber is then tallied giving log volumes by diameter and length combinations.

Dimension: Vernacular for lumber which is 2″ thick or greater in nominal thickness and is utilized in applications where serviceability and structural strength is the primary concern. Differentiated from studs, in that dimension is commonly produced in longer lengths (10–20′), wider widths (up to 12″), and has more stringent strength requirements and less wane allowance.

Earlywood: The portion of the annual growth ring that is formed early in the growing season (also called springwood). It is both softer and weaker than the latewood portion.

Edgings: The tapered and wane edge of a flitch that is sawed off via ripping (normally by a piece of equipment called an edger) to produce lumber.

End grain: Having the characteristics of a cut surface which is perpendicular to the wood fibres. Rough and difficult to make smooth.

Feed speed: The speed (normally measured in distance or number of pieces per an allotted time) at which logs or lumber is fed through processing equipment.

Feed-rolls: Rollers or wheels which propel logs or lumber through saws or other processing equipment.

Fibreboard: A sheet, generally a panel, that is formed from refined wood fibres that are pressed together in varying densities, and with different amounts of glues and additives to achieve different properties.

Fines: Refers to wood particles which are classified as being too small for the end use intended. Normally used in conjunction with specifications for wood chips.

Firmwood: Wood fibre that is suitable for making chips or other solid wood products such as lumber or veneer. Firmwood fibre.

Fishtail veneer: Veneer recovered from the outer portion of the log, generally during the round-up process, with a 'fishtail' appearance and thus yielding half length sheets after trimming.

Flitch: A sawn piece of a log or cant that has not yet had the edgings removed and thus is tapered lengthwise and is made up of edges that are entirely of wane.

Green: Wood cut from a live tree and unseasoned.

Hardwood: Trees that are botanically classified as Angiospermae. Also called broadleaved.

Heartwood: Wood that is technically dead (it only serves a structural function for the tree) and resides from the pith to the sapwood.

Hog: A piece of equipment that normally utilizes rotating hammers to beat chunks of wood and tree residue into smaller conveyable chunks.

Hog fuel: The product of a hog as utilized for mulch, raw material for making alcohol, burning, etc.

Insulation board: A low density fibreboard utilized for thermal and sound insulating qualities (commonly as ceiling panels), having a specific gravity of less than 0.50, and as low as 0.02 (commonly around 0.26).

Joule: A metric unit of energy. One joule equals 0.2388 calories or 0.0009481 BTU.

Kerf: The volume of wood that is removed by a saw when cutting through wood.

Kiln: A chamber used for drying wood in a controlled manner.

Knot indicator: An imprint on the surface of a log that indicates the past presence of a limb.

Kraft paper: A process for making paper whereby the lignin is removed from the wood pulp by subjecting the wood fibre to heated chemicals. This process yields long fibres, thus strong paper.

Latewood: The portion of the annual growth ring that is formed late in the growing season (also called summerwood). It is both harder and stronger than the earlywood portion.

Load cell: A sensor that measures static tension or compression, and is used to accurately ascertain the weight of something. Commonly mounted under log bunks or in log and lumber handling equipment to measure the weight of logs or lumber.

Lumber recovery factor (LRF): The ratio of lumber recovered, as measured in board feet, relative to the volume of logs used to produce the lumber, as measured in cubic feet or cubic metres.

Lump sum: The sale of standing timber based on the consolidated value of all timber to be harvested within the sale area, e.g. purchaser pays $45,000 for all the trees marked for cut in the sale area.

Mechanical paper process: Produced from wood pulp which is made from grinding the fibres apart instead of chemical separation. This process yields more wood pulp than chemical processes, such as the Kraft process, but weaker paper.

Medium density fibreboard (MDF): Fibreboard with a specific gravity of greater than 0.5 and less than 0.8.

Mensuration: The process of measuring.

Moisture content: The amount of moisture contained in wood fibre. Normally reflected as a percentage of the bone-dry weight of the wood. Occasionally reflected as the total weight including moisture (generally when referring to wood for combustion).

Neiloid: A frustum of a cone having a decreasing rate of taper, thus having concave sides.

Nominal measure: Existing in name only, e.g. a board which has nominal dimensions of 1″ thick by 8″ wide is actually 0.75″ thick and 7.25″ wide.

Optimizer: Computerized controls that use shape images of logs, cants, flitches or boards to set milling equipment to get optimal recovery.

Oriented strand board (OSB): A sheet or panel made up of long chips or strands, alternatively layered and pressed together with glue, heat and pressure.

Overrun: Refers to lumber recovery when the output of products in board feet exceeds the input of logs in board feet, and reflected in the percentage that output exceeds input, e.g. 115 mbf of lumber produced from 100 mbf of logs is 15% overrun.

Overs: Wood residue which is classified as being too large for the end use intended. Normally used in specifications for wood chips.

Paraboloid: A frustum of a cone having an increasing rate of taper, thus having convex sides. Often described as being 'bullet-shaped'.

Particleboard: A panel made of particles (not processed fibres) of wood which are bonded together using heat pressure and glues. Particles are generally sawdust and shavings.

Peeler: A log which is intended for rotary peeling.

Pith: A soft, often dark core appearing at the centre of the annual rings on cross-cut logs or lumber ends.

Plug: A piece of wood pressed into a panel or board to cover a blemish. Often used in plywood to cover loose knots or blemishes.

Primary log breakdown: The machine which initially processes a log into a cant.

Primary product: The main manufactured product of a mill, e.g. lumber, plywood.

Recovery factor: The ratio of wood product volume produced per log volume consumed. If it takes $1\,m^3$ of logs to make $0.5\,m^3$ of lumber, the recovery factor is said to be 0.5 (or 50%).

Residuals: General term for chips, shavings, sawdust, bark and other products that are residual to making primary products.

Residue: Similar to residuals, as the term wood residue normally refers to chips, shavings, sawdust, bark, etc., but can be a primary product as is the case with whole log chips.

Revised Scribner: The Scribner board foot volume chart, which was revised in 1972 and used in California, Oregon, Washington and Alaska. Slightly different volumes (always $\pm$ 10 bf) than the original Scribner table that is used elsewhere.

Rough: When used in reference to lumber, as in rough lumber, it means not surfaced. When used in reference to a log, it means covered with large knots.

Round up: The point where a tapered and often elliptical peeler block becomes a cylinder and thus produces a continuous ribbon of full width veneer.

Roundwood: Normally used to indicate wood in the log form with bark, but also is occasionally used to indicate manufactured products such as poles or house logs.

Rule: Because in many parts of the world a 'log rule' means a product output scaling method, the use of this term is limited strictly to regulations alone or only when referring to product output scaled logs, e.g. Scribner log rule. The term is not used in conjunction with cubic scaling methods, e.g. USFS Cubic Log Scale (see: Scale).

Sapwood: Often lighter coloured wood which occurs in the outer perimeter of the log between the cambium layer and the heartwood. Sapwood is living.

Sawdust: The particles that are removed by the saw from wood during the sawing process.

Sawyer: The person who saws a log into a cant, flitches or boards.

Scale: 1. The official method of measuring a log, as in a rule or a standard (Alberta Cubic Metre Log Scale); 2. The act of measuring log volume; 3. The volume itself.

Scaling cylinder: A hypothetical cylinder which has the small-end diameter and the length of a log segment. Referred to by some log scaling rules.

Scanner: A machine which uses mechanical means, photo-cells, or multipoint lasers to measure size and/or shape of a log, cant, flitch or board for determining volume or interfacing with milling equipment via an optimizer.

Seasonality: Vernacular for changes to the amount of moisture contained in the bole of a tree in response to changes of the season.

Second cut: Refers to logs or crosscuts that are not the first log or crosscut closest to the ground, but the second or more one up from the butt.

Secondary breakdown: Sawmilling equipment that processes flitches, cants and lumber, as in trim-saws, edgers, gang-saws, re-saws, etc.

Segment (log): An individually measured or segregated section of a log, e.g. a log which has a nominal length of 32′ (9.8 m) is often measured and accounted for as two individual 16′ long (4.9 m) log segments under many scaling methods.

Shake: A split in the wood fibre which occurs along the annual growth rings.

Shavings: A thin curled particle of wood generated by the rotating heads of a planer during the surfacing process of lumber.

Shop lumber: Also called factory lumber. Utilized as stock to extract cuttings for further manufacture, e.g. components of doors, windows and furniture.

Skip: An area of a surfaced board that has not been planed. Generally caused by an undersized dimension, causing the planning knifes to miss the surface of the lumber, and thus creating a rough area.

Slab: The outer perimeter of a log, which is removed during the milling process, and contains the rounded (wane) portion.

Slabber: Rotating chipping heads that grind off the slab(s) instead of sawing them off.

Sleeper: Also called a tie. Used as a support underneath railroad tracks, both on the ground and on bridges.

Slice: A process whereby wood fibre is cut via a knife rather than a saw. Commonly used for veneer, but occasionally used for lumber as well.

Slope of grain: Wood grain that developed with a spiral orientation around the tree trunk rather than the normal vertical orientation. Produces lumber with a diagonal grain orientation that tends to be weak and easily warped. Also called twist.

Soft rot: Wood which has been decayed by fungus and is soft.

Softwood: A tree botanically classified as Gymnospermae. Also called a conifer.

Sort: Vernacular for a group of logs that are segregated, based on characteristics (generally species, diameter, length).

Southern pines: Generic term used to describe any one or a combination of the following pines which are native to the southeastern USA, and have similar qualities and characteristics: loblolly pine (*Pinus taeda*), longleaf pine (*Pinus palustris*), shortleaf pine (*Pinus echinata*) and slash pine (*Pinus elliottii*).

Spangle: A multidirectional separation in the wood fibre across the annual growth rings. A log defect that normally transects the pith of a log. Two or more checks which transect the pith of a log.

Specific gravity: Generally, when used for wood, the bone-dry weight of a given wood volume (measured before drying) is indexed against the same volume of water, e.g. if a bone-dry m^3 of wood weighs 500 kg per m^3 it has a specific gravity of 0.5 as water weighs 1000 kg per m^3.

Spike-rolls: Feed-rolls with spikes for better control of logs in processing equipment. Normally found on timber harvesting equipment and debarkers.

Stere: A unit of measure of stacked wood (wood, bark and void space) occupying a space of $1\ m^3$, e.g. a stack of wood $1\ m \times 1\ m \times 1\ m$ is one stere.

Stratification: A grouping of a population with similar characteristics used in statistical sampling to reduce variation, and thus improve the quality of the sample and/or reduce the number of samples needed to obtain a desired level of accuracy.

Stroke de-limber: A machine used to convert tree stems into logs by removing the limbs and cutting the stem into measured log lengths.

Strip veneer: Narrow pieces, or strips, of veneer which are recovered between defects during the clipping process.

Stud: Lumber which is commonly used in a vertical application in timber framed walls. Always 2″ thick and 3–6″ wide in nominal measure. Commonly produced in lengths of 6–10′ (8′ most common).

Stumpage: The price paid for log volume still standing on the stump (the purchaser pays for harvesting and transportation costs).

Surfaced: Processed through a planer to provide a smooth finish and provide exact sizing. Also called planing or dressing.

Sweep: Refers to a consistent rate of bend or curve in the lineal direction of a log or tree.

Target size: The size that a lumber or veneer manufacturer is aiming for, in the rough green state, to have enough extra dimension to account for shrinkage, surfacing, etc.

Tension wood: The wood that resides on the upper side of a leaning tree trunk. It is characterized by narrow annual rings, raised grain, excessive lengthwise shrinkage, and warping.

Thinnings: Small trees which are removed to improve the growing conditions of the residual stand of trees. Can be milled or chipped.

Tie: Also called a sleeper. Used as a support underneath railroad tracks, both on the ground and on bridges.

Ton rate: Selling logs on a value per weight basis.

Tree length: A tree stem from the stump to the merchantable top. A log that is made up of an entire merchantable portion of a tree.

Trim allowance: Unmeasured additional length to allow for end checking, shrinkage, unsquare cuts and uniform sizing.

Trim ends: Short sections of lumber, commonly 2′ (0.61 m) in length and shorter, which are trimmed off lumber to improve the grade or standardize the length.

Twist: See slope of grain.

Unbleached paper: Made from wood pulp which still contains the lignin, and is thus a natural cream or tan colour.

Under-run: Refers to lumber recovery when the output of product in board feet is less than the input of logs in board feet. Reflected in the percentage that output is less than input, e.g. 95 mbf of lumber produced from 100 mbf of logs is a 5% under-run.

Veneer: A thin sheet of wood which can be produced by rotary peeling, slicing or sawing. Commonly a component of plywood or other products.

Veneer recovery factor (VRF): The ratio of veneer or plywood recovered, as measured in square foot $^3/_8$″ or square metre 1 mm, relative to the volume of logs used to produce the lumber, as measured in cubic feet or cubic metres.

Wane: A characteristic and grade defect of lumber which contains the rounded area (and thus a void) from the perimeter of the log.

Waste gate: A gate just behind a lathe which when opened, is used to dispose of waste veneer. Generally, the veneer from the round-up process and defective veneer is sent under the waste gate, and thus to the chipper.

Wet-wood: An abnormal water-soaked heartwood associated with anaerobic bacteria.

XY charger: An optimizer that utilizes a scanner to determine the optimal location on a peeler block to centre the chucks of a veneer lathe, and places the peeler block there to obtain the optimal recovery.

Index

African log scaling methods 38–44
African timber species
Bone-dry weight and volume conversions 250–254
Log weight-to-volume data 235–240
Alberta Cubic Metre Scale 18–22
One diameter regression formula volume chart 209–210
Asian, Australian and Oceania timber species
Bone-dry weight and volume conversions 248–250
Log weight-to-volume data 232–235
Asian log scaling methods 38–47, 197

Band saws 139
Bark
Content in stacked measure 79, 183
Effects on solid wood content of stacked measure 80
Heating value 178–179, 181–183
Products 131, 177–180
Recovery 2, 169, 177–180, 227–230, 240–241
Role in preventing wood degradation 82, 123
Volume and weight 32, 129, 227–230, 241–242
Basic density *see* Specific gravity
BC Firmwood 15–18, 84, 90, 92, 130, 151–152, 164
Board foot (lumber) *see* Lumber measure
Boards (appearance lumber) 137, 146, 176
Lumber recovery from cubically scaled logs 145–146, 150, 152, 156–158
Lumber recovery from product output scaled logs 153, 156–158
Bone dry weight
Bone-dry ton (BDT) 170–173, 177, 180, 243–254
Bone-dry tonne (BDMT) 170–175, 177, 180, 243–254
Bone-dry unit (BDU) 170–177, 180, 243–254
Bone-dry weight and volume conversions of selected tree species 172–173, 243–254
Fibre content (%) 170
Units of measure 169–170
Brereton, Bernard 39
Brereton log scale 38–42
British Standard log scale 196
British Thermal Unit (BTU) 181–183
Butt flare 15, 19, 23, 36–37, 84

Chemi-mechanical paper 180
Chips 117, 120–121, 131, 140, 142, 146–149, 169–171, 174–176
Fines 121, 171, 174, 176
Overs 171, 174
Residual 174–177
Whole log 174
Circular area defect formula 14, 21
Circular saws 139
Clark, Judson F. 67
Computer tomography 82
Conditioning (peeler block) 164
Conversion ratios (commonly used) 193–195
Between log scales 83–93
Cord 74, 78–79, 181–182
Cubage au Réel 33–36
Cubic foot, half segment volume chart 205–206
Cubic metre
Full segment volume chart 211–212
Half segment volume chart 207–208
Cunit 10
Curve sawing 139

Decimal C 52–53, 58–59
Defect
  Effects on log grade 117–121
  Effects on lumber recovery 142
  Effects on plywood and veneer recovery 162–163
  Effects on pulp log grade 121
Defect deductions 8–9
  Alberta Cubic Metre Scale 21–22
  BC Firmwood 16–18
  Brereton 40–42
  British Standard 196
  Cubage au Réel 35
  Doyle Log Rule 64, 67
  Effect on conversion ratios 84, 90
  Effect on plywood and veneer recovery 162–163
  Effects on log grade 17
  Effects on lumber recovery 142
  Effects on weight ratios 129
  German log scaling standard 196
  Hirogoku 197
  Hoppus 44
  International ¼″ 70, 73
  JAS Scale 46–47
  Maine Rule 199
  New Brunswick Board Foot Rule 201
  New Brunswick Cubic Metre 198
  New Zealand 3-D and Mid-girth 37–38
  Newfoundland Board Foot Rule 200
  Norwegian Cubic Log Scale 197
  Ontario Board Foot Rule 202
  Ontario Cubic Method 24–26
  Quebec Cubic Metric 198
  Russian Government Standard 33
  Scribner Long Log Rule 56, 60–61
  Scribner Short Log Rule 51, 54
  Swedish National Board of Forestry Log Scale 28–29
  USFS National Cubic Log Scale 12–14
Density, wood 128–129, 180
Diagram rule 6, 47–48
Diameter 7–8
  Effects on log grade 118
  Effects on log quality 122
  Effects on lumber recovery 140–141
  Effects on plywood and veneer recovery 161–162
  Elliptical 7, 118, 162
  Rounding rules 7
  see also the various log scaling rules
Diameter deduction 14, 18, 22, 26, 29, 35, 38, 42, 51, 54, 60–61, 67, 73
Diameter measurements 7–8
  Alberta Cubic Metre Scale 19
  BC Firmwood 15–16
  Brereton 39–40
  British Standard 196
  Cubage au Réel 34
  Doyle Log Rule 62–63
  German Log Scaling Standard 196
  Hirogoku 197
  Hoppus 43
  International ¼″ 68–69
  JAS Scale 45
  Maine Rule 199
  New Brunswick Board Foot Rule 201
  New Brunswick Cubic Metre Scale 198
  New Zealand 3-D and Mid-girth 36–37
  Newfoundland Board Foot Rule 200
  Norwegian Cubic Log Scale 197
  Ontario Board Foot Rule 202
  Ontario Cubic Method 23–24
  Quebec Cubic Metric 198
  Russian Government Standard 30
  Scribner Long Log Rule 55–56
  Scribner Short Log Rule 49–50
  Swedish National Board of Forestry Log Scale 27
  USFS National Cubic Log Scale 10–11
Dimension lumber 132–134, 136–137, 145, 175–176
  Lumber recovery from cubically scaled logs 145, 148, 150, 152,
  Lumber recovery from product output scaled logs 154–158
Doyle Log Rule 6, 61–67, 76–77
  Lumber recovery 158

East-side Scribner *see* Scribner Short Log Rule
Ethanol alcohol 183
European log scaling methods 26–36, 196–197
European timber species
  Bone-dry weight and volume conversions 254
  Log weight-to-volume data 240–242
European Union Log Scaling Standard 196

Factory lumber (softwood shop) 134, 136–137
Firewood 78
  see also Wood energy
Formula rule 6, 48
Francon system *see* Hoppus
Freshness, log (excessive delay in log delivery) 95
Frozen wood 139

German Log Scaling Standard 196
Goskomlesprom of Russia 29

Grain characteristics
Compression wood 119, 145
Earlywood 119
Effects on log grade 119
Effects on pulp log grade 121
Growth ring density 119
Latewood 119
Tension wood 119
Gross volume determination 8
Alberta Cubic Metre Scale 20–21
Automated measurement systems 80–82
BC Firmwood 16–17
Brereton 40–41
British Standard 196
Cubage au Réel 35
Doyle Log Rule 63–64
German Log Scaling Standard 196
Hirogoku 197
Hoppus 43–44
International ¼″ 69–70
JAS Scale 46–47
Maine Rule 199–200
New Zealand 3-D and Mid-girth 37–38
New Brunswick Cubic Metre Scale 198–199
Newfoundland Board Foot Rule 200
Norwegian Cubic Log Scale 197
Ontario Board Foot Rule 202
Ontario Cubic Method 24–25
Quebec Cubic Metric 198
Russian Government Standard 31, 33
Scribner Long Log Rule 56–58
Scribner Short Log Rule 50–53
Swedish National Board of Forestry Log Scale 28
USFS National Cubic Log Scale 12–13
Guyana 43

Hardwood lumber 136, 138, 145
Lumber recovery from cubically scaled logs 145, 149, 150–152,
Lumber recovery from product output scaled logs 153, 155–158
Heartwood 75, 118–120, 125, 169
Heating values 178–179, 181–183
Hirogoku 197
Hog fuel 169
Holland Rule *see* Maine Rule
Hoppus 42–44
Hoppus, Edward 42
Hoppus superficial feet 43–44
Hoppus ton 43–44
Huber formula 9, 23–24, 30–31, 33, 35, 91

Insulation board 180
International ¼″ 6, 67–73
Lumber recovery 153,159
International Technical Association of Tropical Timber log scale (ATIBT Method) 38–42

Japanese Agricultural Standard (JAS) 44–47
Volume chart 218–219
Japanese Brereton 38–42
Joule 181–183

Kerf 48, 61, 67, 134, 139, 176–177
Kiln 177
Knots
Effects on log grade 118–119
Effects on lumber recovery 142
Effects on plywood and veneer recovery 159, 162
Kraft paper 180

Laser scanning 80–81
Length (log)
Effects on log grade 118
Effects on log quality 121–122
Effects on lumber recovery 141–142
Effects on lumber recovery (cubic scaled logs) 151–152
Effects on lumber recovery (Scribner Long Log Rule) 157
Effects on lumber recovery (Scribner Short Log Rule) 156
Effects on plywood and veneer recovery 162
Scaling length vs. segment length vs. mill length 141
Length deduction 13–14, 18, 22, 29, 35, 38, 41–42, 51, 60, 67, 73
Length measurements 8
Alberta Cubic Metre Scale 19–20
BC Firmwood 16
Brereton 39–40
British Standard 196
Cubage au Réel 34
Doyle Log Rule 62–63
German Log Scaling Standard 196
Hirogoku 197
Hoppus 43
International ¼″ 69
JAS Scale 45–46
Maine Rule 199
New Brunswick Board Foot Rule 201
New Brunswick Cubic Metre Scale 198
New Zealand 3-D and Mid-girth 36

Length measurements *continued*
Newfoundland Board Foot Rule 200
Norwegian Cubic Log Scale 197
Ontario Board Foot Rule 202
Ontario Cubic Method 24
Quebec Cubic Metric 198
Russian Government Standard 30
Scribner Long Log Rule 55–56
Scribner Short Log Rule 50
Swedish National Board of Forestry Log Scale 27
USFS National Cubic Log Scale 11
Limbs (effects on log quality) 122
Log cost 5
Log decks volume measurement
Closed decks 107
Count based 112–113
Deck factors 108–110
Depletion 107–108, 110, 112
Open decks 107–108
Physical measurements 109
Sample measure 111–113
Scaling into a deck 107, 110
Solid wood factor 109
Stacked measure 107–110
Weight measure 112
Log grade 5, 17–18, 27, 33, 82, 117–121
Log scaling
Bias 5
Check scaling 5
Converting between methods 83–93
Definition 5
Log segmenting 11–12
Log shape 8–9
Log yard inventory 107–114
Cutout 108, 113
Determined from deliveries and production 113–114
Log rotation 108
Using log tags 110
see also Log deck volume measurement
Long-butting (cull volume: effects on log quality) 122
Lumber
Finger-joint 140
Nominal sizing 133–138
Target size 134, 136, 139, 176
Surfacing *see* Planing
Trim ends 140
Wane 49, 137, 140
Width yield by log diameter 134–135
Lumber measure 133–138
Board foot 133–134, 136–138
Board foot actual to nominal size 133–134, 136–138
Cubic volume 138
Rough-green vs. finished-dry 133
Lumber recovery
Defined 131
Measured as a percentage 131–132
Lumber recovery factor (LRF) 131–132
To determine mill usage volume 113–114
Effects of saw kerf 139
Trends by log size and lumber type 145–152

Maine Rule 199–200
Mass scaling *see* Weight scale
Mechanical paper 180
Medium density fibreboard (MDF) 180–181
Merchantability factor
Alberta Cubic Metre Scale 22
BC Firmwood 18
Brereton 42
Cubage au Réel 35
Doyle Log Rule 67
International ¼″ 73
Maine Rule 199
New Brunswick Board Foot Rule 201
New Zealand 3-D and Mid-girth 38
Newfoundland Board Foot Rule 200
Ontario Board Foot Rule 202
Ontario Cubic Method 26
Russian Government Standard 33
Scribner Long Log Rule 61
Scribner Short Log Rule 54
Swedish National Board of Forestry Log Scale 29
USFS National Cubic Log Scale 14
Mill usage (Determining volume) 114–116
Mis-sorted logs 122
Moisture content
Effects on heat recovery 181
Effects on measuring chips and residual wood fibre 169
Effects on shrinkage and lumber recovery 143–145
Effects on shrinkage and plywood and veneer recovery 163–164
Effects on specific gravity (basic density) 170
Heartwood 125–127
Logs 125–128
Lumber 134, 137–138
Sapwood 125–127
Shavings (planer) 177
Myanmar 43

New Brunswick Board Foot Rule 201
New Brunswick Cubic Metre Log Scale, 198–199
New Zealand 3-D and Mid-girth 36–38
  Volume chart 214–215
Newfoundland Board Foot Rule 200
North American log scaling methods 10–26, 47–73, 197–202
North American timber species
  Bone-dry weight and volume conversions 172–173, 243–246
  Log weight-to-volume data 126–127, 227–230
Norwegian Cubic Log Scale 197

Oceania log scaling methods 36–47
Ontario Board Foot Rule 202
Ontario Cubic Method 22–26
Optimization, shape sawing (mill) 80–82, 139
Oriented strand board (OSB) 121, 169, 171, 180–181
Overrun 132

Paper 180
Papua New Guinea Brereton 38–42
Particleboard 176, 180–181
Peeler
  Conditioning 164
  Core 161
Percentage deduction 13, 21–22, 42, 51, 67, 70, 73
Philippine Brereton 38–42
Planing
  Allowance 134, 139–140
  Effects of grain distorting 119
  Production of shavings 177
  Volume loss from 133–134, 146–149
Plywood and veneer
  Round-up process 161
  Size control 161
  Target sizing 161
  Volume measurement 159–160
Plywood and veneer recovery 159–167
  Cubically scaled logs 164–166
  Product output scaled logs 164–167
Pulp
  Log grading 120–121
  Log stacked measure guidelines (Swedish) 80
  Pulp log scaling 14, 18, 22, 26, 29, 33, 38, 54
  Recovery from wood fibre 180

Quality control, log manufacturing 121–123
Quarter girth formula *see* Hoppus
Quebec Cubic Metric 197–198

Residues and chip (wood) recovery 169–183
Russian Government Standard 29–33

Sample scaling 93–105
  Allowable sampling error 94–98
  Coefficient of variation (CV) 95–98, 100–101, 103, 105
  Determining sample size needed 94–97
  Level of detail to extrapolate 101–103
  Prior load expansion 104–105
  Probability 94–96
  Sample selection 93
  Simple sample 97–98
  Standard deviation (SD) 95–96
  Stratification 93, 99–101, 105
  t value 94–96
  Three-P 98–99, 105
  Time period of expansion process 102–105
  Weight sample 98
Sapwood 39, 75, 119–120, 125, 169
  see also Moisture content, Sapwood
Sawdust 131–132, 139–140, 146–149, 169, 175–177
Sawing variation 139
Scaling cylinder 26, 48–50, 55–56, 60–61, 68–69, 91, 120, 141–142
Scanners 26, 80–82, 115–116, 160–161
Scribner
  Long Log Rule 7, 55–61
  Long Log Rule lumber recovery 157
  Long Log Rule vs. Short Log Rule comparison chart 222–223
  Revised 49, 52–53
  Short Log Rule 7, 48–55
  Short Log Rule lumber recovery 156
Seasonality 128, 169
Segment length, USFS 11, 50, 69, 203
Shavings 131–132, 146–149, 169, 175, 177
Shrinkage
  Allowance in lumber measure 138–139
  Effects on specific gravity 170
  Lumber 8, 61, 68, 122, 128, 132–134, 138, 143–149, 177
  Plywood and veneer 161, 163, 165
Slab loss 48, 61, 68, 80, 140, 161, 174, 176–177
Slope of grain 9, 13, 17, 22, 35, 51, 60, 67, 73, 119, 139, 145

Smalian formula 9, 12, 16, 19–20, 23–24, 31, 33, 91, 197–198
South and Central American log scaling methods 38–47
South and Central American timber species
  Bone-dry weight and volume conversions 246–248
  Log weight-to-volume data 230–232
Southern pine 74
Specific gravity 128–129, 170–173, 177–181, 193, 227–254
Spin-out *see* Unchuckable
Squared area defect formula 13, 17, 21, 37–38, 41, 46–47, 51, 64, 70
Squareness of bucked ends, effects on log quality 122
Stacked measure 78–80, 181–183
Stain
  Blue stain 119
  Effects on log grade 119
Step function 8, 141
Stere 74, 79, 181–183
Studs 118, 121, 132–134, 136–137, 175
  Recovery from cubic scaled logs 145, 147, 150–152
  Recovery from product output scaled logs 154, 156–158
Surface measure, plywood and veneer 160
Surfacing *see* Planing
Swedish National Board of Forestry Log Scale 26–29
  Volume chart 213
Swedish Timber Measurement Council 26–27
Sweep and crook
  Deduction 13–14, 17, 21–22, 38, 41, 51, 60, 67, 73
  Effects on log quality 120, 123
  Effects on lumber recovery 141–142
  Measured via a scanner 82

Taper, log 8
  Alberta taper regression formula 20
  Effects on converting between scaling methods 90–91
  Effects on log grade 121
  Effects on lumber recovery 142–143
  Effects on plywood and veneer recovery 162–163
  Regional taper factors in Russia 32
  see also the various log scaling rules
Taper distribution
  Alberta Cubic Metre Scale 18, 20
  BC Firmwood 16
  Brereton 40
  British Standard 196
  Cubage au Réel 35
  Doyle Log Rule 63
  German Log Scaling Standard 196
  Hirogoku 197
  Hoppus 43
  International ¼″ 68–69
  JAS Scale 46
  Maine Rule 199
  New Brunswick Board Foot Rule 200
  New Brunswick Cubic Metre Scale 198
  New Zealand 3-D and Mid-girth 36–37
  Newfoundland Board Foot Rule 200
  Norwegian Cubic Log Scale 197
  Ontario Board Foot Rule 202
  Ontario Cubic Method 24
  Quebec Cubic Metric 198
  Russian Government Standard 31–32
  Scribner Long Log Rule 56–57
  Scribner Short Log Rule 50
  Swedish National Board of Forestry Log Scale 26, 28
  USFS National Cubic Log Scale 11–12
  USFS segment 204
Target size and size control 139
Tree length logs 18, 23–24, 97–98
Trim allowance 8, 11, 14, 16, 19–20, 22, 24, 33, 45–46, 50, 55–57, 62–63, 69, 203
  Effects on log quality control 122
  Effects on plywood and veneer recovery 162
Twist *see* Slope of grain

Unchuckable 162
Under-run 132
United States Forest Service (USFS) 10, 48, 68
  National Forest Cubic Scaling Handbook 10
Unnecessary log damage 122–123
USFS National Cubic Log Scale 10–14, 90, 92, 109, 151–152, 164

Veneer
  Clipping 161
  Rotary peeled 159, 162
  Sliced 159

Washington and Oregon mill survey, Scribner to BC cubic metre comparison 224
Waste gate 161

Water (weight-to-volume ratio) 125, 170
Weight scale 6, 73–78
Weight-to-length ratios 76–78
Weight-to-volume ratios
Common world timber species (in the log form) 227–242
Logs 74–77, 125–130
West-side Scribner *see* Scribner Long Log Rule
Wet wood 119
Wood energy 177–179, 181–183

X-ray imaging 83
XY charger 161

Yukon territory 15

Printed and bound by CPI Group (UK) Ltd, Croydon, CR0 4YY

16/09/2026

14963433-0002